The Evolution of Society

Erik W. Aslaksen

The Evolution of Society

An Information-Processing Perspective

Springer

Erik W. Aslaksen
Allambie Heights, NSW, Australia

ISBN 978-3-031-23053-0   ISBN 978-3-031-23054-7 (eBook)
https://doi.org/10.1007/978-3-031-23054-7

This Springer imprint is published by the registered company Springer Nature Switzerland AG
The registered company address is: Gewerbestrasse 11, 6330 Cham, Switzerland

Acknowledgements

I acknowledge the traditional custodians of the land on which I live—the Gayamaygal people of the Eora nation—and pay my respect to their elders past and present. The many valuable comments and corrections to the draft manuscript provided by Dr. Herma Buttner FRSN are greatly appreciated, as are the discussions with Dr. Donald Hector FRSN regarding the issues raised in Chap. 13. It is also a pleasure to acknowledge the excellent support of the Springer editorial and production staff and, as always, the unreserved support of my wife, Elfi, which made it all possible.

Symbols

α	Identity partition
β	Level of stress
γ	Strength of the identity
ϑ	The set of assertions in Θ
Θ	Identity (a subset of memory)
K	Commonality
μ	Information flow
ν	Strength of an assertion
σ	Predicate in an assertion
Q	The union of individual ϑs
r	Restraint
S	Subject in an assertion
s^{σ}	An assertion with subject S and predicate σ
$s(\nu)$	Number of assertions with ν arguments
u	Mental processing capacity
w	Number of information items in Θ
$X(S^{\sigma})$	Set of all arguments in Θ associated with the assertion S^{σ}
z	Number of assertions in Θ

Contents

Chapter 1
Introduction

1.1 Motivation and Background

This book is concerned with applying the top-down approach of systems engineering to the development of a contribution to our understanding of society and the forces that drive its evolution. It is motivated by my assertion that we are currently at a critical point in this evolution and that a failure to recognise this could have unfortunate consequences. But for such recognition to develop and become prevalent, the public discourse needs to be underpinned by a common view and understanding of society. A view that transcends that of any particular society or any particular current issues, and that, while it acknowledges the important roles played by belief and culture, does not reflect any particular one, and allows us to step back and see society as the basic feature of human existence. The current state of the public discourse, as I observe it here in Australia and also in other parts of the West, is so concerned with the symptoms of a malfunctioning society that a consideration of the underlying causes hardly finds any space, and it is dominated by a commercial culture; it has become mainly a means of promoting our consumer society.

The next section sets out my justification for the assertion that we are at a critical point, but let me first explain the background to this work—how it developed out of a decades long practice of engineering in industry. My first job after graduating from the Federal Swiss Institute of Technology (ETH) in 1962 was with Bell Telephone Laboratories in Allentown PA, developing microwave ferrite devices for both terrestrial long-haul networks and for satellite communications. Despite being at the device level, this involvement with the telephone network gave me my first introduction to complex systems and to the methodology for dealing with them—systems engineering, as described in the seminal report by Hans Bode (Bode, 1967). My interest in, and application of, systems engineering continued, and when I came to Australia in 1982, I had the opportunity, together with Prof W.R. Belcher, at what was then the New South Wales Institute of Technology and is now the University of Technology

E. W. Aslaksen, *The Evolution of Society*,
https://doi.org/10.1007/978-3-031-23054-7_1

Sydney (UTS), to develop what was at the time one of the first—possibly the first—complete, two-semester undergraduate subject in systems engineering (Aslaksen, 1992).

A central feature of systems engineering is its holistic approach to engineering; it sees engineering embedded in a wider context, first, an industrial context, and then in a social context. The purpose of engineering is to meet needs expressed by society, thus extending the role of the engineer well beyond purely technical problem-solving (Aslaksen, 1996). In many of the major engineering projects I have been involved with, including mines, railways, and mobile communications systems, there was a significant social, or non-technical, component, and so the interaction between engineers and society led to my involvement with the forum on Philosophy, Engineering, and Technology (fPET). This involvement is reflected in the prominence given to engineering in Chap. 8. From there it was an obvious step to apply the general system approach, as detailed in (Aslaksen, 2013), to understanding society itself, which has been the focus of my research over the last five years or so. It is reflected in several publications (Aslaksen, 2018/20/21), of which this work is an edited, integrated, and greatly extended version.

In particular, the organisation of and additions to the material in this book reflects the progression of my focus, from the interaction between individuals, in the form of the social bond, and the influence of technology on this interaction, to the society resulting from this interaction and its evolution. This raised the issue of the stability of this evolution, and the most recent publication (Aslaksen, 2021) was an attempt to quantify this stability by considering the description of society in economic terms. The focus of the present book is now squarely on presenting an understanding of the evolution of society that can address the special situation we currently find ourselves in.

There are several approaches to gaining an understanding of a new and complex problem situation, and the one that forms the backbone of the development in this book is the system approach (Aslaksen, 2018), as introduced in Sect. 1.3. Another approach is to look for a similar problem situation in the past and identify elements of the solution that could be applicable in the current situation. This approach appears in many guises, and we can characterise them by what we might call the distance between the current and the similar problem situation. A useful measure of this distance is the difference in professional specialisation of the people involved in solving the two problems. For example, if the new problem is the development of a high-speed bearing, the useful earlier similar problems and their solutions are most likely to be found by searching the literature and patents in the area of high-speed bearings. The area of expertise is the same in the current and the earlier problems, and the distance is small. An example of this approach is the well-known TRIZ methodology (TRIZ).

In this book we apply this approach in the form of a number of *analogies* between society and physics and chemistry. This is obviously an approach with a large distance, and that brings with it its own problems. In his famous lecture series, *What is Life?*, Erwin Schrödinger made the connection between physics and biology, and recognised the dilemma poised by spanning such a wide distance between two areas of knowledge: "I can see no other escape from this dilemma (lest our true aim be lost forever) than that some of us should venture to embark on a synthesis of facts and theories, albeit with second-hand and incomplete knowledge of some of them—and at the risk of making fools of ourselves. So much for my apology." (Schrödinger, 1944). It is inevitable that sociologists will find some of the descriptions of features of society overly simplified or not according to current scholarship, but it is my intention that the insight provided into what is essentially a very complex, adaptive, self-organising system will more than compensate for these blemishes. Anyway, so much for my apology.

However, in addition to the distance between the two areas, there is an additional significant difference: The evolution of society is a process, and the structures we shall consider are of interest because of how they reflect this process, and the interactions determining the structures are those driving this process. In the case of chemistry, the structure of a molecule is permanent, and is determined by forces rather than by processes. That is, we shall be drawing on similarities between a dynamic and a static phenomenon, and this requires both some mental agility and careful formulation. Hopefully, the formulation is adequate to avoid any misunderstandings.

From this brief description it should be evident that I am not a social scientist, nor is it my intention to become one; my purpose is to present an engineer's view of society. However, this view is not developed in a vacuum; it relies on, and is related to, many parts of social science, and Sect. 1.4 attempts to locate the present work within that body of knowledge.

1.2 The Current Situation

My assertion that we are currently at a critical point in the evolution of society is based on several arguments. The first argument arises from a view of the evolution of society as the current stage in a general process of evolution, to be developed in more detail in Chap. 2.

Within this stage, society has evolved through another sequence of stages—for example, families, clans, fiefdoms, kingdoms, and nation states—stages of increasing complexity and extent of what we shall call *common infrastructure*. It consists of the institutions and organisations that provide the services classified as the common good, such as education, health, defence, law and order, etc. The current stage in this sequence of stages is that the elements are nations; the world's population is contained in something around 200 more or less well-defined and stable nations, and in nations the common infrastructure can be represented by the total Government expenditure. As the society evolves, the Government expenditure, as a proportion of GDP, increases, as shown in Fig. 1.1 for Sweden (red line) and Australia (blue line), so that for a mature nation-state, this proportion is in the range 18–26% (Sweden was chosen for comparison because of the availability of statistical data and for its progressive social policies, see also Sect. 3.4.2.)

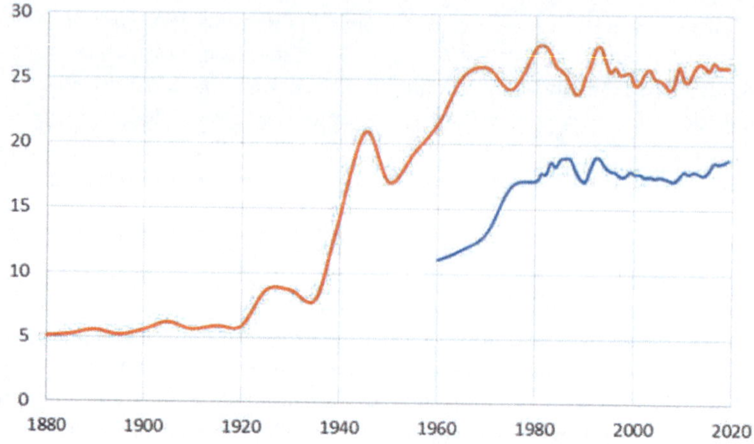

Fig. 1.1 Total Government expenditure as a percentage of GDP, for Sweden (red curve) and Australia (blue curve)

The location of a developed nation within this range is mainly a question of political ideology and of the understanding of the relative roles of individual and society, and their symbiotic relationship. This understanding, which will appear in various guises throughout the book, is central to the characterisation of the state of a society.

One possibility is then that we are currently in the final step in the evolution of society based on nations as the elements in the evolution of society, in which these nations form the last new entity—a world society in the form of a united nations. But in the sequence of stages, the motivation for the grouping of elements into a new, more complex entity was the net benefit it brought to the elements in competition with other similar entities, and for this last step, discounting a Star Wars scenario, there would be no other entity to compete with, so that this motivation is missing, and hence my argument that this is a special situation. This difficulty is reflected in the lack of success in progressing in this direction—first the League of Nations and now the UN. As shown in Table 1.1, the annual expenditure of the UN is only 0.06% of world GDP, so compared with the about 20% for a developed society like a nation, we are hardly even at the beginning of creating such a world society.

Table 1.1 United Nations expenditure in $US for the year 2018, as reported in https://unsceb.org/total-expenses, accessed 20.06.2022

Expense category	Expenditure
Humanitarian assistance	19,228,144,556
Development assistance	16,844,407,512
Peace operations	9,866,214,744
Global agenda and specialised assistance	6,837,061,899
Total	52,775,828,711

The second argument for why we are at a special point in the evolution of society has its origin in about 1450 AD. Up until that time, the world's population had developed as largely disconnected societies in various parts of the world—Europe, the Americas, China, India, Africa, and so on. But at that time, for reasons that are well known, one part of the world—what we now identify as the West—was able to embark on a quest to conquer, colonise, or subdue the rest of the world, mainly by the use of force and often brutally, and to extract a great deal of wealth that spurred the accelerated development of science and technology in the West. That chapter is now at an end, but the result is that a relatively small part of the world—say, 1 billion people out of a world population of 7 billion—is enjoying a standard of living several times that of the rest of the world. So, we have a polarisation on a world scale never seen before.

This polarisation is the main reason why a possible second option for the next step in the evolution of society—that the nations form groups of nations, with these groups developing into entities, each with its own common infrastructure, and competing on a level playing field—is unlikely, at least in the short term. The West seems bent on maintaining its hegemonic position, and a main strategy is to portray any existing or emerging group as an enemy rather than as a competitor and partner in a world society.

The third argument for why we are at a special stage of evolution is that it is only in the last fifty years or so that we have realised the fact that what we call Nature, or our physical environment, is no longer something that exists independently of us; and it is getting to the point that this Nature is increasingly manufactured by us; it is becoming a product of, and integrated into, our society, just like the built environment. This new, integrated society of tomorrow and into the future will be our product—a product of what will, in Chap. 3, be identified as the *collective intelligence*—without any divine intervention. There are several pointers to this—global warming, pandemics, and the flow of refugees and immigrants—and one might expect that they would serve as catalysts for unified action, but the experience to date has not been encouraging.

With this view of the evolution of society as a production process, we can draw an analogy with industrial production processes and the quality revolution that took place during the 50s and 60s. Basically, it was the realisation that the most cost-effective approach to ensuring a high-quality product, as evidenced for example by a low failure rate, was not to perform extensive testing of the product, with associated corrective work or rejection of defective product, but by identifying what part of the process was the cause of the defects, and then fixing that. The analogy is that society, at every level—state, nation, international—has many defects, and instead of just focusing on fixing those defects, we should focus on the process that produced that society and determine how it can be improved so that it will, in the future, provide societies with less defects.

My concern is that there is little understanding, in the general public, of the urgency of this reorientation of our thinking about society's problems. When we look at the last 200 years, we marvel at the acceleration in the development of society in terms of such characteristics as wealth, education, communication, power generation, and the creation of knowledge, but somehow, we fail to consider what this means for the future. Can this acceleration possibly continue? If not, we cannot simply apply a brake to the current development format; we need to come up with a completely different format for the evolution of society, and it is this required break with the past that makes the current situation so special.

1.3 The Approach to Describing Society and Its Evolution

1.3.1 Society

The word "society" is used to describe a wide range of associations between the people in an identifiable group. Some of them very loose and playing only a minor part in the lives of the people involved, others describing a strong interaction that constitute a significant aspect of the existence of the members. Among the former are such societies as most professional societies, special interest or hobby societies, and sporting associations, among the latter are religious societies, such as the Society of Jesus or the Society of Friends. All of these are characterised by the fact that the members belong because they want to belong; membership is an expression of the individual's will. The societies we are interested in, which are generally nations or, in some cases, the world's population as a single society, are of a different type. In these societies, people are members by virtue of being resident within the geographical area that defines the society; the extent to which an individual participates in the activities of the society varies considerably, and in some cases, there is no definite requirement to participate in any of them at all. Throughout the book there will be references to the extensive literature on society in order to examine aspects of the concept, but as the basis of most of the work presented, a society is simply the collection of all the people who recognise themselves as being members of that collection.

It might be useful to compare this initial, high-level definition of society with a definition by an influential author who we shall encounter a couple of times later on, Jürgen Habermas. In his book, *The Theory of Communicative Action*, he states:

> I use the term culture for the stock of knowledge from which participants in communication supply themselves with interpretations as they come to an understanding about something in the world. I use the term society for the legitimate orders through which participants regulate their memberships in social groups and thereby secure solidarity. By personality I understand the competences that make a subject capable of speaking and acting, that put him in a position to take part in processes of reaching understanding and thereby to assert his own identity. (Habermas, 1984, v2, 184).

So, for him society is what we identify as the *structure* of society, and what he calls culture is close to what we shall call society's *belief system*, and his 'personality' is close to the processes we include as *intelligence*, both defined in our model in Chap. 4. With a few explicit exceptions, in this book we shall take the representation of the concept 'society' to be either as 'nation' or as 'world'; which one of these is applicable will be easily understood from the context.

1.3.2 Society as a System

A society is clearly a very complex entity, but complex entities are encountered frequently in engineering, with examples in both hardware and software, and within such entities as the telephone system, the Internet, and the Space Shuttle. To handle such complexity, the systems engineering methodology was developed, starting in the 1930s, and it is based on the recognition that complexity is a characteristic of the description of something, not of the thing itself. To demonstrate this, consider such a common item as a car. To the car dealer, a car is described by a few variables, such as make, model, year, and market value. To the owner, it may even be described by a single variable—the cost or owning and operating. But to a fleet owner the cost is complex, comprised of capital cost, operating cost, and maintenance cost, and to the mechanic, the car is a complex entity, consisting of numerous interacting parts, grouped e.g., by functionality. The point is that the complexity of any item depends on our relationship to it; how we need to describe it in order to establish that relationship. And one mode of description that allows us to handle complexity in a well-defined manner is *the system concept*; that is, a description in terms of interacting parts, with the interaction being the crucial component that distinguishes a system from a collection of parts. Formally, a system is defined as consisting of three sets,

- a set of *elements*;
- a set of *interactions* between the elements; and
- a set of *external interactions*.

The system methodology, as a general methodology for handling complexity in any endeavour, and described in detail in (Aslaksen, 2013), is to develop the description in a top-down manner, with the first level of breakdown being into "areas of interest", or *views*, and then to develop each view in a step-wise fashion, by breaking it down into further levels of increasing detail, or decreasing level of *abstraction*, each level consisting of a set of *elements* and the *interactions* between them. The meaning of abstraction here is essentially the same as the one used by Georg W.F. Hegel in his essay *Who Thinks Abstractly* (Hegel, 1966) and developed further by Andrew Feenberg as "instrumentalization theory" (Feenberg, 2013). In the example of the car, the views could include cost, performance, and availability, with cost broken down into cost element, performance into functional elements, and availability into reliability/maintainability elements.

Our initial, or *primary*, view of society as a system is in terms of its active components—its human members—and we define *society* as consisting of two sets:

- the set of individual members—the *individuals*; and
- the set of their public actions—the *interactions* between them.

For the time being we are neglecting the external interactions. In addition, there is a further physical component—the *belongings*—which we treat as the internal environment in which the individuals and their interactions exist and produced by them; this will become clearer as the description of society is developed in more detail. The belongings include artefacts, such as tools and utensils, equipment and devices, clothing, artworks, means of transportation, and buildings of all sorts; domesticated animals; physical infrastructure, such as arable land, waterways, railways, roads and bridges; and the physical means of storing and transmitting information. They are identified as belongings because they belong to a particular society; they form part of the characterisation of that society.

The public actions of the individual members are those actions that establish some form of relationship with at least one other member. Obvious examples are those actions performed in public; a more subtle example would be someone sitting alone at home reading a book—the relationship is with the author. The opposite are private actions, such as sleeping or brushing teeth; they form part of the description of the individual. Public actions have a physical content, as in writing a letter or shaking hands, but also, as we shall discuss in more detail in Sect. 5.1, an information content, as in the information contained in the letter or the acknowledgement expressed in the handshake.

Regarding this approach, there are a few points that need to be briefly stated. The first is that the *partitioning* of an entity—also called its *breakdown*—into interacting elements is always a matter of choice, and while there are certain rules that can assist in making a description more useful, what is considered useful depends on the intention of the person developing the description; there is no fixed relationship between an entity and its description as a system. This is what makes systems engineering somewhat of an art.

The second point is that the choice of elements, i.e., the choice of partitioning of the entity to be described as a system, introduces a *structure* of the view. In the case of the above view of society as a system, the individuals are initially, at the highest level of abstraction, all identical, so that the system is *unstructured*.

As a brief digression, two very different structures have been prominent in the social and political science literature. The first emphasizes the importance of the individual as a social actor, with its interpersonal relations and motivations of personal conduct, so that the structure consists of the different *roles* individuals take on, such as friend, enemy, competitor, lover, etc. The second, identified with the work of Karl Marx, emphasizes the importance of *classes* within society, in which the individuals are members (German "Träger") due to their common circumstances. The significant interactions are between these classes, in the form of conflicts and class struggle, leading to structural determinism as the dynamics of society. In this sense, the Marxist view of society represents a system approach to handling the complexity of society.

The third point is that, in any given view, some or all of the relationships between the elements on a particular level can be expressed in terms of a *model*, and the characteristics of the elements become *model parameters*. The level of the model corresponds to the level of the system breakdown.

A fourth point is that there will generally be relationships between different views of the same entity, and such a relationship can be either vertical or horizontal. A vertical relationship exists when one view, the *secondary view*, refers to the elements of another view, the *primary view*, as the basis for its own elements. In the example of the car the cost view would be subordinate to the component view, in that the cost would first be broken down into the cost of the physical components, and then into various cost types for each component. In the case of society, the primary view is the one defined above, with individuals as the elements; two secondary views are defined in Chap. 3. A horizontal relationship is a relationship between two secondary views and arises simply from the fact that they relate to the same entity.

It follows that the level of breakdown of a secondary view cannot exceed that of the primary view. Furthermore, by focusing on one view, we are ignoring, or *hiding*, the complexity of the other views, just as a model on one level hides the complexity that would be revealed on lower levels. The extent to which such hiding is justifiable with regard to the questions being asked or the issues being investigated is always an important issue in applying the system methodology.

There are a few further aspects of our description of society as a system that need to be noted, and the first is that only in the most primitive of societies is the interaction between individuals limited to direct, person-to-person interaction via speech, gestures, or physical contact. In all other societies there are additional forms of interaction that involve a *medium*, such as pen and paper, a telephone, a gun, etc.; in short, some *application of technology*. These media could become additional elements in the system, with their own descriptions of their behaviour, and contribute to the behaviour of the system, i.e., of the society. However, in the following, we shall consider these applications of technology as extensions of the human interaction capabilities, so that the human becomes a *hybrid* of purely human characteristics and technological characteristics. An important consequence of this approach is that identical individuals implies that the interactions are also identical; conversely, any difference in interaction media implies different individuals and hence a structuring of the society.

Secondly, and again in all but the utopian case of a completely homogeneous society, the interactions result in a structuring of the society, i.e., the system consists of identifiable subsystems, and depending on in which of these subsystems an individual finds itself, it displays different properties. That is, one individual is able to perform many roles (e.g., wife, mother, club member, worker, etc.). These roles are not the ones mentioned above; they do not structure society, but are defined by the structure of society, as defined by the division of labour.

Thirdly, the choice, as far as describing a particular aspect of society is concerned, lies in choosing a suitable description of the behaviour of the individual. That is, selecting a subset of the set of the individual's interactions with the "outside world", but without any concern for the internal structure of the individual, or, in other words, adopting the view of a psychologist or sociologist rather than that of a biologist or neuroscientist. This treatment of the individual is fundamental to understanding society as a system of individuals, and to emphasize this point, I will give an analogy with systems engineering. Consider a very well-known element of many technical systems: an M6 $\times$ 20 bolt with hexagonal head. As an individual element, it is defined in great detail in a standard, such as a DIN standard, where all its physical characteristics, including its dimensions, the tolerances on these dimensions, its material composition and grain structure, and its surface finish are specified, together with some functional requirements, such as load capacity. This element could be used in numerous applications (i.e., systems), from a sewing machine to a space craft, and within each of these to perform various functions, such as that of a fastener, or to adjust the tension of a spring, or the position of a lever, and so on. In the description of the behaviour (functionality) of any one of these systems, none of the individual characteristics of the bolt listed above appear; the element is described solely in terms of its functionality, its purpose as an element of that particular system. Reflected onto our treatment of society as a system, this means that the individual as a stand-alone entity is of little interest; it is only the behaviour of the individual embedded in the particular view of society that is relevant. This can create a barrier to acceptance of the idea of a system view for some people, for example, philosophers, who like to view humans mainly based on ideas going back to Plato and Aristotle; persons who did not have (and could not have) any idea of what modern society would look like. Or, to put it more bluntly, the value of an individual, as a single entity, is that of an animal. Everything that we think of as human characteristics only exist in the context of other individuals, i.e., in the context of society. And it is interesting to note that this idea, expressed as "I am because of who we all are", is at the core of an African philosophy, dating back to pre-colonial times, called Ubuntu (Mugumbate & Nyanguru, 2013), to which we will return in Chapt. 10.

Fourthly, and perhaps most importantly, society displays a particular kind of dynamics, resulting from the fact that the behaviour of an individual is not time-independent, nor even a known function of time, as is the case with most technical systems; the behaviour of an individual is a function of its previous experience. Starting at birth, the experience accumulated through interaction with parents, siblings, friends, teachers, colleagues, etc. forms the behaviour of the individual. That is, the *ability* to take adaptive action as a result of sensory inputs is an intrinsic feature of the species. But the *basis* on which the action is taken is the accumulated experience. Similarly, the nature of the interaction changes from generation to generation, both in its form and its intensity, and this change is driven by technology. Direct, face-to-face interaction through increased mobility, and other forms of interaction mediated by information technology, but also through the increase in education, as enabled by the surplus generated by applications of technology. Again, the *ability* to communicate—the ability to speak and the development of language—is an intrinsic feature of the species. But the *utilisation* of this ability is being greatly enhanced by technology, so that one measure of the evolution of society is the development and application of technology.

The final aspect of the description of society as a system of interacting individuals has the character of a dichotomy. On the one hand, we treat the individual as a standalone entity, separate from society, with its own characteristics and ability to interact. On the other hand, this individual is one of the constituents of society; as such it has no existence independent of it. We shall meet this dichotomy in several places in later chapters, each calling for a slightly different mental picture, but at its core it is something we all experience every day: I certainly recognise myself as an independent individual, but I am constantly made aware of how that individuality is circumscribed by society. How well that tension is handled can be seen as a central characteristic of society.

1.3.3 Information, Action, and Will

The actions performed by the members of society are based on their physical circumstances and on the information available to them at the time. Leaving the physical circumstances aside for the moment, the information available to a member of society is a combination of information stored from previous experience and information received at the time, and the processing of the received information depends on the stored information and the member's processing power. The latter may vary somewhat from person to person, just as other physical characteristics do, but within the degree of accuracy of our high-level approach, we shall assume that the processing power is the same for all persons. It then follows that the result of the processing depends on the stored information only; that is, when two persons arrive at different interpretations of the same received information and take different adaptive actions, it is due to differences in their stored information, and not to any inherent differences in processing power.

An important corollary to this view of the individual is that what an individual takes to be reality is a result of this information exchange between the individual and its environment, both social and physical. The realisation that much of what we take to be reality—to be true—depends on the information to which we have been exposed, i.e., is socially constructed, will be called upon in several places throughout this book and will be central to developing our understanding of the evolution of society.

The assumption of individuals as identical processors implies a simplification in connection with the old controversy about nature versus nurture; that is, to what extent the behaviour and mental capabilities of a person are determined by inheritance. There are two aspects to this issue: One is that we all inherit traits characteristic of our species, such as processing of sensory inputs, memory management, reasoning, and speech. There is no doubt a certain variation about a mean value of these traits and capabilities, but whether they show any systematic variation (race, gender) or are completely random is irrelevant for us, as we operate only with averages. The second aspect concerns additional behavioural traits, such as a predisposition to physical aggression, or the interpretation of a difference in looks or behaviour as a threat, or more ethical traits, such as truthfulness, altruism, and compassion. Are they inherited or are they a result of the environment in which a person grows up, i.e., of information received? If we, for a moment, use an analogy between the processor and a computer, then it is difficult to distinguish whether the variation in the output (behaviour) resulting from a certain input (situation) is due to a difference in the firmware (mental processing capability) or in the software (experience) the computer uses to produce the output. The simplification is to assume that the mental processing capability is the same for each person, and that a person's behaviour, including the extent to which it exercises the processing capability, is determined only by its experience, from which it follows that the content of the individual's memory is determined by the individual's interaction with society, and is something that has to be created anew in every individual, hence the crucial role of education, as will be discussed in Chap. 9. The interaction is with persons that represent the history of society over a preceding period equal to, in principle, the age of the oldest living person. This time, and the time it takes to create the initial memory content, are two irreducible time constants that enter into the dynamics of society.

Furthermore, humans, viewed as organisms, have not changed significantly over the last 10,000 years or so; we look the same, our physical capabilities are much the same, and so is the processing power of the brain. Combining this insight with the foregoing, it follows that the behaviour of society is completely determined by the information contained within society. Everything created by humans—by human action—could be described in terms of the information that led to these actions, and so we end up with two complementary descriptions of society: either in physical terms—things and actions—or in terms of information. And, correspondingly, our treatment of society and its evolution will be based on two secondary views of society—an *information view* and an *action view*—defined in Chap. 3, and then developed in subsequent chapters, partly in the form of models.

Of these two views, the information view is the main view, with the actions as consequences of information. This understanding is central to all the following developments in this book, so that even when we consider only actions and their consequences, it is always with an implicit recognition of the causal information. The individual's evaluation of information, i.e., its intelligence, detailed in Chap. 4, can lead to adaptive action, but between the two is an enabling process that acts as a gate—the finest intelligence in the world has no effect on the evolution of society if it is not used to generate adaptive actions. That enabling process is the *will*. Our sensors may perceive changes to our environment, but to then engage the processes and convert that input into adaptive behaviour requires an effort, an exertion of will-power. Will is the force driving evolution; intelligence is the process steering evolution, and we shall return to this in Subsect. 7.2.2.

1.3.4 Evolution of Society

In Chap. 2, the evolution of society is presented as the current stage in a general process of evolution, and it is under that perspective that the development of our understanding of the evolution of society will progress through the rest of the book. And, in particular, and without drawing too long a bow, we can discern an analogy between how the interaction between individuals evolved into ever larger societies with complex structures, culminating in nations, and how now the interaction between nations is evolving into a world society with its own structures. In the former case, the structuring led, in many cases, to violent struggles between the structural elements within nations; in the latter case, it is not difficult to see the potential for similar struggles, just with enormously increased destructive capability, and some aspects of this possibility are discussed in Chap. 13.

However, the evolution of human society has been the subject of study for many centuries, often under the title of sociocultural evolution and focused on the process of structural reorganisation. During the Enlightenment and well into the nineteenth century the dominant view of social or cultural evolution was as a steady progression in a unilineal fashion through a set of stages, from a state of barbarism to our modern (Western) industrial society. Different societies were just seen as being at different stages of this evolution, and various theories were developed within this view. Some of the well-known authors include Hegel, August Comte, Herbert Spencer, and Adam Smith. From the second half of the nineteenth century onward, the field work of anthropologists started to provide an increasing body of empirical data, and while this was first fitted into the prevailing theoretical framework, it soon became apparent the simple unilineal progression was at best a high-level approximation, and the interests of sociologists turned from society in general to particular societies, both less developed indigenous ones and various more developed, industrial societies, by considering specific, local conditions, such as natural resources and the level of knowledge. It was also inevitable that, as the twentieth century progressed, there would be an increasing emphasis on the influence of technology and on society as a complex, self-organising system. Already the importance of the division of labour (as a result of technology) and of the means of production (and associated class structure), as described by Émile Durkheim and Karl Marx, respectively, and reflected in the critical assessment of Max Weber, showed the direction in which the understanding of social evolution was developing, and these themes (technology and system) were developed further by such authors as Lewis Henry Morgan, Leslie White, Talcott Parsons, Gerhard Lenski, Nicholas Luhmann, and Edward Goldsmith. I must emphasize that this book is very different to that work, nor does it aim to contribute directly to sociocultural theories. The current work is focused on the interaction between individuals as one of the two basic components of any society (the other component being the individual) and on the current state of that interaction under the influence of applications of technology within the framework of our political/economic system. However, it does connect with the socio-cultural body of knowledge in the two areas of technology and systems, and I shall refer to it where relevant (and where I am aware of it). And a first reference can be made already here, in that our separation of evolution of the individual (genetic change) from the evolution of society has a counterpart in the Dual Inheritance Theory, as it is set out in the book *Not by Genes Alone*, by Peter J. Richerson and Robert Boyd (2004). The evolution of society is the cultural evolution, and the authors emphasize that it is more dynamic, rapid, and influential on human society than any genetic change of the individual. They also give the following definition of culture: "*Culture is information capable of affecting individuals' behaviour that they acquire from other members of their species through teaching, imitation, and other forms of social transmission.*" (p. 5), and this is almost exactly what we shall call *identity* in the next chapter; the only difference being that the identity is not only (although mostly) acquired through social interaction, but also through observation and introspection. The book gives a compelling picture of the cooperation of genetic and cultural processes in the evolution of humanity, and gives a detailed account, supported by numerous examples, of the operation of cultural evolution over the last 10,000 years or so, to which we shall refer in a couple of places further on.

Throughout this book, 'evolution' in the context of society will obviously be a major focus; how to characterise it, how to measure it, and how to determine what drives it. Two chapters are dedicated to this, Chaps. 2 and 10, but these and related issues also appear in numerous other places in various degrees of detail. But if we, at this early stage of our development, take a step back and ask: What is it that distinguishes a society from a collection of individuals, it is the interaction between the individuals, and the evolution of society is the evolution of this interaction. At this high level of abstraction, the interaction, as such, is not measurable, and so we look for variables that can define it and its evolution in strength, in complexity, and in an increasing number of detailed variables as society evolves. There is a danger of focusing on these variables as if they described something *sui generis*, and to lose sight of the fact that society is just us—what we are and what we do. A similar situation arises with technology; we hear how we are being overwhelmed by technology and how *it* is changing our lives. But technology has no life or agency of its own; it is nothing but what we decide it to be and do. Let us keep this perspective in mind.

1.4 Location Within the Body of Knowledge

The body of knowledge related to society as a subject is vast, and the question that will arise whenever anyone considers the present work is "where in this body of knowledge is it located?". Besides the obvious links to technology, as underpinning the existence of modern societies, and to engineering through the systems engineering approach, the first place to look is in the part of the body of knowledge specifically devoted to the study of societies and the relationships among individuals within them—social science. Social science is composed of numerous branches (see e.g. the listing in https://en.wikipedia.org/wiki/Outline_of_social_science, accessed on 20.06.2022), and of those the present work has strong links to communication studies, economics, political science, and education. But above all this work, and in particular Chaps. 4 and 5, is an attempt to provide a theoretical framework within which all these branches can be accommodated, and as such it is located within Social Theory.

When presenting work on social theory to the sociological community, which is contained almost completely within academia, there appears to be less interest in the originality of the ideas and propositions put forward than in the work's conformance to, and location within, the existing literature, documented by extensive excerpts and references. While it is, of course, unthinkable, one cannot quite avoid a sneaking suspicion that a main purpose of publication, and thereby criterion for acceptance, has become the mutual recognition within the community—a perversion of what

Thomas Kuhn identified as maintaining a paradigm. There are certainly exceptions, e.g., as noted by David Harris,

> The titans of social theory meet, discuss and publish their work at an apparently 'pure' level, where arguments are pursued for their own sake, following internally-set agendas of relevance and interest, with no apparent ulterior motives, no particular anxieties to justify their 'quality' to exterior bodies, and no obvious desire to maintain institutional, national or subject matter boundaries. https://www.arasite.org/dhintro1.htm (accessed 20.06.2022),

but that is (unfortunately) not relevant in the current case.

Accordingly, as a virtual genuflection to the powers that be, this section attempts to locate the current work within the existing framework of social theory, and a first question is: Can it be considered to be within this framework at all? It is certainly about society, and it is theoretical, but if we refer to a definition of social theory, e.g., as in Wikipedia,

> Social theories are analytical frameworks, or paradigms, that are used to study and interpret social phenomena. As a tool used by social scientists, social theories relate to historical debates over the validity and reliability of different methodologies (e.g. positivism and antipositivism), the primacy of either structure or agency, as well as the relationship between contingency and necessity. Social theory in an informal nature, or *authorship based outside of academic social and political science, may be referred to as "social criticism" or "social commentary", or "cultural criticism"*

then there is certainly some doubt. However, in another definition, by Austin Harrington,

> Social theory can be defined as the study of scientific ways of thinking about social life. It encompasses *ideas about how societies change and develop*, about methods of explaining social behaviour, about power and social structure, gender and ethnicity, modernity and 'civilisation', revolutions and utopias, concepts and problems in social life (Harrington, 2005),

and in one by Anthony Elliott,

> In contemporary social theory, certain core themes take precedence over others, themes such as the nature of social life, *the relationship between self and society*, the structure of social institutions, the role and possibility of social transformation, as well as themes such as gender, race and class (Elliott, 2009),

the italicised items seem to indicate some overlap between social theory and the view of society presented in this paper. And, referring to the introduction by Peter Kivisto to the book *Social Theory: Roots & Branches* (Kivisto, 2013), where social theories are characterised as "lenses into aspects of social realty", and "Sociology is a big tent discipline. It relies on a wide range of research methodologies as well as a broad spectrum of theoretical approaches.", I am hopeful that the view presented here will be accepted as a contribution to social theory (even though it is authored by an engineer).

If, as a start, social theory is divided into three main areas,

- **Symbolic interactionism**, which considers the symbols and details of everyday life, what these symbols mean, and how people interact with each other. Symbolic interactionism traces its origins to Max Weber's assertion that individuals act according to their interpretation of the meaning of their world.
- **Structural functionalism** addresses society as a whole in terms of the function of its constituent elements; namely norms, customs, traditions, and institutions—a complex system whose parts work together to promote solidarity and stability. Major contributors to this approach include Herbert Spencer, Émile Durkheim and Talcott Parsons.
- **Conflict theory**, which originated primarily out of Karl Marx's writings on class struggles, focuses on the negative, conflicted, and ever-changing nature of society, with social order maintained by political and economic domination.

then the present work must, by default, be located in the area of structural functionalism, as it has little, if anything, in common with the other two areas. But this classification is not the most significant characterisation; much more important would be a classification in terms of the level of abstraction, in which case this view and model of society sit at a level above all the existing social theories, and therefore not readily identified as belonging to any one area. With reference to the paper by Gerhard Lenski, *Rethinking Macrosocial Theory* (Lenski, 1988), this is the top layer of a multilayer theory. If the view is expanded to the next level of detail of description, relationships to the work of many of the well-known sociologists become evident. In particular, as the view is of society as an information-processing system, the relationship to the *sociology of knowledge* can be developed in considerable detail, referring to the work of Karl Marx, Émile Durkheim, Max Scheler, Kurt Mannheim, Norbert Elias, Niklas Luhmann, and to the book by Peter L. Berger and Thomas Luckmann, *The Social Construction of Reality*. In the following chapters, this relationship is indicated for completeness, where appropriate.

1.5 Overview of the Book

1.5.1 Structure

The purpose of the book can be summarised as developing an understanding of society and its evolution that will allow us to perform a critical evaluation of the current state of society, adopting the approach outlined in Sect. 1.3, and this is reflected in the structure shown in Fig. 1.2.

It must be emphasized that this block diagram is a grouping of the material according to a few high-level themes; it does not imply that these blocks or the chapters contained in them are stand-alone products. There is significant information flowing from earlier chapters to later chapters, and for a full understanding of the propositions and arguments presented the chapters need to be read in sequence.

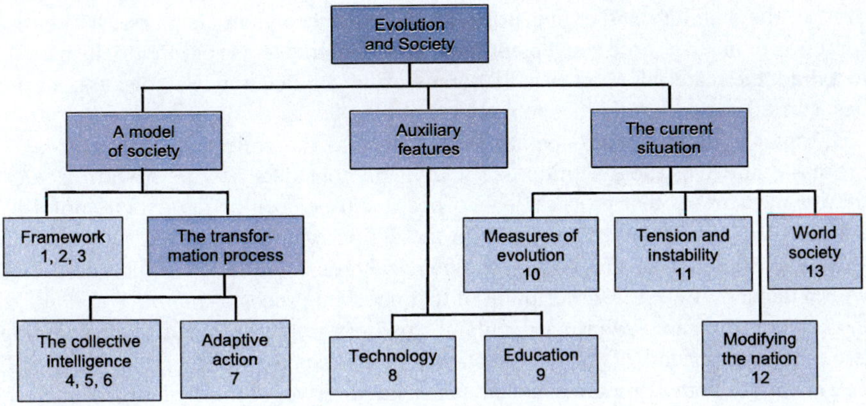

Fig. 1.2 Organisation of the contents in terms of the main subjects and the allocation of chapters to these main subjects

1.5.2 *The Chapters*

Chapter 2 presents a general picture of evolution, within which the evolution of society is a special case, and a central concept in this picture, and one which we shall utilise in several places further on, is that of a *generalised energy* that encompasses both the normal use of the concept, as in physics, and human activity. Society is created by the interaction of its members, and so the evolution of society can be viewed in terms of the evolution of this interaction, which is, essentially, the evolution of the technology that mediates this interaction.

Chapter 3 defines and develops the *two views* of society introduced in Sect. 1.3.2 and shows how both these views can be considered to have a fundamental two-component structure—two classes of information in the information view, and two cycles in the action view; all four linked to each other. However, the structures of the two views differ considerably, and the reason is that, while in the information view it is normal to have a top-level model in terms of identical individuals, this is not useful in the action view, because actions are associated with actors; that is, with people in particular roles. A given person will usually act in several roles, and the interactions are between such roles (according to the division of labour). Hence, while the information view is developed in terms of a single model, the action view consists of several models, each relating different sets of roles.

An important concept introduced here is the *collective intelligence* as the process that drives the evolution of society, and it can be seen as a process that links the two views, in that the action view describes the conditions under which the collective intelligence transforms society's *belief system* in the information view.

The two views are detailed in the next two chapters—the information view in Chap. 4 and the action view in Chap. 5. The approach taken in both chapters is to develop the associated models in as parsimonious a fashion as reasonable, removing most of the ancillary and explanatory material to subsequent chapters. One reason for this is to make it more convenient for anyone wanting to use or refer to the models to extract the material, another is that many of the explanations and discussions are best carried out with reference to models in both views.

Chapter 6 addresses the operating conditions of the collective intelligence, the process that drives the evolution of society, and considers how to provide a quantitative measure of that process. This process can be seen as the overarching link between the two views because, while the action of this process is to transform society's belief system, the process, as a physical process, relies on support described within the action view. A description of that physical process—*adaptive action*—is the subject of Chap. 7. Further details of this linkage are provided by considering two main components of society—technology in Chap. 8 and education in Chap. 9. Technology is both a product of society and a major driver of the evolution of society, and education is a main component of what we can think of as the *metabolism* of society—maintaining the state of society.

The remaining chapters are concerned with the implications of the view of society developed here. Chapter 10 considers various means of obtaining quantitative data for defining the state of society; in particular, through existing economic data. Chapter 11 is concerned with the structuring of society; what causes it, how to categorise it, and the new features of society introduced by it, such as power, tension, and stability. Chapter 12 contains two suggestions for changes to aspects of current developed societies that would move them closer to the ideal presented in this book. Chapter 13 changes the focus, which until then has been mainly on society in the form of a nation, to a world society. Such a future society might take on various forms, including a society with nations as the individuals, but there are several significant difficulties to be overcome to form any world society at all. A possible piecewise approach is put forward as a subject for further discussion.

Chapter 2
Evolution

2.1 A General Process

At a sufficiently high level of abstraction, we can see the evolution of society as the current stage in a general process of evolution. A possible identification of the stages of the general process is shown in Table 2.1, and, leaving out the first stage, which is still under some discussion, the evolution progressed through several stages, each one conceptually consisting of two phases: a *growth phase*, in which the density of elements increased, and with it the intensity of their interaction, until, in an *interaction phase*, new entities were formed.

If we introduce a *generalised concept of energy*, the formation of entities, whether existing or new, involved the conversion of *free energy* into *bound energy*. In the case of atoms and molecules, the generalised energy is just the normal definition of energy in physics, with free energy as kinetic energy and bound energy as the binding energy of the constituents of the entities. In the case of life forms, the energy is in the form of activities (as in processes), with free energy represented by the activities of entities on their own, and bound energy represented by the activities of entities as components of a greater entity.

This generalised concept of energy will be referred to throughout the book, and so it is appropriate to already here point out how it must be distinguished from another use of energy as a central feature of society—in the work of Herbert Spencer, e.g., as explained in *First Principles* (Spencer, 1899). For Spencer, this energy was the non-human energy sustaining the operation of society and exploited by it in overcoming the forces restraining its development, such as the energy in flowing water, in wind, and in fire. Our concept relates to the *interaction* of the elements making up a society, and the generalisation allows it to represent actual physical energy, as in Stages 2 and 3 of Table 2.1, as well as activities associated with the exchange and processing of information, in Stage 7.

In each stage, the complexity of the new entities, as in the number of elements in the entities, increased, and, if we think of the proportion of free energy to total energy as *temperature*, the temperature decreased from stage to stage. But also, if

E. W. Aslaksen, *The Evolution of Society*,
https://doi.org/10.1007/978-3-031-23054-7_2

Table 2.1 Evolution as a succession of stages, each consisting of two phases, with the transition from the first to the second resulting in the creation of a new type of entity through interaction, i.e., through the conversion of free into bound energy

Stage	Entities evolving in the growth phase	Entities arising in the interaction phase	Nature of interaction
1	Free energy	Particles	Nuclear force
2	Particles	Atoms	Electromagnetic force
3	Atoms	Molecules	Covalent binding
4	Molecules	Cells	Exchange of matter
5	Cells	Organisms	Functional diversification
6	Organisms	Animals	Central control
7	Genus homo	Societies	Exchange of information

What is not shown, is the decrease in the time between the appearance of new entities

we, again in a generalisation of the particle case, think of two elements interacting in a collision to form a new compound element, the energy must be great enough to overcome the repulsive potential and form a permanent bond, but not so great that the bond is disrupted. That is, there is a temperature range in which an entity can exist, and it decreases with increasing complexity. We are familiar with this in the form of homeostasis and the very narrow range of internal temperature required for the existence of the cells in our own bodies, and the general relationship between complexity, temperature, and temperature range is illustrated in Fig. 2.1.

If we consider a particular entity in the growth phase, we will call the totality of other entities with which it can interact its *environment*, and, with an appeal to imagination, the general view can be extended by introducing the concept of *survival* as a characteristic of the relationship of the entity with its environment. In the picture of evolution illustrated in Table 2.1, we can basically distinguish three different cases: In the case of a particle (atom or molecule), the relationship is a *passive* one, with

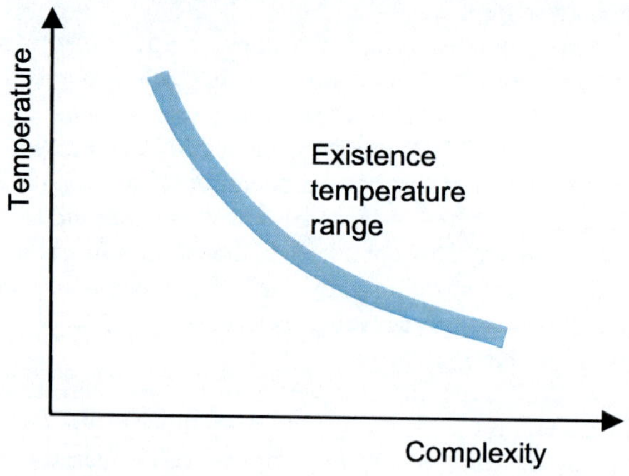

Fig. 2.1 The temperature range in which a collection of bodies of a given complexity can exist

the survival of the particle a function of the environment. As an example, an iron atom in a humid environment will, sooner or later, be converted to an iron oxide molecule; the survival rate will depend on whether the atom is alone or in a lump of iron (a "society" of iron atoms). In the case of a cell, the interaction is an *active* one, with the cell being able to selectively interact with the environment to maintain its defences against the environment's deleterious influences and thus improve its survival rate. The third case, which is a further case of an active relationship, is that of an organism—in particular humans, but also some animals, where the individual is able to modify its environment to improve its survival rate; e.g., by using fire to stay warm or to ward of dangerous predators, to build a shelter against the elements, to engage in agriculture, to extract energy from flowing water, modify plants and animals by breeding and genetic technology to improve their value to us, and so on (The fourth case, still to come, is where the humans modify themselves by genetic engineering to adapt to the environment, e.g., by reducing body size and weight for space flight.)

It is important to note that the step from passive to active taking place in Stage 4 in Table 2.1 involves a complete change in this general process of evolution. It is no longer by interaction and combination; it is of course not the case that a lung, a heart, a kidney, etc. collide in space and form an animal. The process is one of sexual *reproduction* subject to mutations, but the outcome of the process can still usefully be characterised in terms of bound energy and complexity.

As a slight digression, the concept of a generalised energy can be illustrated by introducing an analogy between humans and water molecules. At high temperatures, the water molecules move around freely, colliding infrequently, but not forming any lasting relationships. This is the very early phase of hominin existence as individuals. As the temperature drops, at some point the molecules condense into a body of water, where their freedom of movement is restricted, and the interaction with other molecules is much greater. This is the phase of nomadic societies. If the temperature drops further, the water freezes and becomes a block of ice, restricting the freedom of the molecules to move even further, but enabling them to engage in collaborative movements with their fellow molecules in the form of vibrations. This is the phase of settled societies that started with the agrarian revolution. And the block of ice now has the new property of rigidity, of being able to resist external attempts to change its shape. As long as the temperature stays below 0 °C, the block of ice remains unchanged (ignoring sublimation). But once the temperature reaches 0 °C, any increase in the free energy results in the melting of a part of the ice. If the energy is subsequently lowered the water will freeze again, but now the ice has a somewhat different shape. And, of course, if the energy is increased too much, the ice block turns into water, and the original shape is lost. Thus, for a smooth evolution of society, the fluctuations in free energy must be just right; too small, and society ossifies; too large, and the results are catastrophic. This is sometimes expressed by saying that society evolves "on the edge of chaos", although there is nothing chaotic about the phase transition of water.

The purpose of presenting this general view of evolution is firstly, to suggest that evolution is not a process that progresses towards a goal, in the sense that each stage is getting closer to it. Evolution is a never-ending process with *survival* as its enduring guiding principle; it is only that what survives and the criteria for achieving it are increasing in complexity. Secondly, and as a corollary, to counteract the common view that humans are the final product of evolution. We are just the building blocks of the next level of complexity, societies. And thirdly, to emphasize that society is a living organism, subject to the same issues as all organisms—health, sickness, and death.

2.2 Evolution of Organisms

Restricting our interest in evolution to the third case—the evolution of organisms, from single-cell organisms, such as slime mould, to its currently most splendid result—us, there are a number of parameters that correlate with this evolution. We have already noted complexity (as in the number of interacting components) and the generalised concepts of free and bound energy; the free energy is represented by the activities performed by the individual bodies without regard to or involvement of other bodies; it is an expression of the capabilities of the individual body. The bound energy arises from the relationships established by the interaction between the bodies, in the form of an exchange of information, or communication, and the scale is given by the strength of the interaction. As both energies are now represented by dissipative processes, collections of living bodies, or organisms, cannot be isolated systems, but must be immersed in an environment from which they can extract energy (metabolism).

At some point in the evolution, organisms develop the ability to interact indirectly; that is, not through direct chemical interaction, but by using signs (dance/bees), chemical markers (pheromones/ants), or sounds (most animals), and it reaches its culmination in *homo sapiens* with language as the means of interaction. The collection is now a society, and its existence requires a balance between the kinetic and structural energies: basically, a balance between the benefit of being a member of society and the associated loss of individual freedom and opportunity, both *as perceived by the individual*. Membership requires an agreement to abide by certain rules and regulations—a contract between the individual and society, with the benefits being provided by one part of the common infrastructure and the contractual obligation being realised by the other part of the common infrastructure—the legal system and law enforcement.

A further aspect of complexity is the structural complexity—the number of different relationships between the elements (cells) of the organism. In terms of the organism as a process, this is an expression of the division of labour and of the efficient organisation of the process, and there is a clear correlation with the degree of central management in the form of a central nervous system and a brain. And again, in a highly simplified manner, we can distinguish three stages of this evolution: In the first stage, the function of the central control system is to determine appropriate physical responses to sensory inputs, in the form of actions (movement, generating sound, etc.). This may be envisioned as a *look-up table*; if the sensory inputs are represented by a high-dimensionality vector, then to each point in this space, there corresponds a resultant physical action. The look-up table is inherited, and while the signals received, processed, and generated represent information, there is no *thought* involved, and the issue of consciousness is not relevant here.

In the second stage, sensory inputs result in one of three actions. One is, as in the first stage, an immediate physical action according to an inherited look-up table. The second is that the input is evaluated based on previous inputs, as stored in a *memory*, and the appropriate (or best) response is selected as that providing the greatest probability of survival. The third action is that the input information is stored in memory, without any immediate physical action. It is the ability store information as *thoughts* that is the core of consciousness and identity; when new inputs arrive and they are evaluated against this previous experience, the resultant action reflects that experience—it is unique to the individual organism.

In the third stage, the ability to process information has developed beyond processing sensory inputs; it now has the ability and capacity to recall items of information from memory and process them with regard to their relationships and implications. As a result of this processing—*thinking*, new items of information can be created; the new information is an emergent property of the system, created by letting items of information interact in the brain, so in addition to the *activities domain* of actions and processes, the organism now has a *mental domain* which includes information items not directly related to physical experience.

This is clearly a very simple and limited description of the evolution of organisms, but it allows us to realise that organisms can be described in two complementary ways: either through their actions, which can be externally observed, or through their information processing. In the first stage, and to a large extent also in the second, the information processing can be deduced from the action. For example, if the organism suddenly jumps away, we can deduce that the processing identified the immediately preceding sensory inputs as constituting a threat. In the third stage, the relationship between information and action is much more complex, and while action is still determined by information, there is no longer any simple relationship between this information and the information received as sensory input.

It might be tempting to compare this duality of description—in terms of action or in terms of information—to the duality of description in physics, where e.g., an electron or a photon can be described as either a particle or a wave. But, while we in both cases will use the description that results in the best understanding of a particular situation, the two cases are fundamentally different. In the case of physics, the two modes of description refer to the same physical entity—the electron or photon, whereas in the case of organisms, the action and the information are both real entities: related parts of the organism.

2.3 Evolution of Society

2.3.1 It Is not a Random Process

We are currently located in the last stage in the sequence shown in Table 2.1—the evolution of the genus *homo* into its various species, of which only our species—*homo sapiens*, or humans—has been present in the period of interest in the context of this book, which we may limit to the last 10,000 years or so. In this period the physical features of the species have not changed significantly; in particular, the brain and its inherent capability and capacity for processing information have remained largely unchanged. What has changed, however, is the level of interaction between humans, and the associated creation of societies. We may therefore think of this period as a separate stage of evolution—the evolution of the genus society into a number of species in the form of different societies—where each new species presented a new environment for the individual, and to be successful, the individual had to adapt to the demands of the stronger and more complex interaction. This included changed behaviour and such new concepts as compassion and solidarity, and a very interesting question, but one that we cannot pursue in this book, is whether this social behaviour is something the newborn child must learn in each generation or whether some of this acquired behaviour becomes a heritable trait through the process of *epigenetics*. That is, epigenetics would be to the evolution of society what genetics is to the evolution of organisms, and it is this evolution of the genus society that is the subject of this book.

From the outset we need to realise that, while we have placed the evolution of society within a general process of evolution, it is principally different to all preceding stages of evolution in that, in those stages, changes resulted from random events—collisions in the case of atoms and molecules, mutations in the case of organisms—over which the participants had no control or influence. In the case of society, the changes are the result of actions by the members; changes that are, for all practical purposes, unpredictable. Not only in the sense of their content, but also in the sense of their timing or rate of occurrence, even in a statistical sense, which results in a completely different dynamic behaviour. If we look back at society, evolution is very evident; it has taken us from the cave to where we are today. Considering human history over the last ten thousand years or so, there can be little doubt that the "richness" of life, as measured by the diversity of our concepts and experiences, both on the average and in total, has increased exponentially in historic times, despite many setbacks in the form of wars and ideological subjugation. In most of the world, the opportunities for self-fulfilment available to the average person through material well-being and the associated free (or non-working) time, education, and an intellectually stimulating environment have led to societies that are again promoting those same factors, while attending to the various issues, moral and otherwise, accompanying that development. While it may appear that "the good life" has become synonymous with material success; it must not be overlooked that the access to and participation in all forms of art and political, scientific, and religious discourse, as well as a greater understanding of our environment through education and widespread dissemination of information have greatly increased the non-material content of the average person's life.

As we note earlier, this evolution is not towards a given goal; it is simply an ongoing process with survival as its guiding criterion, and it is no more possible to predict where humanity and society will be a thousand years from now than it would have been to predict, 100 million years ago, the appearance of man. If we describe the state of society in terms of a set of variables, **S**, we can describe the history of society in the same terms, but we cannot determine the future values of these variables based on this history, due basically to what is the unpredictability of human actions. But we can identify the process that drives this evolution—the *transformation process*, which we shall develop in some detail in Chaps. 4–7, and it will allow us to say something about the *direction* is which we are currently heading, which immediately raises the question: is there a preferred or 'good' direction? Most attempts to provide an answer to this question presuppose that it is to be sought in something external to us, by recourse to a divine or supernatural authority. But if we understand that we are now the masters of evolution, the answer must be sought in an understanding of the characteristics of that mastery—of the process that drives evolution, and that *the basic criterion is survival.*

A similar view of evolution of society, although in a somewhat different context, was expounded some time ago by Sir Peter Medawar in a series of BBC lectures in 1959 (Medawar, 1959), and as a biologist and immunologist (and Nobel laureate) he used the evolution of homo sapiens and, in particular, of the human brain as the basis for his argument. For this purpose, he divided the evolution of the brain into four stages: In the first stage, the brain was an organ that responded only to external stimuli by reactions that were already present in the brain. That is, a certain stimulus, which he called an elective stimulus, elected its corresponding reaction, but the brain would not react to stimuli that did not fall in this group. In the second stage, the brain began to be able to accept instructive stimuli; stimuli that contained information about how it should be processed. The development in these two stages depended entirely on a genetic heredity, whereas in the third stage, a non-genetic system of heredity evolved that allowed brains to do more than merely receive instructions; it made it possible for them to be handed on. The fourth stage is the systematic change in the nature of the instructions passed on from generation to generation; an evolution that has been progressing at an accelerating pace for the last couple of centuries. The conclusion Medawar draws from this argument is that social change is not governed by any laws other than laws which have at some time been the subject of human decisions or acts of mind, and the mechanism that supports this change process is the non-genetic heredity mediated through the transfer of information from one generation to the next.

There can no longer be any doubt about the fact that the development of humanity and our environment is driven by us. We are no longer the pawns in Nature's game of the survival of the fittest in a distribution created by random mutations; we run the game. Adam Smith's "invisible hand" is no longer invisible; it is our hand.

2.3.2 Survival

When we speak of 'survival', it means on the one hand that something persists in time; on the other hand, that it does so despite something that is detrimental to its persistence. As individuals we speak of surviving an ordeal or a challenge, but when we speak of the survival of humanity, the species *homo sapiens*, or human *society*, it must relate to something other than the survival of individuals, which is severely limited. Maybe it is one or more characteristics of individuals that survives, such as the genetic code? But we can already see the possibility of modifying the genetic code; for example, to eliminate a disease or to make people smaller in order to reduce their ecological footprint and make space travel more economical. So, by this definition, the species would become extinct, but there clearly is something that persists. Another characteristic used to define a species is the ability to produce fertile offspring. But we could conceive of a future where all offspring is produced in a factory environment, so again the species would become extinct, but something persists.

We should also consider that we are likely to become increasingly characterised by a combination of what we might consider purely human characteristics and those of applications of technology; that is, we will become hybrids or cyborgs. That development has already started; the behaviour of a person in our society today is significantly affected by technology. And embedding applications of technology into the human body is well under way; currently only replications of human organs and parts, but additional and/or supporting capability will follow. So, if we want to say that we, humans, are surviving through all of this, what is it that survives? What survives in all the foregoing examples is *the continuity of the evolution of society*, whereby society we understand all humans and their artefacts, the interactions between these, and the environment in which they operate. This continuity means that, at any point in time, the changes taking place in society in the next time increment are completely determined by the previous history of society. We can express this symbolically by the equation

$$\frac{d\mathbf{S}(t)}{dt} = \int_0^t f(\mathbf{S}; t)\mathbf{S}(t)dt \tag{2.1}$$

where $\mathbf{S}$ is the set of variables describing society, and $f(\mathbf{S}; t)$ is a function expressing both a weighting of the variables in determining the future development and the fact that recent developments are more significant than ancient ones. That is, $f(\mathbf{S}; t)$ is a rapidly decreasing function as we go back in time, which is why the choice of the point $t = 0$ is unimportant; starting, say, 10,000 years ago should be sufficient.

We now see that what survives is, in the case of humans, not a particular physical structure; that is, a particular ordering of physical elements in space, such as the molecules in the genome, or relationships between humans, such as a particular social structure, but a particular ordering of events in time, as expressed by Eq. 2.1.

And the meaning of 'survival' in the context of humans is much more complex than what we might at first think of as the most immediate manifestation: the survival of the individual. We certainly have a strong personal survival instinct; we will fight anyone that tries to eliminate us, we will go to great pains and expense to avoid succumbing to an illness and do our utmost to survive disasters and persecution. But on the other hand, there are numerous examples of where we are willing to give up our lives for our children, for our country, for a religious belief, and so on. And then there is the very deep desire to survive death in some form; again, in our children, but also in works that will remind posterity of our existence, such as a handprint on a cave wall, a poem, a scientific discovery, a world record, actions that will gain us a place in history, or a great monument, such as the Taj Mahal or the pyramids. But it is even more subtle than this; for those of us who cannot on our own produce something that will endure, there is a desire to be part of something that will. An example of this is the dedication and sacrifice of those who built the great cathedrals.

What we have, then, is an inherent will to survive, but that, just as with so many other aspects of what it is to be human, the form of this will to survive has developed as society developed. With an increasing understanding of the meaning of society and our place in it has come an evolving understanding of the meaning of survival; both concepts have become more closely related and more complex. Considering that more than 99% of all species that have existed are now extinct, a simple-minded extrapolation of that process would indicate that the days of homo sapiens are also numbered. However, as already noted, we will soon have the ability to alter our genetic makeup (in addition to the ability of altering the "environment" in which society exists), and so generate new subspecies as changed circumstances demand. And while that process carries its own risks in the form of unintended consequences, we might avoid the straight-forward extinction of the previous 99% of species.

2.3.3 The Time Factor

Sometime in the early days of the genus homo, humans became aware that they were different from other living organisms. They no longer saw themselves as just another creature living in and as a part of Nature, but increasingly saw the rest of Nature as the environment in which their individual lives took place and which could, to some extent, be manipulated and exploited to their advantage and from which they could obtain some measure of independence. This recognition and understanding were a result of the shift in the balance of capabilities from the physical to the mental. And besides the increasing ability to develop technology and apply it to modify the conditions of existence, it was accompanied by a shift in the balance between physical and mental activity. Existence had no longer just, or even mainly, a physical content, but also a mental content, and thoughts and mental images could take on a reality in a person's mind that was divorced from any physical reality. A consequence of this was that the fear of death, which all animals display, but which they presumably accept as the natural end to individual existence, could now be sublimated by a mental construct in which the individual continued to exist in some form in a mental realm, in contradiction to all physical evidence. The preoccupation with and embellishment of this construct in the form of beliefs, rituals, and religious systems have been central to the evolution of society and an inseparable part of our story of how we ended up where we are today.

In the book *The Death of Forever*, Darryl Reanney explores this issue from various points of view (Reanney, 1991). He gives a very good account of the importance of death, whether consciously or unconsciously, in forming our character, and gives many examples of how our desire to transcend death has manifested itself in works ranging from scratches on a cave wall to the pyramids. It shows how all art includes a desire to overcome the transient nature of existence. Now, "transient" means "lasting only for a short time", and this brings us to a consideration of the nature of time. Reanney sees time as the greatest barrier that Nature has erected between the structure of the human mind and reality, and death, as a marker in time, can only be transcended by a pure, universal consciousness that is freed of time. He believes that, to the highly evolved mind, which has filtered out ego noise, reality appears as a timeless continuum. Without entering into the issue of a pure consciousness, Reanney's treatment of time and its relation to the cycle of human life, both the short-term circadian cycles and the long-term one ending in death, is thought-provoking, but then, in the end it seems to miss what to me would be the proper conclusion: Rather than reality being a timeless continuum; that is, something existing unchanged in time, it is time that has no reality. Time is a concept we have invented as an arbitrary measure of change; the reality is change. In a universe where nothing changes, not a single elementary particle moves, there could be no concept of time. Of course, there would be no humans, either. The concept of time depends on the ability to perceive change, which is again contingent on memory. So, in a universe without organisms with memory there could not be any concept of time, either. We are so conditioned to thinking and speaking of time as an actual physical parameter; for example, we talk of something being a function of time. And when we say something happened later, we think in terms of elapsed time, rather than after something else had changed.

This is relevant to our investigation because the timeframes of the stages of evolution are very different. In the last century, the life span of humans has not increased dramatically when measured in normal time, but if we would use the rate of change of our environment as our 'time' scale, we would say that we are now living very much longer than a century ago. Without providing an exact quantitative definition of 'change', it is likely that more than 99 percent of all change to society in historical times have taken place since the introduction of the efficient steam engine in 1770s. We now have two related processes: changes to society from applications of technology and the associated issues arising from these applications, and the process of developing restrictions to resolve these issues, and in a qualitative manner, we can illustrate their relationship as shown in Fig. 2.2.

The vertical distance between the curves, ε, is a measure of the issues outstanding and needing to be resolved, and we see that despite the time required to resolve issues, Δ, is decreasing, it is not decreasing fast enough to prevent the number of outstanding issues, ε, to increase. In short, society's processes for responding to

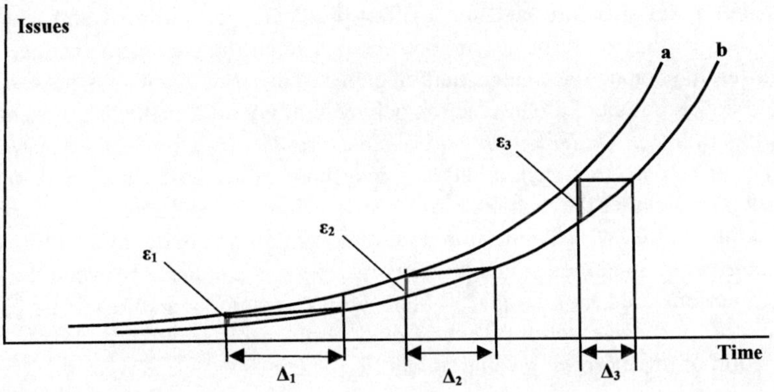

Fig. 2.2 The curve labelled **a** shows the issues that have arisen from the introduction of technology-based applications prior to any point in time, and the curve labelled **b** shows the issues that have been resolved

change are inadequate, and we seem to be facing a potential run-away situation. This problem was identified quite some time ago by William F. Ogburn in his book *Social Change* with respect to culture and original nature (Ogburn, 1922). He considered the adjustment that was required between different parts of culture (or society, in our current terminology); in particular; the adjustment of the non-material culture to changes is the material culture, where the latter included both the natural environment and industry and its products, and showed that there was generally a delay between the change and the required adjustment, which he called the hypothesis of *cultural lag*. He discussed the reasons for this lag, and in doing so, in many respects anticipated the hypothesis of Thomas Kuhn in *The Structure of Scientific Revolutions* (Kuhn, 1962). And in the Summary of Part IV, he considered an outcome which we shall also take up in the later part of this book:

> It is thinkable that the piling up of these cultural lags may reach such a point that they may be changed in a somewhat wholesale fashion. In such a case, the word revolution probably describes what happens. There may be other limiting factors to such a course of development, and our analysis is not sufficiently comprehensive and accurate to make definitive prediction. But certain trends at the present time seem unmistakable.

When comparing the two works of Ogburn and Kuhn, we see that there is a significant difference. Kuhn's concept of a paradigm is a belief system within a branch of science, and its revolutionary change as a result of mounting contradictory evidence takes place within this same environment; any consideration of its effect on the rest of society is secondary. Ogburn's investigation of social change includes this same effect, and identifies many of the same reasons for it, but the 'revolutionary' change takes place in another part of the social structure, particularly in government through such actions as new legislation or changes to the educational system. The main part of Ogburn's cultural lag is the additional lag resulting from this transition across the social structure, and most often predicated on political and economic factors; factors that are of minor importance to a scientific paradigm change.

A different perspective on the situation depicted in Fig. 2.2, and one that resonates with our development of the interaction between individuals, is the one presented by Habermas in his essay *Technology and Science as Ideology* (Habermas, 1970). He identifies an underlying tension between two types of actions: a purposive rational action, which includes the relationship between technology and nature, and a communicative interaction, which aims at mutual understanding and, by implication, of the establishment of an agreed social framework. There is a balance between these two types of actions, and he sees the problem with our time as a shift in this balance towards the purposive action. That is really only a different formulation of the explanation of the increasing value of ε in Fig. 2.2.

2.3.4 The Importance of Information Exchange

Societies, as groups of humans related through mutual benefit in the form of defence and food, existed long before 10,000 years ago; it would have been such bands of nomads that spread *homo sapiens* out over the world and made it the sole surviving species of the genus. And one factor in their success is likely to have been a superior ability to interact via speech, in addition to the interaction via physical means (body language, force). This was preceded by the development of the physical ability to produce an increasing number of sounds that could be associated with such sentiments as happiness, pain, anger, fear, etc., and the further development of sound-combinations to form words and then words into sentences—what we think of as a language. The timing of this development is somewhat uncertain, as speech, in contrast to symbols, leave no trace, but by 10,000 years ago languages were present in all groupings of humans in the world. And that is what makes it a suitable starting point for our study of the evolution of society, as from this point on, that evolution, which is basically the evolution of the interaction between society's members, took off at an exponential rate.

That evolution was driven, in large part, by the development of technology related to the reproduction and transmission of information, and some of the major milestones are shown in Table 2.2. Not only the amount of information exchanged per unit time, but also the spatial reach of the interaction increased, changing the relationship between the interaction and the structural features of society. For example, in the first and much of the second of the three major periods indicated in Table 2.2 (i.e., face-to-face interaction only), there appears to be a clear relationship between settlement size and density, as has been demonstrated by Roland Fletcher in his study of settlement, both ancient and more recent (Fletcher, 1995), and in Chap. 5 we shall show that this relationship is determined by the characteristics of the face-to-face interaction.

In the second of the major periods the reproduction and storage of information progressed greatly, but the transmission was still dependent on the transport of

Table 2.2 The evolution of information technology as a main driver of the evolution of the information exchange within societies

Start date	Information technology	Capability and impact
> 10000 BC	Spoken language (not technology dependent	Time and location restriction
3200 BC	Symbolic writing, clay tablets	Time and location restrictions overcome, but dependence on technology
2000 BC	Alphabet, writing with ink on substrate (papyrus, vellum, paper)	Improved productivity, proliferation of written languages
200	Printing, woodcut and etchings	Reproduction of drawings, limited benefit for text
1040	Movable type	Improved productivity for text
1450	Printing presses	Greatly improved productivity; books and newspapers
1844	Telegraph	Time/distance limitation eliminated
1876	Telephone	No user training required
1920	Radio	Mass distribution of sound signals (voice/instruments)
1927	Television	Second sense (sight) added
1980	Mobile telecommunications, fibre technology, IT	Great increase in service offerings and decrease in cost

matter (letters, books, drawings, etc.), even though that transport was also very much improved through applications of technology (roads and bridges, steam-powered ships and railways). However, these changes in information exchange resulted in larger societies, structured by the intensity of the interaction, and one possible depiction of this process is shown in Table 2.3.

However, if we want to think of the sequence of societies in Table 2.3 as a sequence of new species of the genus *society* arising as a result of man-made "mutations", as we suggested earlier, we have to recognise a principal difference between this process and the process of evolution of organisms. When a new species arises, as for example, when *homo sapiens* arose as a species of the genus *homo*, its members had nothing

Table 2.3 The progression of societies and their constituent elements over time

Elements	Society
Individuals	Family
Families	Clan
Clans	Fiefdom
Fiefdoms	Kingdom
Kingdoms	Nation
Nations	World society

further to do with any member of the family *hominidae* out of which they arose through mutations. But when, e.g., a fiefdom arises out of clans, it incorporates the clans as structural elements; they will change their functionality somewhat, but they retain their existence and distinctiveness. This is a process more akin to the development of single cells into multi-cellular organisms, with the increasing integration of the cells into the organism and their differentiation to form the increasingly complex structure of the organism is reflected in the integration of the individual into society and adaption to its increasingly complex structure. But just as the cell retained its importance as the building block of organisms, the same is true of the individual and of the sequence of structural components arising in the course of the evolution of society. Both families and nation-states retain their identity and importance as they adapt to that evolution. For example, in a nation the family is still an important structural component, but features that were originally characteristic of the family, such as education, have been ceded to the nation, or the state. That is, some of the *sovereignty* of the element is exchanged for membership in the society, as we shall come back to in Chap. 11. The idea of a structure-less society is utopian, but can, as we will do in the following, be employed as a useful abstraction for certain purposes.

The great leap in the information exchange came with the representation of information as electromagnetic signals, and this was the beginning of the third period of Table 2.2—the one we are presently in and which we often characterise as the modern period. As a result, the final stage in the structuring of society—the identification of the whole world as a structural unit—was initiated and is in ongoing development today.

The evolution of society is driven by human intelligence, both by its ability to identify and describe a desired change, and by its ability to realise this description as an actual change to society. There are many measures that characterise this evolution, such as the size of individual societies, the division of labour, the level of education and economic activity, but the one that occupies a particular position in the picture of a society is the use and development of technology. This is because of the strong positive feed-back effect of technology on the development of society and, indeed, on the smallest elements of society—the individuals—and it has led to what is a paradoxical situation. On the one hand, the behaviour of the individuals as stand-alone entities has very little relevance as the elements of a modern society, much as the study of a single ant under the microscope would give little indication of its purpose and behaviour as a member of an ant colony. Intrinsic human characteristics are so intertwined with technological artefacts and their operation that their separation, in terms of the behaviour of the individual in society, becomes artificial. But, on the other hand, the features and evolution of society are completely determined by exactly the intrinsic abilities and behaviour of the individual; society is what it is because we made it so. Technology has no mind or will of its own.

The solution to this conundrum lies in, on the one hand, developing a model of individual behavior—effectively a simple model of intelligence, which is the subject of Chap. 4—that reflects this intertwining by introducing the concept of the individual's *identity*—the things that matter to the individual, the things the individual is willing to make a sacrifice for—and on the other hand by recognising that this identity is a social product, formed by the ongoing and technology-mediated immersion in society through interactions with other individuals, which is introduced in Chap. 3 and then developed further in Chap. 5. The concept of absolute values, unrelated to the society in which they are to apply, is no more relevant than the above concept of the stand-alone individual.

2.3.5 Measures of Evolution

Evolution is often thought of as a gradual change, as "a gradual process of change and development", or more specifically for organisms as "the process by which the physical characteristics of types of creatures change over time, new types of creatures develop, and others disappear" (Cambridge Dictionary). The description as 'gradual' is a matter of the level of resolution at which the evolution is observed; at a high level of resolution the evolution appears as a myriad of very small steps, at a medium level of resolution it appears as continuous, and at a low level of resolution it appears as a few large steps. But in any case, to describe the evolution of something, we first must define the entity that is evolving, and then define those characteristics that we want to consider as measures of its evolution. For example, if we chose to consider the evolution of hominini, we might define the characteristic of interest as brain volume, and thus consider two hominini as evolutionary different if and only if they have different brain volumes. Using our nomenclature from Sect. 1.3, this is a particular *view* of the evolution of hominini, hiding all of their other characteristics, and the *significance* of this view of evolution is then a matter for discussion.

A common approach to defining the entity that is evolving is to define it in terms of its physical components, as these are the ones most immediately apparent to us, and then select the characteristic(s) of interest with reference to these components. In the case of the societies we are interested in, which, with a few explicit exceptions, we shall take to be understood either as "nation" or as "world", we chose the definition in term of three components already introduced in Sect. 1.3.1:

- the individual members—the *individuals*.
- their *actions*; and
- their *belongings*.

The belongings include artefacts, such as tools and utensils, equipment and devices, clothing, artworks, means of transportation, and dwellings; domesticated animals; physical infrastructure, such as arable land, waterways, railways, roads and bridges; and, most importantly, the repositories of knowledge, contained in libraries and other data storage media. They are identified as belongings because they belong to a particular society; they form part of the characterisation of that society.

In the following, we will sometimes distinguish two types of actions—private actions and public actions. Private actions are actions that do not require one or more other individuals for their performance. Obvious examples are such actions as sleeping or brushing teeth, but more subtle examples would be writing a book or doing calculations; it is only publishing the book and making the results of the calculations available for use that would be public actions. The public actions are what we shall call *interactions*; they involve or affect other persons. The distinction between the two types of actions will become sharper as we progress.

The concept of an *individual*, as used in this essay, has a dual definition, much as the dual description of a photon as a particle or as a wave in quantum mechanics (but recall our previous comment on this analogy). On the one hand we define an individual as an average member of society; that is, as a human whose characteristics, such as intelligence, physical capabilities, and life cycle, take on values that are the average of these characteristics over all members of society (although we will, later in this essay, limit this to members over a certain age). This is similar to the use of *per capita* parameter values in economics, and to this extent, the individuals are all identical. But the information contained within them is different for each one, and we are able to speak about the value of a characteristic of this information of a particular individual. For example, as we shall see later, the interaction between two individuals will depend on how much they have in common, which implies that there are things they do not have in common, so that, to this extent, they are not identical. But when we then take the average over all pairs, what two individuals have in common is independent of which two individuals are involved in the interaction. We might express this by saying that behind the individual sits a person who is obscured because of the averaging process.

To approach the problem of defining measures of the evolution of society, we must answer a fundamental question: What do we understand by evolution in the context of society? Or, in terms of our identification of society as a genus: How do we distinguish species of this genus; what makes them different? Consider a nation as an example of a society. If the population increases by 10%, or the GDP increases by 15%, we would not say that this nation has evolved into a different nation—it has only grown. Similarly, if it has a greater quantity of belongings it has grown, but it is not necessarily a different species. Of course, we could ask: Is the concept of a nation really a species of the genus society, in the sense that all nations are examples of the same species and equivalent as far as evolution is concerned? Probably not, but more of that in Chap. 10, which is dedicated to a more in-depth examination of measures of the evolution of society.

Also, as the human, i.e., the species *homo sapiens*, has not developed significantly over the last 10,000 years, that is not where we should be looking for measures of evolution of society.

The clue to finding an answer to the question lies in recognising that the primary driver of the evolution of human society is the interaction between its members, and therefore the primary measures of social evolution must relate to this interaction—to its strength, timeliness, and quality. The evolution of society is not primarily a matter of economic growth, as in GDP or in the level of technology, or even of changes to such parameters as income and wealth inequality; these are measures of secondary effects. As such, if suitably defined, they may be valuable *indicators* of changes to the primary measures; an example of such a relationship is that between interaction and technology given earlier in this chapter. The interaction between the members of society is not something that can be characterised by a single variable, and even a consideration of its basic features will require all of Chap. 5. And the resulting understanding of how to measure evolution of society may turn out to be different to what we expect at this point. But in the next chapter we first take a step back (or up, in terms of abstraction) by recognising that society can be viewed under two complementary perspectives: in terms of actions and processes, and in terms of information. These two views of society and its evolution can be developed separately up to a certain level of detail, but it is in understanding the interaction between them that we develop a deeper understanding of the dynamics of society.

As a subject of study, the evolution of society is located generally within the social sciences, as the branch of science devoted to the study of human societies and encompassing, in addition to economics, sociology, anthropology, archaeology, linguistics, political science, and psychology, and the evolution of society may be measured by any combination of features from these disciplines. And if the concept of society is extended beyond human society to the interaction between members of any species, there is a large body of work concerned with social characteristics that can be employed in comparative studies over a broad range of taxa. A prominent characteristic is social complexity, and a relatively recent article by Bergman and Beehner (2015) provides a review of the literature and promotes a particular definition of social complexity. They propose that "social complexity should be measured as the number of differentiated relationships that individuals have". To investigate the extent to which this definition can be usefully applied to society and, in particular, to our model of society as an information-processing system, requires a discussion of differentiated relationships and their correlation with social cognition, as emphasized by Bergman and Beehner, and this discussion is taken up in Chap. 10.

Another approach to characterising the evolution of society is the recent work undertaken by a large group of scientists and reported in the Proceedings of the National Academy of Sciences under the title *Quantitative historical analysis uncovers a single dimension of complexity that structures global variation in human social organisations* (Turchin et al., 2017). Data was collected from more than four hundred identifiable polities on 51 variables, and with a time resolution of 100 years, and recorded in a public database, Seshat. These variables were aggregated into nine "complexity characteristics" (CCs) as follows:

CC1: polity population
CC2: extent of polity territory
CC3: size of the largest urban centre
CC4: hierarchical measures (administration levels and settlement types)
CC5: measures of professions, bureaucracy, and examination systems
CC6: infrastructure and facilities involved in the functioning of the polity
CC7: information systems (writing, record-keeping)
CC8: literature on specialised topics
CC9: monetary instruments, economic complexity.

After undergoing a selection process to improve its significance, the data was subjected to a Principal Component Analysis (PCA), with the result that one of the

PCs is by far the dominant one, providing a single measure of polity complexity. This allows informative graphic displays of the temporal evolution of single (or groups of) polities, as well as comparisons between different polities (or groups of polities).

Impressive as this work is, the measure of social evolution presented represents a particular point of view—what might be called a historicist point of view. Not only the time period considered—from the Agricultural to the Industrial revolution (more or less inclusive)—emphasizes society's more distant past, but the variables relate to things, to artefacts, to the physical evidence that archaeologists and anthropologists collect. If we instead relate complexity of society to its level of understanding of its environment, of Nature, then the increase in the last 200 years completely dwarfs that of the previous 10,000 years. And the same is true if we look at society as a system of interacting individuals and measure the complexity as the intensity of the interaction. The preoccupation of parts of the academic community with the more distant past, and the veneration of figures such as Plato and Aristotle, is symptomatic of an escape from the bewildering complexity and dynamics of today's society into the comforting certainty and stability of the past, and of a preoccupation with the acquisition of knowledge rather than with the application of it to solving current problems in society.

However, having said this, the view of social complexity as a measure of the evolution of society could be a useful approach to examining where the world society is currently heading, but it will require a means of characterising changes to social complexity in more quantitative terms. A main feature of modern society is its preoccupation with the economy and its various elements, such as trade, the share market, employment, real estate, and taxation; it is the central issue in every election campaign. And it is correspondingly well characterised and measured, with data ranging from global GDP to expenditure on education in Burkina Faso, and from income and wealth distribution to details of household expenditure. In the same way as various measurements of the human organism, such as temperature, heart rate, blood pressure, and analyses of blood and urine samples are used to determine the state of the organism—its health, we might be able to use all the economic data to determine the health of society and how it is developing. This idea is also taken up in Chap. 10.

Chapter 3
The Two Views

3.1 Choice of Views

As noted in Chap. 1, when we want to apply the system methodology to developing an understanding of society, the first step is to identify which aspects of society we are interested in. This is analogous to viewing an object; when we choose a particular perspective, we can see the features of the object facing us clearly, but the features on the far side are hidden to us, and features on the sides are only partly visible. And if the object has many intricate features, we may have to focus on them separately. The complete description of the object requires us to combine all these perspectives, and we may employ advanced 3D graphic displays to improve our understanding of what the object looks like.

In the case of society, our choice of views arises from the realisation that the actions performed by the members of society are based on their physical circumstances and on the information available to them at the time, as already noted in Sect. 1.3.3. Leaving the physical circumstances aside for the moment, the information available to a member of society is a combination of information stored in memory from previous experience and information received through the senses at the time, and the processing of the received information depends on the stored information and the member's processing power. The latter may vary somewhat from person to person, just as other physical characteristics do, but within the degree of accuracy of our high-level approach, we shall assume that the processing power is the same for all persons. It then follows that the result of the processing depends on the stored information only; that is, when two persons arrive at different interpretations of the same received information and take different *adaptive actions*, it is due to differences in their stored information, and not to any inherent differences in processing power.

We now have two complementary views of society—in terms of information and in terms of actions and processes. In accordance with the nomenclature introduced in Sect. 1.3.2, these two views are secondary views, as they both refer to the primary view of society in terms of individuals and interactions. In the *information view*,

E. W. Aslaksen, *The Evolution of Society*,
https://doi.org/10.1007/978-3-031-23054-7_3

the individuals are described in terms of their ability to process and store information—*homo cogitans*—and the interactions are exchanges of information between individuals; in the *action view*, the individuals are producers of artefacts—*homo faber*—and the interactions are in the form of exchanges of goods or their representation as money. These two views are closely coupled through their horizontal relationship, but we need to realise that they are also very different in several respects. One, the description of society within the action view is in terms of readily observable and measurable quantities, such as money, hours of work, attained education level, and so on, whereas the description in terms of information and thought processes is much less developed. This reflects the difference between thought and action—the information view is more abstract and general in nature. You can see what a person is doing; it is much more difficult to determine what the person is thinking.

Two, the action view has matured into a number of different aspects (or sub-views) of society that we can readily observe, such as an economic view, a political view, a technology view, an education view, and so on. The action view is to some extent an overarching view connecting these secondary views. Whereas the information view, as it is presented in this book, is relatively novel and has yet to attract significant attention from the academic community.

Three, in the context of this book, the roles of the two views are quite different. The focus of the book is to develop the information view to gain additional understanding of society and its evolution; the action view is only a means of relating the existing knowledge of the physical world to the information view; it is not something we have to develop. And fourthly, as we develop in the next section, the information view contains the primary driver of society's evolution—the operation of the collective intelligence—whereas the action view only describes the conditions for and the results of this operation.

The two views form a framework for the development of our understanding of society and its evolution throughout this book, but we will sometimes broaden the development by considering the two views to be situated in two domains—the information view in the *information domain*, and the action view in the *physical domain*. That is, we can speak of an element or an item within a domain without necessarily linking it to the associated view of society. For example, a particular item or type of information, such as the information in the social media, exists quite independently of the information view of society, and we may consider it and study it as such. Its treatment in the literature does not involve our view; obviously, as that view is our construct.

3.2 Structure of the Two Views

Referring to the discussion in Sect. 2.4 about what constitutes evolution of society, and to some extent pre-empting the later consideration, in Chap. 10, of measures of evolution, it is very useful for the further development of the two views to partition them into two parts, according to whether they relate mostly to the maintenance and growth of a given society or mostly to the transformation of one society into a different one—i.e., a new species, even though this is by no means a sharp distinction. Starting with the information view, the information to which an individual is exposed can be defined as belonging to one of three classes:

Class 1 The inputs that we either are not consciously aware of or do not actively engage with; perceptions of our environment when we are not focusing on anything specific; they are the inputs we associate with being awake. While some of these inputs are stored, and we might recall them later if prompted to do so, they do not result in any further processing, but they do take up some of the person's processing capacity, in the form of distractions.

Class 2 The factual and uncontroversial inputs of everyday life, mostly connected with family life, work, education, and entertainment; inputs that, when evaluated against the contents of memory, result in the updating of memory without creating any conflicts or inconsistencies, and in predictable actions.

Class 3 Inputs that relate to "the things that matter" (Aslaksen, 2018) or, with reference to (Axelrod, 1997), "culture", which includes beliefs, attitudes, behaviour, art, and social norms. The evaluation of these inputs may result in a more or less direct and immediate adaptive (i.e., behaviour-changing) action, such as participating in a protest march, avoiding eating meat, writing a letter to the newspaper, etc., but may also raise conflicts with existing information that then need to be resolved.

In this view and its associated models (Chaps. 4 and 5), "information" is used in the widest sense, as a concept that includes data, knowledge, understanding, belief, instinct, hope, wish, feeling, and the like. An important perspective on the distinction between these three classes of the information flow is provided by the concept of *attention*—focused mental engagement with a particular information item. Class 1 does not require attention, whereas classes 2 and 3 do. Because of this, the concept of attention becomes a key factor in the discussion in Chap. 5, but for the time being, we will only consider the subdivision of information into Class 2 and Class 3. It will sometimes also be useful to recognise that the information can be classified along a different "dimension", according to whether it pertains to a process, i.e., to how to do something, such as extracting the square root of a number (which we might call an algorithm), or to a thing, such as "the Thirty Years War started in 1618". Under the perspective of education, the former requires practice, whereas the latter does not.

The subdivision of information leads to a subdivision of the actions and processes resulting from the information. The information in class 2 results in ongoing and new applications of technology and associated products, as well as new technology and new information in class 2, in what we shall call the *economic cycle*, while the information in class 3 results in new beliefs and associated adaptive actions, in what we shall call the *transformation cycle*. The transformation cycle is the one evolving society, with the economic cycle providing the environment in which it takes place, and they are called cycles because of the feed-back loop existing between information and action.

A similar division is the distinction between material transformations and transformations of ideological forms expressed by Karl Marx in the Preface to *A Contribution to the Critique of Political Economy* (Marx, 1859):

> The totality of these relations of production constitutes the economic structure of society, the real foundation, on which arises a legal and political superstructure and to which correspond definite forms of social consciousness. The mode of production of material life conditions the general process of social, political and intellectual life. It is not the consciousness of men that determines their existence, but their social existence that determines their consciousness. At a certain stage of development, the material productive forces of society come into conflict with the existing relations of production or -- this merely expresses the same thing in legal terms -- with the property relations within the framework of which they have operated hitherto. From forms of development of the productive forces these relations turn into their fetters. Then begins an era of social revolution. The changes in the economic foundation lead sooner or later to the transformation of the whole immense superstructure. In studying such transformations it is always necessary to distinguish between the material transformation of the economic conditions of production, which can be determined with the precision of natural science, and the legal, political, religious, artistic or philosophic -- in short, ideological forms in which men become conscious of this conflict and fight it out.

Marx considers the *social conditions* as primary, whereas the view presented in this book is that the *consciousness* drives the changes to society. In his essay *Probleme einer Soziologie des Wissens*, Scheler (1924) distinguished between *cultural data* (ideas, beliefs, values) and *real factors* (data about the structure of society), corresponding approximately to our class 3 and class 2. In Karl Mannheim's *Ideologie und Utopie* (Mannheim, 1929) a distinction is made between *exact sciences* and *cultural sciences*, and in Herbert Marcuse's *One-Dimensional Man* (Marcuse, 1964) the second dimension is essentially the processing of class 3 information. The distinction is also expressed in various works on education, such as the *specialised* and *general* components in Émile Durkheim's *Education and Sociology* (Durkheim, 1970), and in *What is an educational process*, by Peters (1967), his *principles* correspond approximately to our class 3 information, as will be discussed in more detail in Chap. 9.

This subdivision of both information and actions into two groups is perhaps the most significant and, potentially, the most controversial aspect of our approach to understanding society, as many people would say it is impossible to separate the two; any action will have some influence on the evolution of society (e.g., the butterfly effect). Nevertheless, this division will prove to be very useful; we only need to be careful to recognise its limitations. What it says is that most of our factual knowledge, such as the value of π, the number of planets of the Sun, names of rivers and cities, how to drive a car, and how to make steel, while very important for sustaining society, does not determine the direction of change of society. As an analogy, if one person shoots another person, the gun, as the item of technology, and thus defined by class 2 information, is an enabler of the action, but the action is determined by the person's processing of the relevant available information (which would be mainly class 3). One could take the view that if the person did not have a gun, the event would not have taken place, and that therefore the gun, i.e., the technology, is the determining factor. That is not our view, nor is it that of our legal system; it is the person that is punished,

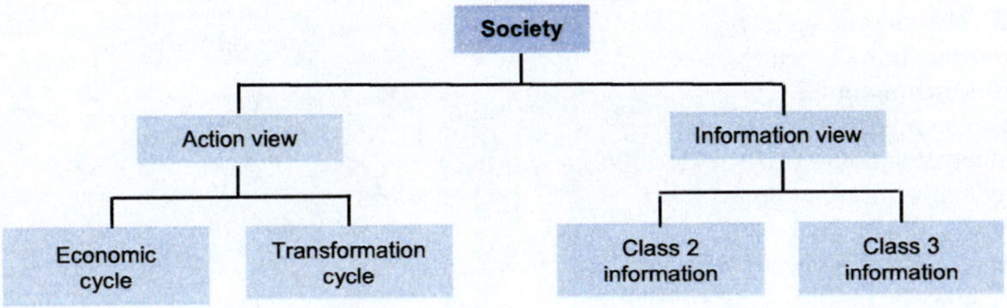

Fig. 3.1 The high-level structure of the view of society presented in this book

not the gun. That is not to say, of course, that one should not be concerned about the availability of the gun, as is true of many applications of technology, but, again, any restrictions or procedures are the result of human decisions, i.e., of information processing, and mainly the processing of information in class 3.

Irrespective of the class of information, we introduce a common quantitative measure through the concept of an *information item*. Information items are the kind of statements and messages we would use in conversation with other persons, such as "I see we are expecting rain tomorrow", "the value of π is 3.1416", "I believe the world was created five thousand years ago", "you should not smoke", or an instruction about how to do something; and the fact that they vary greatly in complexity, potential impact, truth value, etc. is not relevant to our use of the concept. It is an average or a fiction, just as the average person is a fiction, and we will not need any knowledge of its nature, as we will only use it as a relative measure of quantity. In the cognitive and social sciences, information items are often referred to as *cognitions* or *mental states*. The term "information item" was chosen here for two reasons: One, in order to disassociate it from any further level of detail that may be attached to the use of the other two terms, and two, in order to allude to the association of our model with information technology and the brain as a computer.

At this point, the structure of our view of society is as shown in Fig. 3.1. However, it is again emphasized that, although this structure looks very symmetric in the two views, they are quite different, and are treated quite differently in this book.

3.3 The Information View

3.3.1 The Main Components

In the information view, the state of society is described in terms of a *body of information*, and the evolution of society in terms of the changes to this body of information resulting from society's ability to generate and process information. The generation and processing of information is performed by a distributed processing system consisting of the individual members of society interacting with each other in the form of *information exchanges*, and with their environment in the form of *discovery*,

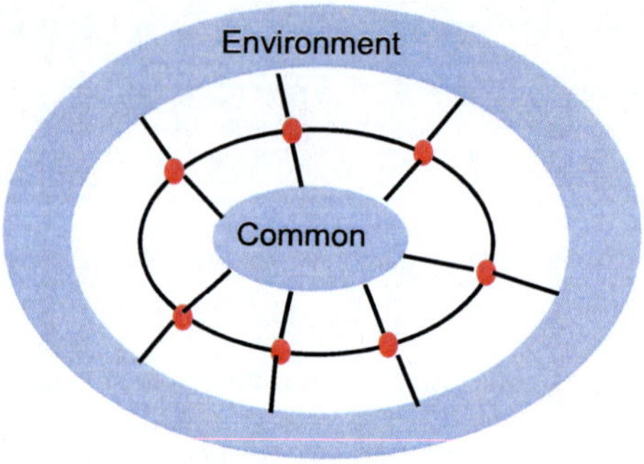

Fig. 3.2 The view of society as an information-processing system, consisting of the individual members and their interactions with each other, with a common information repository, and with their environment. The simple network structure, with nearest-neighbour interactions, is for illustration only

both increasingly supported or mediated by applications of technology. The body of information is contained in the individual members and in a common repository, resulting in a view of society as illustrated in Fig. 3.2.

The process implied by Fig. 3.2 operates on all three classes of information. However, as we are primarily interested in the evolution of society, we shall be focusing on class 3 information, and with that restriction, the process is the operation of what we shall identify as *the collective intelligence*; a consensus-building process that determines society's belief system. It is a process of reflexive interaction and *evaluation*; evaluation of new information on the basis of existing information, leading to either rejection or acceptance, with the latter resulting in an updating of the stored information. As such, this collective evaluation process can be seen as *society's immune system*.

A superficial consideration of collective intelligence might perceive a contradiction, in that, on the one hand, the process supresses fluctuations and seems to lead to a consensus and what one might think would be an optimised steady state of society. Whereas, on the other hand, it is exactly fluctuations, in the form of ideas and actions, that are essential for the process to function. This contradiction only arises if we confuse collective intelligence with a means of achieving conformity and fail to understand the function of collective intelligence in controlling the evolution of society as an ongoing process. What guides this control process of assessment and adaptive action; is it random or is it goal-oriented, in which case, what is the goal? In Sect. 2.1 we put forward the belief (and it is a belief, in the sense that can only be made plausible, but not proven) that the goal has always been the same: *survival*. It is only that what survives, and our strategy for achieving it, have developed and become very much more complex and sophisticated, in line with the increasing complexity of society. One example of this, which is also detailed in Richerson and Boyd (2004, 169–187) as "the demographic transition", is the reduction in fertility rate in affluent societies. Survival is no longer seen in the quantity of offspring, but in their education and professional standing, and the cost of attaining that leads to a reduction in the number of children.

Society does not evolve towards a goal—a fixed state, some future ideal society, a heavenly kingdom—as much as some would like this to be the case. And when we speak of progress, it must be in the sense of change, not in the sense of getting closer to a goal. What we can observe is the evolution of society over historical times, and with whatever measure we use to characterise the state of society, say, **S**, we can determine d**S**/dt. In terms of the wording of the contradiction, the consensus is not about any end state, but about how to ensure the stability of the evolution of society over the next time increment.

But that does not allow us to determine the future course of **S**(t), except perhaps over a short time interval. Rather than being concerned about the state of society and its evolution, we should focus on what drives the evolution. Discounting any divine or supernatural force, the driving force can only be found in us. It is our intelligence, the ability to perceive our environment and to take goal-oriented adaptive action, and the goal is the survival of the evolution process itself.

With this, we have arrived at what is an underlying premise of the present work: The evolution of society is good to the extent that it is determined by the exercise of the collective intelligence. It is, essentially, an affirmation of our trust in ourselves, in the human species, and with this, The Good Society becomes the society that provides the conditions for the exercise of the collective intelligence.

The concept of the collective intelligence will feature prominently throughout the whole development of the evolution of society documented in this book, and it is important that we have a clear understanding of its nature and the role it plays in the evolution of society. The set **S** of variables defining the state of society at a given point in time was introduced above; the choice and number of variables in this set will depend on the level of detail and on what aspects of society are considered important in the particular context. And while, for a given society, **S** will change with time, so that we are tempted to write **S**(t), this is not what we normally associate with this notation; it is not a function. If the present time is $t = t_0$, then $\mathbf{S}(t < t_0)$ is whatever the values of the variables were at time t, and $\mathbf{S}(t > t_0)$ is whatever the values will be at time t, but which cannot be *determined* now. We may *predict* what those values will be, but with a confidence that decreases with increasing value of $t - t_0$.

Correspondingly, if we write d**S**/dt, this is not a function of time that, when integrated, yields **S** as a function of time; it is simply a notation for the effect of the process that changes **S**. This process is the transformation process introduced in Sect. 2.3, and the effect has a direction and a magnitude. In our view of society, both are determined by the operation of the collective intelligence and, as we just stated, this is a consensus-building process that dampens the short-term fluctuations due to extreme minority beliefs. But this still does not answer the question: Is there a "best" long-term direction of the evolution of society? It is the central belief underlying the work presented in this book that the best direction, in the sense of greatest probability of survival, will result from the unimpeded operation of the collective intelligence. That is, the collective intelligence dampens the short-term fluctuations *and* determines the long-term direction of society's evolution, both to the extent that the operating conditions approach their ideal values.

The knowledge that guides the long-term direction is embedded in our brains; it is a feature of our brains, just as we have two arms and two legs, and as such it must be present in our DNA. It evolved in parallel with the evolution of the physical complexity of organisms, and where it failed to keep up with the latter, the resulting species was doomed to extinction. A single cell in a liquid environment can detect the presence of a harmful substance and move in the direction of decreasing concentration. An ant can detect a moving object, such as a finger, as a threat and move away from it. An animal can detect danger through processing various sensory signals and take appropriate action. A human can also detect physical danger, but in addition can identify ideas that threaten society, because now survival is more than just the survival of the individual, it is also the survival of society. *The embedded knowledge includes the knowledge that humanity's future is as a society.*

With our focus on the evolution of society over the last 10,000 years, there has been little or no genetic change, i.e., change to the DNA sequence. But there can well have been heritable phenotype changes that affect gene activity and expression—*epigenetic* changes—and it is tempting to speculate that such changes have taken place—and are taking place—to adapt our behaviour to the demands of an increasingly complex society. The question then arises as to the speed of this adaption, or, conversely, to the time constant involved in the change process and its relation to the speed of the evolution of society. The relationship between change and adaptation was illustrated in Fig. 2.2. The time constant must be a decreasing function of the interaction strength, and while the form of this function is not known, it again points us to the essential role of the interaction and its quality. Given the importance of education to the interaction, it may well be that this time constant is related to the one involved in the intergenerational cycle, as will be discussed in Sect. 9.2.

As an aside, if this knowledge is contained in our DNA, it must be located somewhere in the genome. And as the brain develops in the embryo, this knowledge must be transferred into a configuration of neurons. Understanding this process, as well as the details of how information is represented in the brain, is still a work in progress, but an early and readable account was provided the book *Genome*, by Ridley (1999). In Chap. 7 of that book, he describes investigations, based mainly in linguistics, that would locate this information on chromosome 7. If that is true or not, the idea that some basic aspects of behaviour are part of our inherited nature is central to our argument for the significance of the collective intelligence.

Charles Darwin, in *The Descent of Man*, argued strongly for the evolution of the social and moral "instincts" as part of the evolution of species up to and including humans. In the case of animals, he gave the following argument, with survival as the underlying selection criterion:

> It has often been assumed that animals were in the first place rendered social, and that they feel as a consequence uncomfortable when separated from each other, and comfortable whilst together; but it is a more probable view that these sensations were first developed, in order that those animals which would profit by living in society, should be induced to live together, in the same manner as the sense of hunger and the pleasure of eating were, no doubt, first acquired in order to induce animals to eat. The feeling of pleasure from society is probably an extension of the parental or filial affections, since the social instinct seems to be developed by the young remaining for a long time with their parents; and this extension may be attributed in part to habit, but chiefly to natural selection. With those animals which were benefited by living in close association, the individuals which took the greatest pleasure in society would best escape various dangers, whilst those that cared least for their comrades,

> and lived solitary, would perish in greater numbers. With respect to the origin of the parental and filial affections, which apparently lie at the base of the social instincts, we know not the steps by which they have been gained; but we may infer that it has been to a large extent through natural selection. (Charles Darwin, *The Descent of Man*, Chap. 4)

A realisation of this basic knowledge, or "instincts", has surfaced in various places and in various forms, and at least as long ago as when the Chinese philosopher Mengzi identified it as "sprouts" (Ivanhoe & Van Norden, 2001, p. 112). It was taken up again by Lyndon Storey in his book *Humanity and Sovereignty*, where he identifies this knowledge as "potentials" (Storey, 2006). The point that both of these authors, as well as many others, make, is that this basic embedded knowledge, whatever it is called, is indeed potential; for it to determine our behaviour it needs to be practised, and this requires a sustained effort. In our view, this is driven by the operation of the collective intelligence, which means that no individual or segment of society has a greater influence than any other on this process. It is the ultimate manifestation of equality and of the true meaning of 'democracy'. This concept of an ideal state of society defined by the operating condition of the collective intelligence can be related to Mannheim's concept of a utopia (Mannheim, 1929) as a future state that involves a fundamental break with the ideologies associated with the existing social order. We shall return to Mannheim's work when we consider ideology in more detail in Sect. 11.4, but we should also note Mannheim's view of democratization—the increasing involvement of the whole population in the political process—as an important aspect of the way forward.

Thus, in the first instance, the measure of the operation of the transformation process is a measure of how far removed it is from the ideal state, and the concept of the direction of the change reduces to either towards or away from the ideal state, masking any details of the resulting change in **S**. The changes are simply registered as *fluctuations* in the state of society; recent examples of such fluctuations are the two world wars. Fluctuations play a crucial role in the cybernetic perspective of control or self-regulation of adaptive systems, see e.g. (Buckley, 1968, p. 494).

Possible choices and measures of **S** are defined and discussed in Chap. 10, the operating conditions of the collective intelligence are the subject of Chap. 6, and the completion of the transformation process through adaptive action is the subject of Chap. 7.

The concept of collective intelligence has, in various contexts and formulations, been around for a long time, going back at least to the ancient Greeks. It finds expression in Plato's Protagoras and the myth of how Zeus ordered political wisdom to be evenly distributed among men (Plato, *Dialogues*); in Aristotle's Politics, where in the opening of Chap. XI of Book 3 he says:

> The principle that the multitude ought to be supreme rather than the few best is one that is maintained, and, though not free from difficulty, yet seems to contain an element of truth. For the many, of whom each individual is but an ordinary person, when they meet together may very likely be better than the few good, if regarded not individually but collectively, just as a feast to which many contribute is better than a dinner provided out of a single purse. (Aristotle, *Politics*)

And, again, in Chap. XV, "a multitude is a better judge of many things than any individual."

Cicero ascribes the following to Cato The Elder:

> Cato used to say that the government of Rome was superior to that of other states; because in them the great men were mere isolated individuals, who regulated their constitutions according to their own *ipse dixits,* their own laws, and their own ordinances. Such was Minos in Crete, Lycurgus in Sparta; and in Athens, which experienced so many revolutions, first Theseus, then Draco, then Solon, then Clisthenes, afterwards many others; and lastly, to support the Athenian state in its exhaustion and prostration, that great and wise man, Demetrius Phalereus.
>
> Our Roman constitution, on the contrary, did not spring from the genius of an individual, but of many; and it was established, not in the lifetime of a man, but in the course of ages and centuries. For (added he) there never yet existed a genius so vast and comprehensive as to allow nothing to escape its attention, and all the geniuses in the world united in a single mind, could never, within the limits of a single life, exert a foresight sufficiently extensive to embrace and harmonize all, without the aid of experience and practice. (Cicero, *De re publica*)

The concept of a collective intelligence is the foundation of our normative commitment to democracy as the preferred form of government, and it has also found its expression in a number of fields and related concepts, as described in Malone and Bernstein (2015). The Introduction to that book contains both definitions of collective intelligence and many references to the relevant literature. In recent times, the focus has been on collective intelligence as a factor in the performance of groups (Woolley et al., 2010), and in Pentland (2014) it is explored in the context of what the author calls Social Physics, an approach to understanding social processes and dynamics using Big Data. (In contrast with the reliance on Big Data in Social Physics, our approach could be said to rely on No Data.)

The relevance of the concept to politics and to the evolution of society is indisputable—after all, it is what got us from the cave to where we are today—but it is not so well supported by empirical research (for the obvious reasons of the magnitude and complexity of the task). In this regard, we can note that the book (Malone & Bernstein, 2015), with the title of Handbook of Collective Intelligence, does not actually consider the presence of collective intelligence in the political process, or to what extent it supports democracy. An author that has been much concerned with collective intelligence, but under the name of deliberation in the public sphere, is Jürgen Habermas, and in Habermas (2006) he quotes the definition of the deliberative paradigm as follows:

> The deliberative paradigm offers as its main empirical point of reference a democratic process, which is supposed to generate legitimacy through a procedure of opinion and will formation that grants (a) publicity and transparency for the deliberative process, (b) inclusion and equal opportunity for participation, and (c) a justified presumption for reasonable outcomes (mainly in view of the impact of arguments on rational changes in preference). (Bohman, 1996; Bohman & Rehg, 1997)

The relationship between collective intelligence and democracy is treated in numerous articles and constitutes part of the body of knowledge dedicated to the various forms of political systems and assessments of their relative efficiency; many references can be found in Farelly and Shalizi (2012), and we shall return to this in Chap. 7. That is not an issue of direct relevance to our considerations here, for the following reason: I consider society as a system made up of individuals and the interactions between them, with the emergent behaviour determined by these two sets. In particular, the structure of the interactions will have a major impact on the evolution of the system, and without entering into any arguments about this, the work presented here takes as given, i.e., as an axiom, that the free, unrestricted exchange of information yields the best result, and that any restriction or manipulation of the information exchange will result in a detrimental change to the process of evolution. However, a recent book by Hélène Landemore, *Democratic Reason: Politics, Collective Intelligence and the Rule of the Many* (Landemore, 2017), deserves special mention. Leaving aside such justifications as justice, equality, fairness, and consent, she presents a strong argument for democracy on epistemic, or outcome-based, grounds. The history and current status of both explicit and implicit objections to democracy are given careful consideration and countered in a convincing fashion, and her treatment of the two democratic processes—deliberation and agglomeration, or majority rule—will be referenced again in Chap. 7, where we return to this under the heading of participatory and deliberative democracy in our discussion of the role of political parties.

The concept of *the individual*, as shown in Fig. 3.2 and as used throughout this view of society, has a dual definition, as already noted in Sect. 2.4. On the one hand, the information contained within each individual is particular to the individual, who, when distinguished in this way, becomes a person, whereas as containers and processors of information, individuals are indistinguishable.

Two cases where we need to take account of the individuals as persons present themselves immediately. The first one is reproduction, which involves a particular interaction between male and female individuals based on biology rather than on any exchange of information, and therefore does not seem to fit into our system description. However, we can sidestep this issue by replacing both reproduction and death by a renewal process within the collection of individuals making up society (or a group within society); it becomes part of what we might consider to be *the metabolism of society*. That introduces the second issue—the *lifecycle of the individual*. In our view of society as a system, the individuals are time-independent, and the evolution is an evolution of energy, from free to bound; the individuals are simply the carriers of the process. In the same vein, we can now consider the evolution of society to be an evolution of the knowledge and the interactions, with the physical beings constituting a background on which this evolution takes place. Under this perspective, the early education and integration of children into society becomes part of the *metabolic process*; the process of maintaining the living substance of society. It is somewhat similar to describing the behaviour of a person without considering the fact that, within that person, cells are dying and being created and integrated into the body at a great rate.

The lifecycle of the individual introduces an apparent inconsistency in the concept of the individual as used in this work. On the one hand, the individual is a person who goes through a lifecycle, from birth to death, and in some sections of the work we take this explicitly into account when we consider a particular information process, such as education in Chap. 9. On the other hand, in general and throughout the greater part of this work, the individual is a time-independent processor of information defined by the model developed in the next chapter. This dual meaning of the concept should hopefully not cause any confusion; firstly, because of the preponderance of the latter use, and secondly, because the meaning is evident from the context in which it is used.

If we limit the time period for which our view is intended to the last 10,000 years, this individual has remained unchanged, at least to the level of accuracy with which we are concerned. Consequently, the parameters describing society within the information view, and the changes to these parameters, are completely determined by the interactions shown in Fig. 3.2. The choice of parameters is influenced by the particular model, developed in the next chapter, and the changes to these parameters are the subject of Chaps. 6 and 9. However, already at this point; without any details of the model, but simply as a consequence of our view of society in terms of information, it follows that whatever parameters we develop, they must be in terms of information only. And from our subdivision of information into three classes, an immediate parameter, although not well defined at this point, is the relative importance of these three classes of information.

3.3.2 *Identity*

In Sect. 3.1 we identified three classes of information, and with them two parts of the knowledge base (the storage of class 1 information is, to the extent that it occurs, irrelevant). The smaller part, the class 3 information, which is the part we are mainly interested in, contains the knowledge required to evaluate those inputs that are not only not repetitive, but where some or all of the knowledge entering into the evaluation of the inputs is of a probabilistic nature. That is, it is in the form of more-or-less well-founded *beliefs* (which may include *opinions* and *conclusions*), and we shall further restrict this to beliefs about things that matter to the individual, the things the individual is willing to make a sacrifice for. What an individual considers an item that matters is not determined only by the generally accepted characteristics of this item, such as family relationship, a written constitution, rules and laws, the dogmas of the Church, but by the individual's perception of the item, by additional features that are important to that individual. The relationship with each of these things that matter is peculiar to each individual, and it is something the individual establishes as it goes through its life cycle. This web of relationships is what we shall call the individual's *identity*, and constitutes a subset of the knowledge base, which in all of the following will be denoted by Θ. This identity determines, in effect, what the goal of the individual's "goal-oriented behaviour" is; it is the realisation of those beliefs, making them come true, and thereby it also defines the meaning of "survival" for that individual. Note that "truth" does not enter into this definition of identity, and we may also note that the concept of identity is close to (but somewhat more limited than) what is contained in the use of "ideas" as a catch-all for concepts, beliefs, and values in the paper (Simpkins et al., 2010), to which we shall be referring again later.

The size of the identity, to be defined more precisely below, is usually an increasing function of time (age of the individual) up to a certain time, and then fluctuating and eventually declining. The increase is due to the interaction of the individual with its environment; at first parents and siblings, then teachers and peers, then spouse and children, people in the workplace and people in various organisations (sports, political, etc.), and so on. But the decline is not due to interaction with other people; it is due to the deterioration of the memory function and of the flexibility of the brain to accept and process new information. In the rest of this book, we shall neglect this physical effect, and simply assume that the capability of the brain remains constant over the individual's lifetime.

At this point, we need to put the concept of identity into perspective, because many readers will be familiar with the concept as it is used in other disciplines. Anthropologists may use 'identity' mainly in the sense of 'belonging to'; that is, as consisting of characteristics that define membership of a group, such as ancestry, skin colour, or language group. This is therefore a meaning that is relatively constant over the life cycle of an individual and what we might think of as external in character. It is, therefore, on a first glance quite different to our concept. However, the "belonging to" could also relate to such characteristics as performing certain rituals in response to natural occurrences, such as floods, storms, or draughts, in which case the difference between the two meanings or uses of the word is reduced.

Philosophers take a different approach again; an approach rooted in ontology and what exists. What does it mean for a person to exist? What does it take for a person to persist from one instant in time to another? What are the consequences for identity if all or a part of the brain is removed or transplanted? These, and many other similar questions occupy a considerable body of work, but it would appear that they have little bearing on our meaning of identity as what matters for survival.

Within psychology, a distinction may be made between *ego identity*, the sense of continuity and the issue of persistence; *personal identity*, the behavioural features that distinguish one person from the next; and *social* or *cultural identity*, the collection of social roles that a person is able to play. Within this framework, our use of the concept is clearly close to "personal identity", to the extent that behaviour is an accurate reflection of the survival criteria, an issue we shall return to shortly. However, you will realise that our use of it, as defined above, is a very great simplification. To demonstrate the nature of that simplification, and to be able to refer to some of the detailed features when that proves useful later on, the following is a brief summary of the main characteristics of the concept as used in psychology, drawing on articles by Serafini and Adams (2002) and Schwartz (2001).

The point of departure is the seminal work, *Identity: Youth and crisis*, by Erik H. Erikson, in which he introduces the three levels of identity (ego, personal, and social) and develops a theory of identity formation as the central development task of adolescence, and whose resolution sets the social-cognitive structure of individuality (Erikson, 1968). Then, to measure the state of this process, James E. Marcia

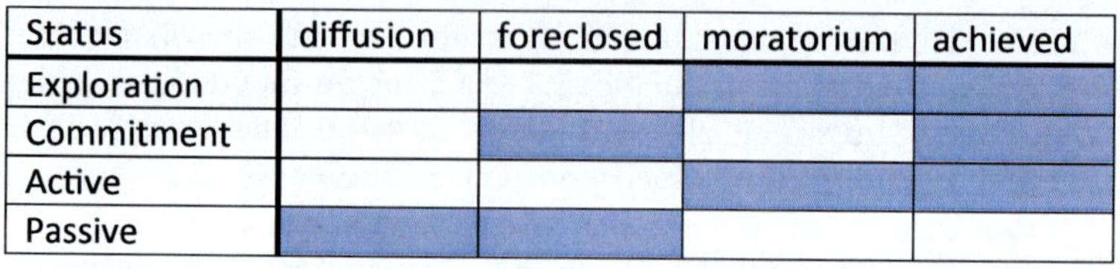

Fig. 3.3 Characterisation of the state of identity

expanded on the theory by identifying four *types* of identity related to two of Erikson's dimensions of formation—exploration and commitment: *identity diffusion*, in which no ideological commitment has been made or is being explored; *identity foreclosed*, in which commitments have been made without any exploration (adopting beliefs and commitment of an authority, such as parents); *identity moratorium*, in which exploration is taking place without making any commitments; and *identity achieved*, in which exploration has been completed and commitments made (Marcia, 1966). Of these four states of identity, the first two may be classified as passive, and the last two as active (Adams & Marshall, 1996), so that we end up with the following characterisation of the state of identity (Fig. 3.3).

We can see that 'identity' is being used with two meanings: to identify an activity or task, and to identify the result of the task, with the latter being measured by the four types. To characterise the task, Gerald R. Adams and Sheila K. Marshall (*op. cit.*) proposed five functions that provide

a. the structure for *understanding* who one is;
b. *meaning and direction* through commitments, values, and goals;
c. a sense of *personal control* and free will;
d. *consistency, coherence, and harmony* between values, beliefs, and commitments; and
e. the ability to *recognise potential* in the form of future possibilities and alternative choices.

A further dimension to the description of identity is the context in which identity becomes observable, such as choice of occupation, political orientation, or religion. Each such content area is called a *domain*, and they can be classified according to a characterising feature. Two structuring of domains have emerged: *dichotomous*, with only two classes (internal and external, or sometimes expressed as ideological and interpersonal), and hierarchical or *clusters*, in which three main clusters are commonly used (psychological, interactional, and socio-structural). Further details of the issue of defining and structuring domains are contained in (Schwartz, 2001); for the purpose of demonstrating how our use of identity relates to the use in psychology, a particular structure, introduced by Adams (e.g. in Adams & Marshall *op. cit.*), will be more appropriate. Adams divided the social context into two levels: the micro and macro contexts. The *micro context* refers to interpersonal exchanges and relations in which personal identity is directly affected by means of dialogue and other forms of direct interaction. The *macro context* refers to more overarching social and cultural contexts, in which social identity is shaped by the invoking of cultural norms, practices, and beliefs. Within this structure, the effects of the macro context are implemented through micro contexts, as in cultural norms being taught to children by their parents. Our approach is, in a sense, the opposite of this; actions on the micro level proliferate to become effects on the macro level, only then, in what might be considered a secondary effect, will there be a reverse process of the effects on the macro level shaping identities on the micro level. As we consider the individual only as an element of society as a system, we do not require any differentiation of the identity. In our view, identity is defined only in relation to its effect within the system; this is a great simplification.

This observation leads us to another relationship between our model and an existing field of knowledge, *memetics*. As the effect of our identity elements is limited to their effect in interactions between individuals, their nature is close to that of *memes*. Memes were first introduced by Richard Dawkins in Chap. 11 of his celebrated book, *The Selfish Gene* (Dawkins, 2006), and were very much in vogue for a couple of decades, but memetics was not able to establish itself as an independent discipline, and its only journal, *Journal of Memetics*, ceased publication in 2005. Memes, which can take on various forms, such as ideas, behaviours, or styles, are envisaged as the counterparts, in culture, to genes in evolutionary biology. But there is an obvious difference, in that genes are the physical carriers of genetic information, whereas memes are basically units of information which are only *represented* in physical form. As a result, there are a number of problems with memetics, as listed e.g., in in the article by Uhlíř and Stella (2012), and we can see to what extent these problems are reflected onto identity elements, in particular, when we restrict our attention to persuasion, as we shall do in Sect. 5.3:

1. The difficulty with defining a carrier of memes (in addition to the content). In the case of identity elements, there is no carrier, only transmission channels.
2. Lack of fidelity in copying. In the case of persuasion, the fidelity is controlled by a central authority (Church, sect, political party, etc.).
3. Natural selection, which should eliminate mutations and leave only a few viable alternatives, but the contrary is observed. In our case, all "variants" are arguments in support of the same subject.
4. The alleged agency of memes. Identity elements have no agency, the agency is the operation of the intelligence.
5. No operationally valid definition of meme as the smallest unit of culture. In the case of identity elements, the issue of "smallest" is resolved by the requirement of no entailment.
6. Meaning of memes; each person has a different interpretation. In the case of persuasion (of an ideology), the scope for interpretation is very limited.

However, concepts and analogies introduced in memetics, such as the analogy between the propagation of ideas and the propagation of a disease (see e.g. Lynch, 1996), may be usefully employed in investigating how the social bond is established throughout a society.

The concept of identity, as it has been introduced here, is a complex concept, and while we will be using it only as a set of items, without any further details, there are many aspects of it that could be explored. One is the *size* of the identity; what is the significance of this parameter? How, and if at all, does it manifest itself in the behaviour of a person? We might think that an opinionated person has a small identity and a broadminded person a large identity, but that is clearly too simplistic, as an opinionated person can have a firm opinion on a large number of subjects (i.e., a large number of assertions and arguments) and hence a large identity, whereas a broadminded person could have only a few, but varied assertions and arguments about the same number of subjects. Even with regard to a single subject, size does not necessarily distinguish between prejudice and broadmindedness; it is the variety of assertions and arguments rather than their numbers that count.

Another aspect is the *dynamics* of the identity; does the size of the identity increase as society evolves? A comparison between 10.000 years ago and today would have to yield an affirmative answer, but the shape of the increase is another matter. Until recently, the increase is likely to have been determined by the increase in knowledge in general, but with the current accelerating growth in knowledge one must ask if there is an upper limit. Regarding a single subject there may be a limit imposed by the ability of the processing function, as already mentioned, but as the identity only takes up a very small part of the storage capacity of the brain, it is difficult to see that the limiting factor could be the storage capacity. It might rather be that there are only so many issues that can be important to us, or that we can be passionate about, and that all others are supressed, and this is effectively the assumption we shall make in our model of the individual in the next chapter. But the validity of that assumption is something I will have to leave to psychologists.

Finally, an interesting question is to what extent the structure of the brain—its allocation of capacity to different functions (e.g., storage and processing)—will adapt to the increasing use of IT in supporting mental functions.

3.3.3 Identity Items

As noted previously, we may think of a body of information stored in an individual as consisting of a number of information items. The information of greatest interest to us is the information of class 3, as it is the information that determines the evolution of society, and as it is stored in the identity, Θ, information items in class 3 shall be called *identity items*. The number of identity items in Θ, i.e., the size of Θ, shall be denoted by w. The concept of an identity item is difficult to define with any precision, as these items take a multitude of forms that we variously refer to as knowledge, concepts, information, data, and beliefs. An initial definition could be to say that an identity item is a *statement*, of the form "I believe that ....", but this then raises the problem of defining the strength of the belief, which may range over such statements as "I believe it is possible that ...." to "I am certain that ....". One approach is to base this weight on the supporting evidence, and there is an extensive

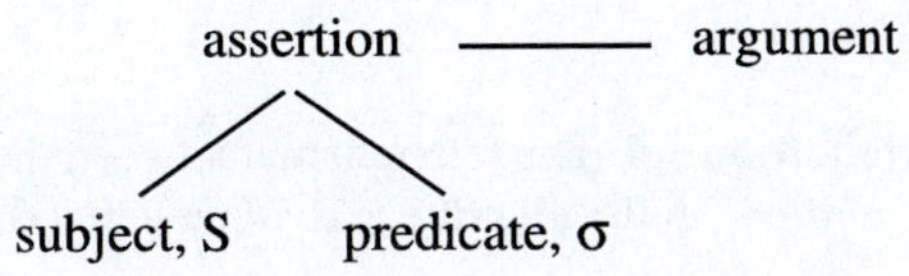

Fig. 3.4 Structure of an identity item

mathematical theory of evidence, in particular the Dempster-Schafer theory, as set out e.g., by Shafer (1976), but that is mainly applicable to the issue of uncertainty in science, with evidence in the sense of determining the probability of the truth of a statement, and of somewhat limited use in the present context. A person can believe something firmly enough to give his or her life for it, without a shred of evidence in that sense. Evidence in support of a belief is a very different concept.

Here we define an identity item as a statement (opinion, belief) with the following structure (Fig. 3.4).

An example of an identity item is the statement "I believe Sydney is the greatest city in the world because its building A is the tallest building in the world", where "I believe" simply identifies the statement as asserting a belief, not a fact, and as all identity items express beliefs, this part is unnecessary (or optional). "Sydney is the greatest city in the world" is the assertion, composed of the subject "Sydney" and the predicate "is the greatest city in the world". The phrase "because its building A is the tallest in the world" is the argument in support of the assertion.

Definition 3.1 Let an assertion be denoted by S^σ, where S is the *subject* and σ the *predicate*, then the set denoted by $X(S^\sigma)$ shall be defined as containing all arguments in Θ in support of S^σ, but such that no argument in this set entails any other argument in the set.

Definition 3.2 The *strength* of an assertion S^σ, denoted by ν, is defined as the number of arguments in $X(S^\sigma)$.

Some of the *w* identity items in Θ are related, in that they have the same assertion; they differ only in their arguments. Let *s*(ν) be the number of assertions with ν arguments, with

$$\sum_{\nu=1}^{w} \nu \cdot s(\nu) = w, \tag{3.1}$$

then we can view *s*(ν) as a distribution of strength, characteristic of the identity. Two extreme distributions are $s(w) = 1$ and $s(1) = w$, and what we might consider a *uniform* distribution, is given by $s(\nu) = w/\nu$.

Definition 3.3 The *strength* of an identity, denoted by γ, shall be defined by the following expression

$$\gamma = \sum_{\nu=1, s\neq 0}^{w} \frac{\nu}{s(\nu)}; \tag{3.2}$$

with $1/w \leq \gamma \leq w$.

Definition 3.4 The relationship between the identity strength, γ, and the number of assertions in the identity, z, shall be defined by the expression

$$\gamma = \frac{w}{z^2}. \tag{3.3}$$

even when the strength distribution may not be a uniform one.

It may at first seem strange that the strength depends on the size of the identity, w, but if we consider the two extreme cases, then, in the case of single assertion with w arguments we would expect the strength of the assertion to increase with the number of arguments supporting it, and in the case of all assertions having only one argument (so $z = w$), the strength becomes increasingly'diluted' with increasing value of w.

To an identity, Θ, with w identity items, there corresponds a *set* of assertions, ϑ, with z members, and in a society of n individuals we can form the union of these n sets,

$$\mathrm{Q} = \bigcup_{i=1}^{n} \vartheta_i. \tag{3.4}$$

Definition 3.5 Each assertion in Q, q_i, $i = 1$ to $i_{\max} \leq n \cdot z$, is characterised by a parameter κ_i, its *commonality*, $1/n \leq \kappa \leq 1$, defined by the requirement that $n \cdot \kappa_i$ indicates the number of identities in which q_i appears.

Associating commonality with assertions rather than with identity items is because what is of primary importance in a society is the commonality of what people believe; the extent to which they have the same reasons for believing it is of secondary importance.

Now pick any one of the items in Q and determine its commonality, continue this process for all the $i_{\max}$ items, and then order them from left to right in order of increasing commonality; we obtain a curve like that shown in Fig. 3.5.

If we do the same for the items of an identity, we get a similar curve, except that the number of assertions along the x-axis is z instead of $i_{\max}$. In particular, the number of assertions with κ equal to 1, or close to 1, will be the same in both cases. So, while α is a global measure of how closely the individuals are integrated into society, and as such an aspect of the degree of social integration, it is also a measure of an individual, in that it defines a subset of Θ with $\alpha \cdot z$ assertions. It is a characteristic of the identity, dividing the identity into two parts: the α-part contains the assertions common to all the individuals, and the $(1 - \alpha)$-part contains the assertions particular to each individual (or to small subsets of individuals) (The α-part is essentially what Durkheim, in his book *The Division of Labor in Society*, calls the *conscience collective* (Durkheim, 1997)).

Definition 3.6 The parameter α, as defined in Fig. 3.5, shall be known as the *identity partition*. In Sect. 6.6.2 it will be identified as one measure of the *essence of society*.

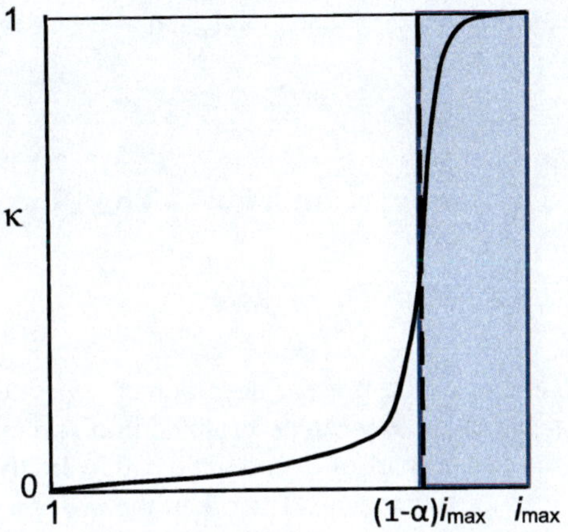

Fig. 3.5 The distribution of the commonality, κ, of the assertions in the society. Approximating this distribution by a rectangle defines the value of the parameter α

Definition 3.7 The α-part of Θ is society's belief system.

Belief, as it is treated in psychology and philosophy, is closely related to ideology, ethics, and religion, and is the subject of a significant body of work. However, works that focus specifically on belief systems are few and far between, and an overview text such as *Social Theory*, by Kivisto (2013), does not contain an entry on belief systems. A few references, although not at the level of abstraction in our model, are *Understanding Beliefs* (Hahn, 1973); *What are Belief Systems*, (Usó-Doménech & Nescolarde-Selva, 2016); and the book *A Theory of Political Obligation* (Gilbert, 2006).

Here we should be reminded that the concept of "the individual" as a representative member of society, which is central to our model of society, is only significant because the members, over which the average is formed, constitute a society, i.e., have a certain level of interaction. The result of this interaction is what is represented by the parameter α, but while α characterises both society and "the individual", the individual member of society has no knowledge of the value of α, nor of the value of κ for any of its identity elements. Hence, the decisions taken by an individual member cannot depend on α; what an individual member is aware of is the distribution of strength, $s(\nu)$, and the strength of the identity.

The strength of the identity, γ, is a numerical parameter, and so we can form the average over all members of society and associate this value with the individual. We then have two parameters characterising the identity of the individual, Θ: α and γ, and it is reasonable to ask if there is a relationship between them. There does not appear to be a direct relationship, in the sense that the value of one determines the value of the other, which is not surprising, as α is a measure of the alignment *between individuals*; i.e., how similar their beliefs are, irrespective of what the beliefs are; whereas γ is a measure of how narrow, or focused, the belief *of the individual* is, irrespective of whether this focus is the same for different individuals. For example, consider two cases, both with $\alpha = ½$: In one case, $z = 2$, $\nu = w/2$, and $\gamma = w/4$; in the other case, $z = w$, $\nu = 1$, and $\gamma = 1/w$; two very different values of γ for the same value of α.

3.4 The Action View—The Physical Domain

3.4.1 The Two Cycles

Central to the present development was the realisation that the activities that take place in society can be grouped into two cycles. One, the *economic cycle*, contains the greater part of the activities, they are the ones that sustain and grow the current 'version' of society. The other, the *transformation cycle*, contains the activities that change one 'version' into the next, and it is this process we understand by 'the evolution of society'. But this then immediately raises the question: what sort of change to society constitutes a transition into a new 'version'? Or, in other words, how do we define a 'version'?

Within the information view, our high-level description of society as an information-processing system introduced in the previous section, the answer is that a version of society is defined by its *belief system*, and thus, the activities of the transformation cycle are those that realise the changes the belief system as changes to the society. Hence, the evolution of society is synonymous with changes to society's belief system, which result form observing the changes to society and processing them through the collective intelligence, and so on—an eternal oscillation between the two views. The purpose of an action, which reflects the preceding internal information processing activity, can be to effect a physical change, to exchange information with other individuals, or a combination of both, but it may sometimes be difficult for an external observer to determine exactly what the purpose is. As a consequence, while in principle and for our purpose it is possible to distinguish between the activities of the two cycles, in practice this is not so, and it is one of the reasons why the division into two cycles is neither obvious nor generally accepted.

Another reason is that, for most people, the activities in the economic cycle dominate by orders of magnitude; they are the activities that make up our daily life and support our existence as physical beings. If we look at the measures used to compare societies, they are mainly such ones as current GDP, growth of GDP, education level, research expenditure, military power, etc., all measures of activities and entities within the economic cycle. If we consider how most of us spend our time, as we shall do briefly in the next subsection, not much is spent on activities in the transformation cycle; at least, not consciously and productively. And if we look at the issues dominating the political agenda, particularly at election time, as was so starkly demonstrated in the most recent federal election in Australia (the nation with the greatest *per capita* wealth in the world), it is all about the economy; not a thought for what sort of a society we would want to be. Under the pressure of providing profitable investment opportunities for the ever-increasing capital, we are being turned into consuming robots.

This focus on the economic cycle was lamented some time ago by Marcuse (1964). Instead of two cycles he talked of two dimensions, and his second dimension is essentially the transformation cycle. For this process to function, there needs to be tension in the first place; it only becomes active when the individual perceives a conflict between the social environment in which it is embedded and its own identity. Acceptance of this conflict as a prerequisite to progress is central to Marcuse's critique of modern society. Referencing Hegel, he states that the power of the negative is the principle which governs the development of concepts, and contradiction becomes the distinguishing quality of reason (Marcuse, 1964, 171). And concept is taken to designate the mental representation of something that is understood, comprehended, known as the result of a process of reflection (Marcuse, 1964, 105). It is this dialectical mode of thought that Marcuse calls *negative thinking*. Negative thinking is the driver of Marcuse's second dimension of the development of society; the driver of the first dimension—*positive thinking*—is the acceptance of the perceived world as the basis for reason. The difference between the two modes of thinking is reflected in the difference between "is" and "ought" (Marcuse, 1964, 132).

Marcuse's concern is summarised in the subtitle to the Introduction to *One-Dimensional Man*—"The Paralysis of Criticism: Society without Opposition". And there should be criticism of a society which he describes as follows: "Its productivity is destructive of the free development of human needs and faculties, its peace maintained by the constant threat of war, its growth dependent on the repression of the real possibilities for pacifying the struggle for existence—individual, national, and international." His analysis of why, in the face of this situation, there is so little criticism, reflects his background as a philosopher sharing many roots with Heidegger, Adorno, and Horkheimer, but at a high level, and in a perhaps overly simplistic view, it is about information; how the control of information can determine our perception of reality and make the irrational seem perfectly rational. The current state of society is made to appear so obviously the logical choice that there is no need to look for any alternative; it's just the way it has to be.

The suppression of the transformation cycle, or the second dimension, is a result of a strong coupling between the two cycles, in that maintaining the status quo of the economic cycle and acceptance of its inequalities always required the suppression of the activities of the transformation cycle. The control of information is nothing new, and religion has always played a major role in effecting it; it was always a central aspect of the catholic church and was exercised with a degree of cruelty that made it a form of terror. The Renaissance was the beginning of a revolt against this control; a process that culminated in the Enlightenment. However, at this time the Industrial Revolution started to realise its own demand for control; a control required for the acceptance of the alienating and degrading conditions of the industrial production process. Today the control is required by the economic cycle in order to maintain a version of itself that has long passed its use-by date.

The interrelation between the two cycles can be viewed under several perspectives. One is from the point of view of the purpose of a change to society. Ever since the appearance of the first life form there has been an incessant process of change, new species arising and others becoming extinct, and it appears to be a feature built into the process of life itself, with new species being created through random mutations and sexual propagation. And now, in the case of *homo sapiens*, a new life form—the society—with its new form of evolution. So, what necessitates change? I believe the answer has been the same ever since life began: to improve the chances of survival. What survives—cell, organism, individual, society, and the criteria for survival—have had to evolve along with the evolution of the life form, but change is an inherent characteristic of life. Just as creation and destruction—birth and death—are inseparable from life, so is change. We might have become somewhat blinded by technology and its applications, but change is not the doubling of the number of transistors on a chip, the increase in GDP, or anything like that; change is change to the life form—the species, which in our case is society. The changes taking place within the economic cycle are of shorter duration and due to stresses within society; we might think of them as tactical changes on the long-term evolution of society. The changes taking place within the transformation cycle are due to the operation of the collective intelligence; we might think of them as strategic changes and as *mutations* of society's belief system. In both processes the criterion is survival; a good change is one that increases the likelihood of survival, a bad change is one that decreases the likelihood of survival. It is possible to see these changes in historical terms, with wars and revolutions as manifestations of the desires for such mutations. As with all mutations, the majority were unsuccessful in establishing lasting changes to society's belief system, but some were successful and were realised as changes and innovations to the structure of society.

Another aspect of the relationship is that many variables, such as the inequality of income and wealth, are common to both processes; the difference being that whereas in the economic cycle they describe the outcome of the process—the state of society, in the transformation cycle they describe the condition under which the process operates.

A third view of the relationship arises as a result of both processes using the same processor—the human brain, and with both competing for the processing capacity.

Finally, in drawing perhaps an overly long bow, we can see an analogy with the work of Noam Chomsky in linguistics and, more specifically, in the structure of grammar and the relationship between this structure and inherent mental states (see e.g., Chomsky, 1980). Those mental structures are the analogue of the individual's innate ability to construct a set of criteria for evaluating change with regard to survival, and our two cycles are the analogue of Chomsky's deep and surface structures, with the deep structure corresponding to the transformation cycle. And with language being the main carrier of information, it would not be unreasonable to expect a relationship between the evolution of language (or speech) and the evolution of society.

Both processes are driven by the exchange of information—in the economic cycle mainly information of class 2, in the transformation cycle information of class 3, but it is the latter interaction that characterises a species and is of interest to us in the current context. That interaction is defined by its *composition* (information quality) and its *intensity* (information flow), and it is the subject of Chap. 5.

3.4.2 Types of Activities

As already noted, the action view is, to a certain extent, a description of a collection of sub-views, and while these more detailed views each encompass the activities within both cycles, they are normally presented without making this distinction. In later chapters we can make the distinction where appropriate. One sub-view is to look at in what activities people spend their time, and to that end we might decide to consider all activities taking place in society to be grouped into five groups, as shown in Fig. 3.6. The interaction between these groups is that the sum of the time spent in each one of them must equal the total available time.

To illustrate the meaning of Fig. 3.6 by an example, let us look at a generalised, and somewhat hypothetical example of a developed nation, but one which is patterned on a real nation, Sweden, for which good statistical data is available (Swedish Bureau

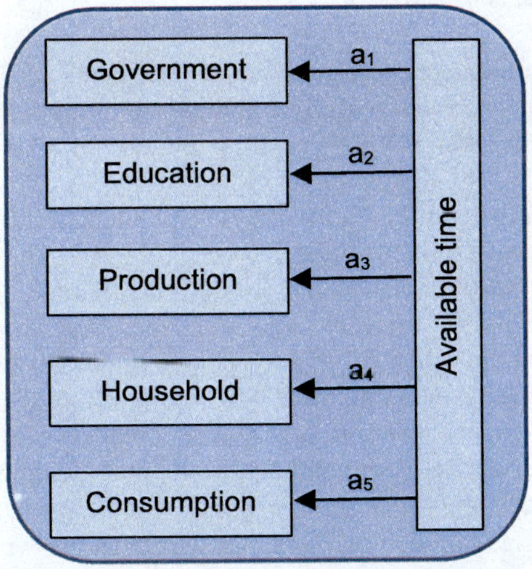

Fig. 3.6 The view of society in terms of the activity groups in which the members of society (i.e. the population) spend their time. The definitions of the five groups are as follows: Government: Legislative, executive, legal and law enforcement institutions, defence, customs, taxation, and other administrative services. Education: Teaching and studying at all levels. Production: Provision of goods and services, incl. research, development, manufacturing, sales, marketing, construction, operation, and maintenance, and incl. government owned operations (transportation, health, water, power, etc.). Household: All the household work that could be performed by paid staff. Consumption: All other activities. The five parameters, a_i, are the proportions of available time spent in each of the five groups over the lifetime of the average person

of Statistics, at www.scb.se/en/finding-statistics/). There, the lifetime of an average person is about 80 years. However, it takes some time after birth before the person accounts for activities and expenditure in its own right, so let us simply discount the first 5 years from this activity view. Of the remaining 75 years, a certain amount is spent sleeping; say, on the average 8 h per day, so that the time available for spending on the five activities identified in Fig. 3.6 is 438,000 h.

Let the working portion of the lifetime be 45 years (from 20 to 65 years of age). From the Swedish data, the time spent working per person in this age range per week is 26 h (to which we add 6 h for commuting) or, over the lifetime, 74,880 h (this excludes household work), and this is distributed over the three activity groups labelled Government, Education, and Production in Fig. 3.6.

Each person might spend an average of 30 h per week, or 1200 h per year, over 15 years (from age 5 to age 20) on education, equating to 18,000 h. If 1 in 13 is engaged in teaching, this equates to another 4680 h on education, or a total of 22,680 h, so a2 = 22,680/438,000 = 0.0518. And if 1 in 5 is engaged in government activities (excluding education), a1 = 12,168/438,000 = 0.0278. As a result, it follows that the time spent on production is 74,880 − 4680 − 12,168 = 58,032 h, and a3 = 0.1325. In this simple example we make no gender distinction, so it might be reasonable to say that each person spends 2 h per day between the ages of 25 and 75 on household work, or a total of 36,500 h, so a4 = 0.0841. We then have that the time remaining for consumption equals 438,000 − 74,880 − 18,000 − 36,500 = 308,620 h, so a5 = 0.7046.

If we now want to consider these activities in the context of evolution, we need to take account of the decrease in lifetime going back in time, at least back to when some data is available, which for Sweden is about 1750. The life expectancy, or expected age at death, for persons at age 5 is shown in Fig. 3.7.

If we now look at the year 1750, the hours available to the average person (e.g., *per capita*) for the five activities in Fig. 3.6 were 204,400. The working portion of the lifetime would have been approximately 27 years (from 13 to 40 years of age), and with a six-day working week at something like 10 h per day, the time spent working over the lifetime was 4680 h. Each person might have spent an average of 15 h per week for 6 years, equating to 4320 h, and teaching would have accounted for about 800 h (i.e., about same ratio of teachers/pupils as in 2000). It is difficult to determine the amount of time spent on household chores, as these were very much integrated into the work (e.g., as in agrarian production for own consumption), but in the absence of anything more definite, we might assume it to have remained basically constant. That is, the gain effected by various household equipment was absorbed by larger houses and higher standards. So, 2 h per day between ages 13 and 40, or 19,656 h. As a result, the time remaining for consumption was about 89,800 h.

From the above (very rough) calculations related to our activity view we only need to keep two results in mind as we progress: Over the period of 250 years, the time available for consumption, i.e., discretionary or "free" time increased by a factor of more than 3, and the time spent on education increased by a factor of more than 4.

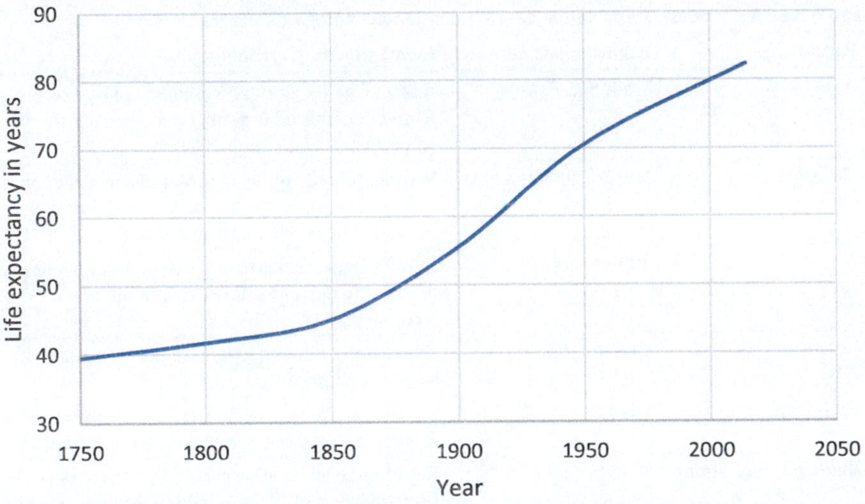

Fig. 3.7 Expected age at death for persons at age 5 in Sweden. This is an approximation due to the uncertainty regarding the extent of infant (i.e., < 5 years of age) death rates in the early years

3.4.3 Technology

As we already discussed briefly in the previous chapter, technology has been one of the major measures of evolution throughout the last 10,000 years, and today its impact on our society is both undisputed and very obvious. Just look around you, and almost everything you see owes its existence to applications of technology. If you are reading this book in a printed version, the paper on which it is printed was produced by a highly sophisticated process that converts woodchips into a fibrous "soup", which is spread out as a thin layer on a moving cloth that drains off most of the moisture and delivers a continuous sheet of felt-like material to a highly sophisticated machine, where it is pressed and dried by passing through numerous hot pairs of rollers at high speed. The steel used in this machine was produced in a complex process using iron ore and coal as its raw materials, and by many individual pieces of equipment, each one itself the product of a long development and improvement process. These raw materials were excavated, crushed, washed, and transported by numerous, high-performance machines, and so on; a chain that goes on almost without limit. Or take an ordinary drinking glass; the numerous applications of technology involved in converting quartz sand into a finished product at a cost of a dollar or less per glass are each based on a huge body of knowledge. Or the incredible precision required in the manufacturing and operation of the weaving machines that produce the cloth we wear. The involvement of technology in our daily existence has become so ubiquitous that it is no longer given much thought, in the same way that we take our natural environment for granted; as something that is simply there.

Table 3.1 A condensed view of the development of technology (Aslaksen, 2013a)

Period	Approximate duration	Development of technology
Ancient	Until 500 BC	Knowledge transmitted verbally and by example. Resources limited to timber and stone, with some metals and simple hand tools
Classical	500 BC to 400 AD	Written records of designs and of scientific input appear. Resources include bricks and concrete, iron/steel becomes more available
Medieval	400 to 1400	Slow increase in knowledge base; improvement of existing designs. Increase in mining technology
Renaissance	1400 to 1650	Expansion of knowledge base due to upsurge in science and printing
Enlightenment	1650 to 1750	Further interaction with science, improvement in fabrication methods (precision, standardisation)
Industrial Revolution	1750 to 1850	Rapid increase in all aspects of technology. Formalisation of the technology through education (textbooks)
Production	1850 to 1980	Very great expansion of the technology; in particular, of the resource base in the form of standard construction elements
Information	1980 and ongoing	An increasing proportion of technology is becoming related to software

As a result of this immediate presence of technology in our daily lives, it is possible to view society simply in terms of its technology, and to view the evolution of society as the evolution of technology; this is what we do when we speak of the stone age, the bronze age, and the iron age. To see how the development of technology fits into the evolution of society, Table 3.1 sets out a more detailed time frame and notes some of the most important characteristics of this development (an even more detailed description was provided in Aslaksen, 2013a).

Applications of technology have become so embedded in society and of such importance in our daily lives that we can say that technology is now a defining component of our culture, together with such other components as art, religion, and the rule of law. The evolution of society is strongly influenced by the development of technology, and as we have seen that we are now in charge of this evolution, our understanding of technology, the manner in which we control the development of technology, and how we decide to apply technology, become essential factors in assessing how evolution will progress from here on.

In the Introduction to the proceedings of the Nobel conference XXI, Byrne (1985) noted that questions of judgement and values permeate issues relating to the use of science-based technology. These issues are more complicated than merely choosing between "good" and "evil". They involve a societal decision as to what is good, or which of competing goods to pursue. Society and its leaders, then, make decisions that determine whether technology will be used responsibly or not, to which of many responsible uses the technology will be put, and what level of undesirable effect will be tolerated. These are matters that demand the attention of the entire society: business leaders, politicians, scientists, and citizens.

It is not difficult to find examples of how, in the past, technology and its application had a major influence on the evolution of society. For example, the ability to form large-scale communities, as in cities, and thereby enabling closer cooperation, depended on the development of civil and structural technology: water supply (dams, aqueducts, wells), sewerage and drainage (pipes, canals, tunnels), brickwork, etc. Or the ability to separate workplace from place of residence through mechanised transport (railway, ferry, tram, bus, and car). And in the present day there are technologies that have, or will have, significant impact on the further evolution of society, such as genetics, nuclear technology, and renewable energy technology. These obvious examples can also make us overlook some less obvious, but, in the long run, not less significant influences of technology, and we shall return to that in Chap. 8. The fact that many (most?) applications of technology have both a positive and a negative effect on society was described as "the technological dilemma" in Buchanan (1992, Ch. 13), and the author emphasized the importance of a wide participation in the decision-making process to minimise the negative effect.

The involvement and importance of technology in all aspects of society, ranging from the most mundane items of daily life, such as brushing teeth or going to the toilet, to international relations and power struggles, as exemplified by the China-US competition, is so obvious and ubiquitous that we no longer think of technology and its applications as something separate from us as humans. But if you close your eyes to shut out all the objects surrounding you, and just look inward at yourself, there is a being with the ability to sense its environment, to think, to remember, to form opinions and make judgements, to create objects (*homo faber*), and, above all, with the ability to relate to and communicate with other humans, and thereby form societies. In principle, all these abilities exist completely independently of any technology; but we have becomes so impressed by, and dependent on the results of our ability to create things that we have lost sight of the primary purpose of creating them: To support and enhance our abilities as human beings, and that the greatest of these abilities is not the ability to create things, but the ability to communicate and form societies. The progress of humanity has not been primarily in developing and exploiting technology, but in forming societies of increasing complexity and ability.

Having said that, what we need to realise at this point is that technology is so closely interwoven with the evolution of society that any consideration of the manifestations of evolution, or of what drives evolution, or of the interactions between individuals as the elements of evolution, must take account of technology. Technology is an expression of both our manual and intellectual abilities, but has, at the same time, provided the momentum for the development and exploitation of those abilities. And it is this positive feed-back that has resulted in the accelerating pace of evolution of society. Without a good understanding of what technology is and of the interaction between it and society, it is not possible to comprehend how society got to where it is today and to see what our realistic options for the future are. As a recognition of this, a separate chapter, Chap. 8, is devoted to developing a better understanding of technology as a component of society.

Chapter 4
Model of the Individual

In the information view, society is described as a system of interacting individuals, so that there are two main components: the individual, as the information processor, and the individual's interactions, as shown in Fig. 3.2. In this chapter we develop a model of the individual as an information processor; the interaction is the subject of Chap. 5.

4.1 Model Development

4.1.1 First Level

The model is developed in a top-down approach, as introduced in Sect. 1.3.2, and at the highest level (greatest abstraction, least detail) the model consists of three components, as shown in Fig. 4.1. The *environment* is the physical (natural and built) environment in which the individual finds itself, and the flow of information labelled *a* is the result of *discovery* by the individual; targeted action by the individual, such as doing an experiment or making a measurement.

The information flow labelled *b* is the information the individual receives by being embedded in and *interacting* with society, where this block is comprised of all the other individuals in society, plus what was labelled as Common in Fig. 3.2. It is the information we receive by listening to the radio, watching TV, or reading the paper or a book, surfing the Internet, conversing with other people, or just glancing at publicly displayed advertising. It is information created and presented to us by other individuals, which we shall call the *circulating information*, and information extracted from the Common repository.

The action of the individual on society can be the presentation of information in some form, but it can also be a physical action, such as a handshake, giving a present, or moving to another place. Whatever form it takes, we shall, in accordance with our view of society as an information-processing system, characterise it by its

E. W. Aslaksen, *The Evolution of Society*,
https://doi.org/10.1007/978-3-031-23054-7_4

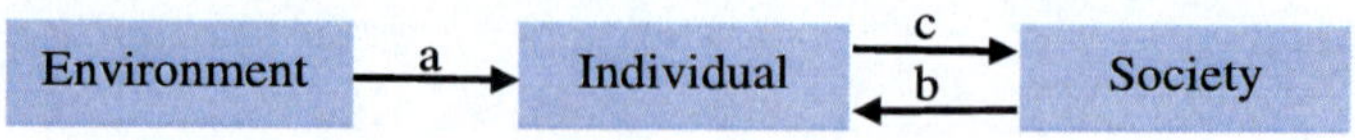

Fig. 4.1 Top level representation of the model as three components interacting by information flows

information content only, and is shown as the information flow labelled *c* in Fig. 4.1. As will be discussed and developed further in later sections, this information flow can be seen as consisting of two components. One is the information intended to influence the identity of other individuals (i.e., class 3 information), the other is information intended to result in adaptive action, i.e., a change in the physical domain.

The representation of the model shown in Fig. 4.1 appears to have an inconsistency, in that it shows a block labelled "Individual" and a block labelled "Society", when "Society" already contains all the individuals. The purpose is to explicitly define the individual's interactions; with that, the further development of the model of society becomes a development of a model of the individual as an information processor.

4.1.2 Second Level

In this next step of developing the model, the individual is decomposed into two components—a Processor and a Memory, as is society—into an Active and a Passive component, as shown in Fig. 4.2. The Active component comprises information produced by all the other individuals in the society; the Passive component comprises all the historical information, as in all forms of documents and electronic data bases, to which the individual has access, i.e., the Common and which we identify as the *collective memory*. This information was, of course, all produced by persons, most now deceased, and while it is subject to a certain amount of attrition (items being lost), the increase in this store of human knowledge is rapidly accelerating. It is true that most of this increase is related to science and technology, but there is also an increase in what we called class 3 information, much of it extending the applicability of previous understanding to the complexity of the current society.

The Processor *evaluates* the incoming information on the basis of the knowledge stored in the Memory, and as a result takes adaptive action. In addition, the evaluation may also result in a change to the information stored in Memory, as indicated by the double-headed arrow.

Figure 4.2 might be construed to mean that all new information is generated by discovery—the information flow labelled a; the rest is just the same information circulating between individuals. But that is not true; it is a basic feature of a system that the system will display behaviours—the so-called emergent behaviours—that are not displayed by any of the elements, and in our case, it means that combining information in the Processor can result in new information. In particular, even in the case of no incoming information, the Processor can work on the knowledge in

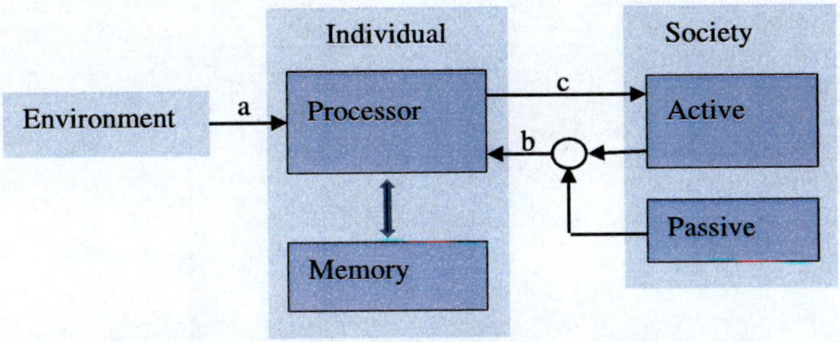

Fig. 4.2 The model developed to the second level. The Active component comprises all the other individuals in the society; the Passive component comprises all the information stored in society, as in all forms of documents and electronic data bases

the Memory, resolving contradictions and finding new correlations and relationships between items of information.

4.1.3 Third Level

At this level, the model is expanded to take account of the earlier subdivision of information into three classes, and we start out by noting a different perspective on the distinction between these three classes of information, provided by the concept of *attention*—focused mental engagement with a particular information item. Class 1 does not require attention, whereas classes 2 and 3 do, and so, as the model is mainly concerned with class 2 and 3 information, the concept of attention provides a measure of the importance of the circulating information from the point of view of the individual.

Of the circulating information items, as well as of those stored in Memory and in the Passive component of society, the overwhelming majority relate to the physical domain. It is the basic information that allows us to function and survive in our environment; it is what children have to learn by experience, such as a flame being hot, water not providing a firm surface, and how to hold a pen. And there is information that we would classify as "facts", such as that the circumference of circle with radius r is $2\pi r$; these are information items that are simply used in carrying out mental tasks, but that require no re-evaluation. For the purpose of studying the evolution of society, we are particularly interested in a small subset of the knowledge base contained in the individual's memory—Θ, the identity introduced in Sect. 3.3.2. This is the knowledge that determines the evolution of society by compelling us to take actions that are intended to change society—what we shall call *adaptive actions*. This subset contains statements about what a person believes in and is willing to sacrifice something for. It consists of information items of class 3 or, basically, information items that are subject to *persuasion*, and we shall call this subset the individual's

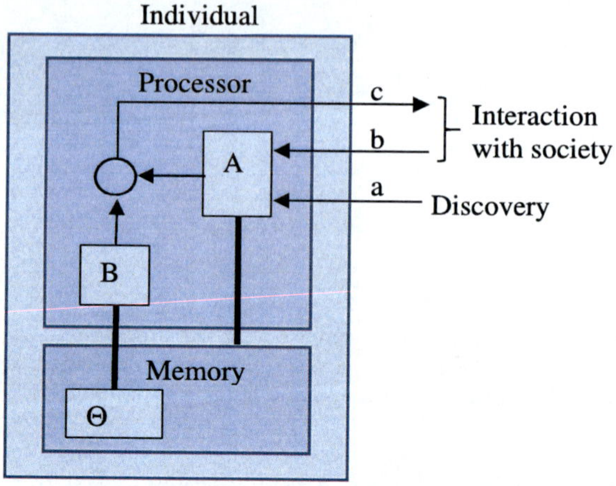

Fig. 4.3 The subdivision of the Memory into the identity, Θ, and the rest (general memory), and the corresponding subdivision of the processing function into two functions, process A and process B

identity. This concept of identity is close to what is identified as *attitude* by Daniel O'Keefe in his book *Persuasion* (O'Keefe, 2002).

In order to reflect this limitation of our perspective, or sharper focus of our view, the level of detail of the model is increased to show the identity as a component of Memory, and the processing associated with it as a separate part of the processor, as illustrated in Fig. 4.3.

Process A is the main process, which *evaluates* the inputs being received from society (other individuals and the common repository of information), what was labelled the circulating information, and from the individual's own process of discovery. In general, the evaluation has one of four outcomes: One is the discarding of any information item that is already in memory. Two is the updating of the knowledge base by incorporating some of the received items and discarding some existing ones; three results in taking action in the form of producing output items that either support or oppose received items, and the fourth is the identification of information items that conflict with any in Θ; in this case Process A simply transfers them to Θ for resolution by Process B. By "conflict with" we shall understand not only direct conflict, as in negation, but also 'question', 'suggest modification', or 'extend'.

Process B is the individual's reflection on the contents of Θ, the "making sense" of the knowledge; a process that can lead to an abrupt change of the identity and may also result in a need to provide the new-found information items as outputs to society. This process includes logical deduction, insight, and revelation; it is often identified as turning knowledge into understanding or into belief, but perhaps the best description of the function of Process B is maintaining a high degree of internal consistency in Θ.

If we consider Process A to be an evaluation process, then the result of the evaluation of inputs in class 1 is that no action of any form is required. Evaluation of inputs in class 2 may result in a judgement in the form of a probability distribution, but this uncertainty is a reflection on the input, not on the evaluation criteria, i.e., the factual knowledge on which the evaluation is based. In both cases, no doubt is left about the evaluation criteria, although they may have been changed, whereas evaluation of inputs in class 3 may raise such doubts. Process B is also an evaluation process, but of the evaluation criteria themselves—the individual's class 3 information, Θ—in light of the doubts raised by Process A about inputs in class 3. From the definition of activity cycles in the previous chapter, we see that Process A is primarily involved in the economic cycle, whereas Process B is concerned with the transformation cycle.

A two-process model of the cognitive function has been around since the work of Amos Tversky and Daniel Kahneman, as reviewed in Kahneman's Nobel Prize lecture in 2002 (Kahneman, 2002). It is suggestive to identify what they call System 2, Reasoning, with Process B, and their System 1, Intuition and Perception, with Process A (refer to Fig. 1 in the lecture) (Other authors, many of which are referenced in Kahneman's lecture, have called System 1 and System 2 slow and fast thinking, respectively). However, this identification is only partially correct, as System 2 incorporates also the processing of the class 2 inputs to Process A, and it misses an essential feature of Process B—the creation of new information, as discussed in Sect. 4.3.

4.1.4 Fourth Level

In most societies today—say, in the form of nation-states, or even in the world society, the direct, face-to-face exchange of information accounts for only a small amount of the circulating information. The majority of information exchanges involve some form of *mediation*, ranging from posted letters to social media, and these media are supported by various applications of *information technology*, from printing presses to mobile phones. The fourth level of model development introduces the concept of a medium explicitly as a component in the information flows, as shown in Fig. 4.4. Medium a would typically be in the form of some sort of instrument, medium c provides the interface to adaptive action in the physical domain, and medium b defines the individual's access to the circulating information.

Each of the media indicated in Fig. 4.4 can be further subdivided into individual *channels*, such as newspapers, radio, or TV; each supported by its own applications of information technology. The subject of channels is taken up again in Sect. 5.2.2.

Associating the media with the individual is our choice; they could have been seen as part of society—a characteristic of the circulating information. The reason for our choice was alluded to already in Subsect. 1.3.2; the increasingly close integration of humans and their information technology makes it reasonable to consider the individual, as seen from society, as a hybrid of purely human characteristics and technological characteristics.

In what in this view would be considered the ideal state of society, the media would be completely transparent, providing *undistorted* transmission of information between individuals, even though different media involve different *limitations* on their information transmission capabilities (e.g., voice only), as compared to face-to-face interaction. That is, the explicit introduction of the media in this fourth level

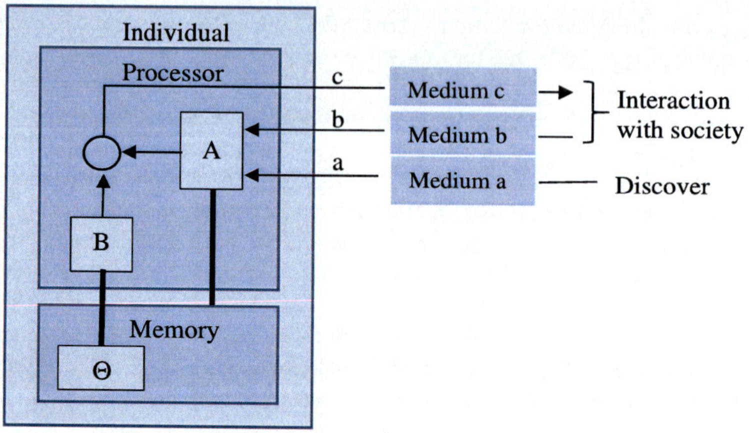

Fig. 4.4 Extension of the model in Fig. 4.3 by the addition of the information exchange media

of the model development allows the model to take into account distortions of the ideal state and be used as a tool to investigate their effects.

4.2 Process Parameters

So far, the development of the model has been purely descriptive, or qualitative; a breakdown of our view of society into interacting components. For the model to be useful, this description needs to become quantitative; the characteristics of the components must be defined in terms of measurable parameters. With our view of society as an information-processing system, the parameters fall naturally into two groups: those related primarily to defining the information, and those related primarily to defining the processing. The former group was the subject of Sect. 3.3.3; the latter group is defined in this section.

The processing of all three classes of information takes place in the individual brains. It appears that the power consumed by the brain is about 12 W, and that this power is relatively constant and independent of any particular brain activity, see e.g. (Jabr, 2012). We shall therefore assume that the energy required to process an information item is proportional to the time required and interpret the components of the processing capacity as the fraction of the total time available for mental processing; essentially, the waking hours, but with a reduction for some activities that occupy us completely and prevent us from carrying out any effective mental processing, such as certain types of manual labour. We then define processing capacities as follows:

Definition 4.1 The individual's total information processing capacity shall be denoted by u_0, and its allocation to the three information classes is defined by u_i, i = 1, 2, 3, with $u_1 + u_2 + u_3 = u_0$. The dimension of u_i is hours per day.

That is, we are assuming that the two processes, A and B, share the processing capacity of the brain.

This allocation is influenced by several factors, including the following:

a. Processing class 1 information requires significantly less energy than processing class 2 information, so that there is an inherent tendency to classify input as being in class 1 (mental laziness or fatigue).
b. We classify information elements as being in class 2, i.e., as being worthy of our attention, only if they promise to bring us some benefit, e.g., as payment for work.
c. The increasing complexity of society and of life in this society is resulting in a need to process a growing amount of class 2 information (as in bureaucracy and "red tape").
d. The information arriving from the environment (i.e., as observations) is inherently in class 2.

To investigate the dynamics of the information processing, the rate at which information arrives at an individual—the information flow—must be defined. Figures 4.1 and 4.3 indicated two information input streams to the individual, labelled a and b, originating in nature and society, respectively. But Fig. 4.2 showed the stream b to consist of two components, one generated by the other individuals in society—what we called the circulating information, and one resulting from an action by the individual—searching the data repository Passive, which we identified as Common in Fig. 3.2. As we are now going to focus on the dynamics of the information, we are more interested in what actions generate information than in where it originates, and so we shall define the following three information flows, measured in information items per unit time:

Definition 4.2 The flow of circulating information input to the individual shall be denoted by μ_a.

Definition 4.3 The flow of information input from the individual's discovery of the environment (stream a in Fig. 4.1) shall be denoted by μ_b.

Definition 4.4 The flow of information generated by the individual's reflection and analysis shall be denoted by μ_c.

Note that these flows are defined as the flow at the input to the individual; that is, after passing through the media in Fig. 4.4.

It is an implicit assumption of the model that all information contained in the individual identities is shared via the circulating information, i.e., included in the output stream labelled c in Figs. 4.1, 4.2, 4.3 and 4.4. That is, the individual's beliefs only have a value in so far as they are made public; the importance of this public role of the individual was emphasized by Hanna Arendt in her discussion of *vita activa* as political action (Arendt, 1958). In terms of our model, this means that, as far as the individual's information output to the circulating information is concerned, medium c in Fig. 4.4 is completely transparent. As a result, we can think of the class 3 elements in the circulating information as a *circulating collection of identity items*, with the total number of identity items in the circulating collection, W, given by $W = n \cdot w$, and within that collection a collection of Z assertions, with $Z = n \cdot z$. This means that the assertions in the α-part of Θ each appear n times in the circulating collection, and that α has the same meaning in the circulating collection as it does in Θ. The continuous processing of the circulating information by all the members of society constitutes an averaging or consensus-building process, which we identified earlier as the operation of the *collective intelligence.*

Definition 4.5 Let the portion of μ_a which catches the attention of the individual be denoted by $x \cdot \mu_a$, and let the fraction of that information which relates to Θ; i.e., is of class 3, be denoted by y; then the parameter μ_1 is defined as the flow of information input relating to Θ, i.e. $\mu_1 = y \cdot x \cdot \mu_a$; this is the flow which is evaluated by Process B.

When we, in the next section, develop an understanding of what determines this flow, and how this influences the identity, we need to keep the following points in mind:

i. The individual has no knowledge of society's belief system and will evaluate the inputs and act accordingly based solely on its own identity.
ii. Only if medium b is transparent are the information items arriving as input to the individual a random selection of the items in the circulating information, and the items in the flow μ_1 a random selection of the identity items in the circulating collection of identity items, W.
iii. In the interaction with the circulating collection of identity items, the size of the identity, w, remains constant. We can observe, in ourselves and in others, that we can only maintain a limited number of things that matter to us, that we are passionate about, that we are willing to sacrifice something for. If we accept new items, develop new passions, they tend to displace some existing ones. There is certainly a significant variation between people in the number of causes they care about, but we shall assume that, from adolescence onward, which is the part of the lifetime in which the individual can influence the evolution of society, the size of the individual's identity is fixed and equal to w. That is, new identity items *replace* existing ones.
iv. The size of the identity is measured in terms of identity items rather than in terms of assertions. The reason for this is that a strong assertion tends to replace weaker ones; in the extreme case of an assertion with strength $\gamma = w$ there is no room for any other assertion.

4.3 Model Dynamics

The assertion associated with an item arriving as part of μ_1 is either identical to one already in Θ, and the probability of this is α, or it is different to any assertion in Θ, and the probability of this is $(1 - \alpha)$. In the first case, the item is either identical to one in Θ (i.e., has the same argument), in which case it is simply ignored, or it provides a new argument for the assertion, which may or may not be accepted. If the item is accepted, the value of ν for that assertion increases by 1 and that of another assertion decreases by 1; however, at the current, high-level development of the model, this process, which results in a change of the strength of the identity, is ignored and the value of ν is the same for all assertions (the uniform distribution discussed earlier). In the second case, the assertion is potentially in conflict with an existing assertion in the $(1 - \alpha)$-part of Θ (by definition, there can be no conflict with the assertions in the α-part), and needs to be processed by Process B, so that there is a flow of potentially conflicting items, $(1 - \alpha)\cdot\mu_1$.

Definition 4.6 The level of unresolved conflicts (and these items are not in Θ), measured as a fraction of w, is perceived by the individual as a *stress*, and shall be denoted by β.

Definition 4.7 The rate at which Process B is able to resolve conflicts will be assumed to be proportional to the stress, and the proportionality factor, which is a property of Process B and has the dimension of an information flow, shall be denoted by μ_2.

That this rate increases with increasing value of β is reasonable; as the stress increases, we put more effort into lowering it. But setting the rate proportional to β is somewhat arbitrary and done just for simplicity, and it ignores any saturation of the capacity of Process B.

With this, the change in the stress is determined by the following equation:

$$w\frac{d\beta}{dt} = \mu_1 \cdot (1 - \alpha) - \mu_2 \cdot \beta; \tag{4.1}$$

As far as the identity, Θ, is concerned, the resolution of a conflict has, basically, one of two outcomes: Either the conflicting item of information is accepted, which deletes one item from the $(1 - \alpha)$-part of Θ and adds one to the α-part, *assigning the same value of* ν *to it as that of the deleted item* (as per our assumption of a uniform distribution, and extending the definition of ν to be either the size of the set X or a value inherited in a replacement), or the conflicting item is rejected. Which of these two outcomes eventuates depends on two things: the *persuasiveness, p* of the conflicting item, and the strength of the assertion in the item to be replaced, ν. The probability of replacement becomes lower the greater the strength, and in the absence of any other information, we shall assume that this influence of the strength is represented by the simplest function, $1/\nu$.

Definition 4.8 The *persuasiveness, p*, which is a characteristic of the argument, shall be defined as the probability of acceptance when $\nu = 1$ (A useful model of persuasion is developed in the doctoral thesis by Mark J. Stoda, *Symbolically seeking a change of mind: a model for comprehending, constructing and critiquing persuasion*, and we shall return to that in Chap. 5).

With this, the probability of replacement is p/ν, and the resulting change to α can be expressed as a function of p and ν:

$$\Delta\alpha = \frac{p}{\nu \cdot z}; \tag{4.2}$$

Furthermore, as previously mentioned, each individual generates new information, both as a result of discovery and as a result of reflection or analysis. The former was shown as the discovery path in Fig. 4.1 and labelled by μ_b in Definition 4.3; the latter, which may be considered an 'internal' flow, was labelled by μ_c in Definition 4.4. This combined flow of new information is important in determining the longer-term evolution of society, in particular if α approaches 1, but it is always true that both μ_b and μ_c are much smaller than μ_a; our regular information processing is concerned overwhelmingly with issues we are familiar with, rather than with new discoveries.

Definition 4.9 The information flow μ_0 is defined as that part of the information flow generated by the individual, $\mu_b + \mu_c$, that is of class 3 and results in a change in Θ.

If the assertion to be replaced is in the $(1 - \alpha)$-part, there is no change to α, but if it is in the α-part, it is deleted, and the replacement added to the $(1 - \alpha)$-part (Note that we are assuming that there is no conflict involved in this replacement process, thus no stress.) Hence, the rate at which α is decreased by this "rejuvenation" process is $\mu_0 \bullet \alpha$, and we obtain the following expression for the change in assertions:

$$z\frac{d\alpha}{dt} = \mu_2 \cdot \frac{p}{\nu} \cdot \beta - \mu_0 \cdot \alpha, \tag{4.3}$$

where the second term implies the transformation of items into assertions in the resolution process.

The two external and two internal sources of information, and the three classes of information, can be summarised in Table 4.1.

The two equations, Eqs. 4.1 and 4.3, form a set of coupled linear first-degree differential equations for the two functions $\alpha(t)$ and $\beta(t)$. Solutions to these equations

Table 4.1 Categorisation of the information flows by source and class

Source		Class		
		1	2	3
Interaction	μ_a	$(1\text{-}x)\ \mu_a$	$x\ \mu_a$	$xy\ \mu_a \equiv \mu_1$
Discovery	μ_b		$\mu_b + \mu_c - \mu_0$	μ_0
Analysis	μ_c			
Process B	μ_2			μ_2

are well known (Kamke, 1943), but without knowledge of the values of the coefficients it is not possible to determine a particular solution. But we can solve the two equations for the steady state, resulting in the following expressions:

$$\alpha = \frac{\pi}{\pi + \omega}; \tag{4.4}$$

and

$$\beta = \frac{\omega}{\pi + \omega} \frac{\mu_1}{\mu_2} \tag{4.5}$$

where $\pi = p/\nu$ and $\omega = \mu_0/\mu_1$.

The two parameters α and β express the main features of our model of society as an information-processing system, so it is important to keep in mind their meanings as well as the limitations imposed on them by the high level of abstraction of the model. The parameter α, the *identity partition,* is defined in terms of an approximation, the *rectangular approximation* illustrated in Fig. 3.5. No estimate of the accuracy of this approximation is given; its justification is simply the well-known effect of belief through repetition, also known as the illusory truth effect, as demonstrated e.g., by advertising, and supported by a significant body of research, to which a recent reference is Hassan and Barber (2021). That article demonstrates the highly non-linear nature of the effect, which further supports the approximation.

Equation 4.4 shows that if $p = 0$, then $\alpha = 0$, which means that the society has disintegrated, and it is "each man for himself". That is, a certain amount of persuasiveness, or, what is more appropriate in terms of π, a certain amount of *acceptance* is required in order to establish the relationship, or bond, between individuals that forms them into a society. But the value of α is not an immediate measure of the state of society in the sense that the greater the value of α the stronger or better the society; $\alpha = 1$ would mean that the society has become completely ossified and incapable of evolving. This is prevented by the presence of ω in the denominator of Eq. 4.4; it is what Big Brother tries to suppress—a measure of *individuality.* The flow of new information, μ_0, is what fuels the evolution of society, so a healthy and vigorous society will have a value of α somewhere closer to the centre of the range 0–1. If this ideal value of α depends on the state of the society is not clear at this point; aspects of this question will be raised in Chaps. 6 and 10.

Another question is the criticality of fluctuations in the value of α away from this ideal value. The function α(π;ω) is displayed in Fig. 4.5. In a modern society the value of ω is much less than 1; the individual (i.e., the average person) receives the overwhelming majority of information items in Θ from society. And if we, in accordance with the previous comment about the value of α, take that value to be somewhere around 0.5, we see that the value of α is very sensitive to the value of π; *i.e.*, to the *persuasiveness* of the arguments in the circulating information, something that obviously has a significant implication for the stability of society.

The quantity βw is the number of unresolved items that are challenging assertions in Θ. But as only items in the $(1 - \alpha)$-part of Θ can be challenged (as there is

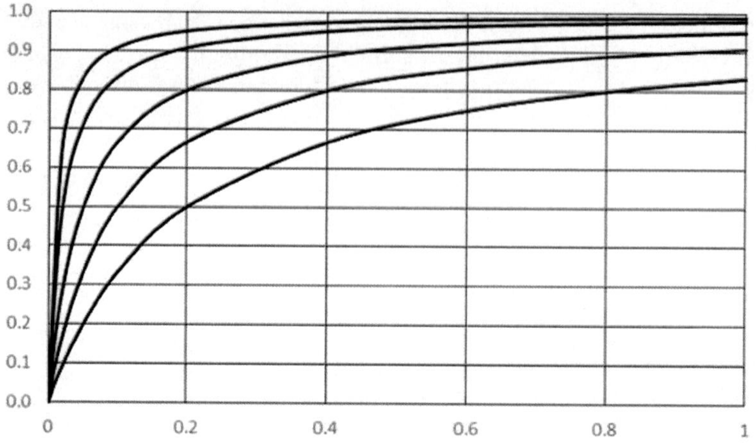

Fig. 4.5 The function $\alpha(\pi;\omega)$ for five values of the quantity ω, from top to bottom 0.01, 0.02, 0.05, 0.1, and 0.2

full agreement on the items in the α-part), we have $\beta \leq (1 - \alpha)/\nu$, with the equality meaning that the identity of the individual is completely challenged. For this extreme situation not to occur, we have the condition

$$\frac{\mu_1}{\mu_2} \leq \frac{1}{\nu}; \tag{4.6}$$

which basically says that the flow of information that challenges the identity of an individual must be less than a certain fraction of the capability of the individual to resolve such challenges (This might remind us of Ashby's Law of Requisite Variety (Ashby, 1956)).

Equation 4.5 shows that the level of conflict within Θ, which we have equated to stress for the individual, depends on the various flows of information. But can these flows take on any value? The processing of these flows is performed by one processor—the human brain, with a finite capacity, and there are therefore three processes sharing this capacity. The first is the initial processing of the input from the circulating information, μ_a. The second is the processing of all the class 2 information flow, and these two make up Process A. The third is the resolution of conflicts, Process B. If we denote the demands these three processes make on the processing capacity as fractions of that capacity by u_i, we have the following equations:

$$u_1 = a_1 \cdot \mu_a; \tag{4.7}$$

$$u_2 = a_2 \cdot (x \cdot \mu_a + \mu_b + \mu_c) \leq 1 - u_1; \text{ and} \tag{4.8}$$

$$u_3 = a_3 \cdot \mu_2 \cdot \beta = a_3 \mu_1 \frac{\omega}{\pi + \omega}; \tag{4.9}$$

where a_i is the time spent on processing one item of information in process i. Equation 4.7 represents a simplification, in that we are taking this process to be a constant background process, depending only on the amount of circulating information, and not on its composition, as already indicated.

The inequality in Eq. 4.8 implies the order in which the three components can draw on the total processing time: first u_1, then u_2, and then u_3, if there is any time left. It also shows that there is an upper limit on the parameter x from Definition 4.5:

$$x \leq \frac{1 - a_1\mu_a - a_2(\mu_b + \mu_c)}{a_2\mu_a}. \tag{4.10}$$

This upper limit turns the individual's attention into the limiting factor in the consumption of information; in the words of Matthew Crawford, "Attention is a resource—a person has only so much of it." (Crawford, 2015). And Eq. 4.10 displays the expected dependence on $(\mu_b + \mu_c)$: the more the individual is immersed in its own projects (e.g., study and research), the lower the available amount of attention that can be allocated to the circulating information.

This concludes the development of the model of the individual in our view of society as an information-processing system, consisting of individuals as information processors and the interaction between them as information flows. At this point, the view of society is a high-level, or a highly abstracted, view; to apply the model and obtain verifiable results, it must be applied to a lower-level, more detailed view of society. This will often be in terms of economic processes, but possibly also in terms of such processes as conflict dynamics and belief propagation, as will be mentioned in later chapters.

Chapter 5
The Interaction

5.1 Types of Interaction

In the primary view of society as an information-processing system consisting of interacting individuals, and with the individuals, as the processors, having remained essentially unchanged over the last 10,000 years or so, it is clearly the interaction that plays the main role, both in our understanding of society and of the evolution of society. Accordingly, the present chapter is concerned with developing a more detailed understanding of the interaction, and we do so by utilising both a top-down and a bottom-up approach.

The approach to understanding society presented in this book is an application of the *top-down* system approach to handling complexity, starting with a high-level description of society as an unstructured system of identical individuals, and with the interaction described in terms of an individual's interaction with (the rest of) society. The book *The Social Bond* (Aslaksen, 2018) took the opposite, or *bottom-up*, approach, starting with the smallest social element—two individuals—and examining the interaction between them. Society would then be described in terms of a multitude of such *binary interactions*.

The two approaches have a great deal in common: In both, the structure of the description is the same: two views—an information view and an action view—and each again subdivided into two parts—two classes of information in the case of the information view, and two cycles in the case of the action view (although this was not made explicit in *The Social Bond*). And such central concepts as identity and collective intelligence appear in both. However, with our aim of understanding the current state of society—increasingly the world society—and the current direction of its evolution, the bottom-up approach would struggle to progress up through all the structural levels to represent the complexity of that society. But firstly, the binary interaction is a component of the interaction between the individual and society—it is something we experience every day—and the relative importance of this component is one possible characteristic of the individual's interaction with society. And secondly, the binary interaction allowed us to investigate aspects of the interaction

E. W. Aslaksen, *The Evolution of Society*,
https://doi.org/10.1007/978-3-031-23054-7_5

that are hidden in the current approach, and because these aspects will arise again in the further development of our top-down approach, we include a discussion of the binary interaction from *The Social Bond* in the next section.

5.2 Binary Interactions

5.2.1 The Basic Model

In Subsect. 3.3.2 we introduced the concept of the identity of an individual as a set of identity items. The identity of individual *a* will now be denoted by W_a, and in order to describe changes to the identity, we need to give it an internal structure, and the following segmentation will prove to be useful ($i = a, b$):

- W_0 The items that are common to the two individuals, resulting from a current background and culture; these items are not affected by the interaction.
- $W_{i,1}$ The items that result from the individual's own experience, but that are subject to being copied to, and removed by, the other individual by the interaction.
- $W_{i,2}$ The items that have been copied to the individual from the other individual in the interaction.

For simplicity, and without affecting our arguments in any significant way, we assume that both identities have the same maximum size, which we normalise to 1, so the segments can take on values that are a fraction of 1. And, because the number of elements in any identity set is a very large number, we may, for convenience, consider these numbers to be continuous functions of time. With this understanding, we can use the differential operator d/dt to denote "the rate of change", even though, in the strict sense, it is always zero.

The reason for this particular segmentation is that we assume that there exists a well-defined and measurable quantity that can be identified as the *strength* of the interaction between two individuals, *a* and *b*, and we shall denote it by μ_{ab}. We want the strength to somehow express the ability of *a* to align the identity of *b* with *a*'s identity, and that alignment can be increased either by eliminating elements of $W_{b,1}$, or by increasing $W_{b,2}$, or a combination of both. Correspondingly, the strength of an interaction has two components; μ_{ab}^+, which relates to the ability of *a* to copy elements from $W_{a,1}$ to $W_{b,2}$, and μ_{ab}^-, which relates to the ability to remove elements from $W_{b,1}$. Note that, as elements of $W_{a,1}$ are added to $W_{b,2}$, the part of $W_{a,1}$ available to be copied shrinks correspondingly, as a given element of $W_{a,1}$ cannot be added to $W_{b,2}$ more than once. The elements that are copied to the other identity are kept track of in a virtual part of the identity, so that it is only the quantity $(W_{i,1} - W_{j,2})$ in either identity that is active in the interaction. We also need to account for the fact that the size of the identity is limited to 1. As it fills up, it takes more effort to add elements, and this is accounted for by multiplying μ^+ by the size of the free space in the identity, $(1 - W_0 - W_{i,1} - W_{i,2})$. Then, considering only the interaction between the two individuals a and b, the four components of the strength are related to the changes in the identities through the following equations, which also effectively provide the definitions of the components:

$$\begin{aligned}
\frac{d}{dt}W_{a,1} &= -\mu_{ba}^{-}(W_{a,1} - W_{b,2});\\
\frac{d}{dt}W_{a,2} &= \mu_{ba}^{+}(W_{b,1} - W_{a,2})(1 - W_0 - W_{a,1} - W_{a,2});\\
\frac{d}{dt}W_{b,1} &= -\mu_{ab}^{-}(W_{b,1} - W_{a,2});\\
\frac{d}{dt}W_{b,2} &= \mu_{ab}^{+}(W_{a,1} - W_{b,2})(1 - W_0 - W_{b,1} - W_{b,2});
\end{aligned} \tag{5.1}$$

In Subsect. 3.3.3 we introduced the parameter α as the proportion of the identity containing the assertions common to all individuals; basically, a measure of the individual's social integration (Definition 3.6). In the case of the binary interaction, α becomes the proportion of the identity common to the two participants in the interaction, and so, in the context of binary interactions, we call α the *alignment*, and define it by the expression

$$\alpha = \frac{W_0 + W_{a,2} + W_{b,2}}{W_0 + W_{a,1} + W_{b,1}}. \tag{5.2}$$

The strength of the interaction, μ_{ab}, is determined by three factors. The first is a characteristic of individual a, and may be described by such terms as charismatic, magnetic, persuasive, or dull, uninspiring, and so on. The second is the communications channel between a and b, the access a has to b, such as through the media, through an organisation (e.g. religious group, political party, union, business affiliation), or through informal groupings based on friendship or common interests, and so on. The third is the nature of the individual b, in the sense of receptiveness to changes in identity. That depends on the extent to which the suggested changes resonate with (or conflict with) the identity items already present in b; something that was identified as "cognitive advantage" by Simpkins et al. in (2010). While that work is concerned with interaction between communities rather than individuals, and with idea networks rather than identities, many of the characteristics identified and discussed there can be reflected onto our binary interactions, e.g., the fact that the interconnectedness of the identity elements is a significant factor. Strong interconnectedness presents less opportunity for new elements to intrude and be accepted. However, we need to realise that most of the literature on network-enabled interaction is focused on the effect this has on cognition, as by Benjamin Smart, Winston Sieck, Paul Smart, and Shane Mueller in *The Extended Mind and Network-Enabled Cognition* (Smart et al., 2008) or by Clark and Chalmers in (1998), whereas we are concerned with a particular subset of this—persuasion rather than cognition, as in conversion of belief. And also, our current basic model of the binary interaction does not allow this third factor to be expressed. We shall return to a more detailed discussion of interaction strength in Sect. 5.2.2.

For any one individual, the two strength coefficients, μ^+ and μ^-, may often have the same, or nearly the same, value, as both depend on the persuasiveness of the individual. But they may not have, as in those cases where the identity items in $W_a - W_b$ do not detract from or interfere with those in $W_a \cap W_b$, and in cases where there is an asymmetry in the effort required to develop the two components of the dependence. Moses was successful in getting the Jews to accept the new god and forsake all others, whereas missionaries often had some success in getting indigenous people to accept the Christian god as an additional god, but could not get them to completely renounce their old ones.

The relationship between two individuals may be symmetrical, $\mu_{ab} \approx \mu_{ba}$, as between two friends, or asymmetrical, with $\mu_{ab} >> \mu_{ba}$, as between Moses and the Jews, or between an emperor and his subjects. In the latter case, such an asymmetry is not reflected in the alignment, α, so that in order to include it in the characterisation of the relationship between the two identities, we need to introduce a further parameter, the *asymmetry*, ε_{ab}, defined as follows:

$$\varepsilon_{ab} = \frac{W_{b,2} - W_{a,2}}{W_{a,2} + W_{b,2}} \tag{5.3}$$

Obviously, $\varepsilon_{ab} = -\varepsilon_{ba}$, and $|\varepsilon| \leq 1$.

Solving the equations in Eq. 5.1 is straight forward by means of a small VisualBasic routine, and for the particular, but important, case where $\mu_{ab} = \mu_{ba}$, three results for the alignment as a function of time are shown in Fig. 5.1. The alignment increases towards its equilibrium value, which for the parameter values chosen is 1, but it is not mainly this value that is of interest to us; it is the initial rate of increase, which we want to interpret as the strength of the interaction. While the general form of the function $\alpha(t)$ is a function increasing monotonically towards its equilibrium value, the derivative is not always a monotonically decreasing function, and so does not allow the initial part of the function to be approximated by an exponential function characterised by its time constant as a single parameter. However, the values of the functions coincide, more or less, at the point half-way between the initial and final values, and so in this case we shall define the *alignment rate*, α_0, as follows:

$$\begin{aligned} &\Delta\alpha = \alpha(t \to \infty) - -\alpha(t = 0) \\ &t_0 : \alpha(t_0) = \alpha(t = 0) + \Delta\alpha/2 \\ &\alpha_0 = \Delta\alpha/2t_0 \end{aligned} \tag{5.4}$$

The three quantities $\Delta\alpha$, t_0, and α_0 are all functions of the seven parameters μ_{ab}^+, μ_{ab}^-, μ_{ba}^+, μ_{ba}^-, $W_{a,1}$, $W_{b,1}$, and W_0. There are some simple relationships, e.g. if μ_{ab}^- and μ_{ba}^- are both non-zero, then $\Delta\alpha = 1 - W_0$, and if $\mu_{ab}^+ = \mu_{ab}^- = \mu_{ba}^+ = \mu_{ba}^- = \mu$, then there is a linear relationship between α_0 and μ, with the initial value of $W_{a,1} = W_{b,1} = W_1$ as parameter, as shown in Fig. 5.2. However, in general there is no

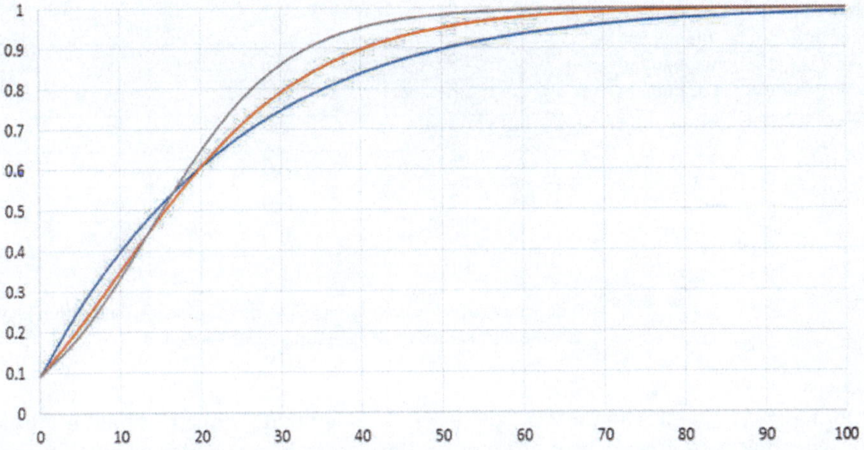

Fig. 5.1 Three alignment functions, all with $W_0 = 0.1$, $W_{a,1}(t = 0) = W_{b,1}(t = 0) = 0.5$. Top function, $\mu^+ = 0.02$, $\mu^- = 0.1$; middle function, $\mu^+ = 0.05$, $\mu^- = 0.05$; bottom function, $\mu^+ = 0.1$, $\mu^- = 0.02$. The horizontal scale is in units of time, and the dimension of μ is per unit time

simple relationship, and it is better to consider the significance of particular cases, as was done in *The Social Bond*.

The subject under consideration may be a limited subject within an area such as religion, politics, social justice, and so on, and the various quantities, such as W_0 and W_1, can then be taken to be specific to this subject. As a hypothetical example of such a situation, consider two friends *a* and *b*, who relate to one another on an equal basis,

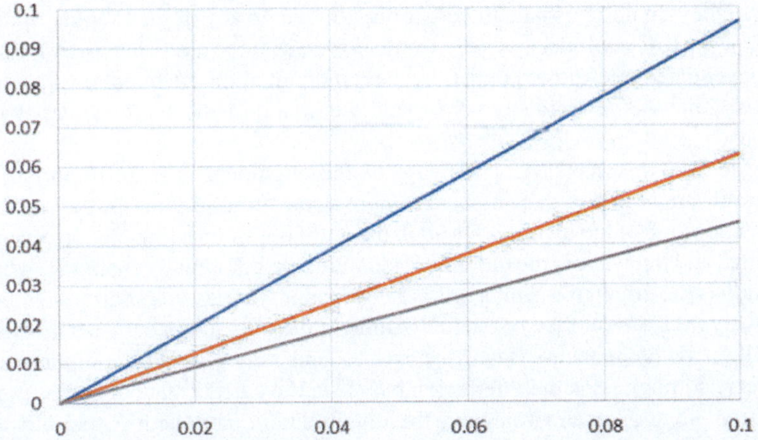

Fig. 5.2 The alignment rate, α_0, as a function of the interaction strength μ, with the initial value of W_1 as a parameter, from the top, $W_1 = 0.2, 0.5, 0.8$, and all with $W_0 = 0.1$. Both α_0 and μ have the dimension per unit time

Table 5.1 For each of the nine evaluations, the upper value is t_0, in hours, and the lower value is the alignment rate, α_0, in per hour

μ	$\mu^+ = \mu^-$	$\mu^+ = 0$	$\mu^- = 0$
0.1	9.64	26.39	12.97
	0.0479	0.0175	0.0173
0.2	4.83	13.21	6.76
	0.0956	0.0349	0.0340
0.4	2.42	6.61	3.40
	0.1906	0.0698	0.0678
	1	1	0.538

The values of μ are in per hour. The three values in the bottom row are the end values of the alignment, $\alpha(t \to \infty)$

e.g., $\mu_{ab} = \mu_{ba}$, and who meet once a week for one hour. At one meeting, *a* raises an issue that is of great concern to him, and on which he has quite strong opinions (many identity items), and discovers that *b* also has strong, but generally different opinions. Thus, on this issue the common ground, W_0, is small, say, $W_0 = 0.1$, and the strong opinions take up more than half the remaining space in the identity that can be allocated to a single issue, so that the initial value of $W_{a,1} = W_{b,1}$ and equals, say, 0.6, which means that the initial value of the agreement on this issue, $\alpha(t = 0) = 0.0769$. This issue dominates their discussions for the next meetings, until they have increased their agreement to, say, $\alpha = 0.5$ (and they might agree to disagree about the remaining opinions).

If we, at first, assume that each one spends an equal amount of effort on persuading the other to accept their opinions and to abandon their own, then $\mu^+ = \mu^- = \mu$. Now assume that each one only tries to show the other one how wrong he is; that is, $\mu^+ = 0$. At the other extreme, the two friends are very concerned not to antagonise each other, so they do not persuade the other one to give up any of his existing opinions; that is, $\mu^- = 0$. The model gives the results for these three cases shown in Table 5.1.

If we equate the effort expended by each individual on the bond to $(\mu^+ + \mu^-) \cdot t_0$, we see that the best strategy is to spend the time on both to accept and abandon opinions.

The reason why the rate of increase of the alignment is more important than the equilibrium value is that in society there is no "equilibrium" in the sense of a final, static value of the identity. Through the activities of observation, introspection, invention, and innovation, the individual identities are continually changing; what we can study is the effect of the interaction between individuals in a short time interval, effectively the $d\alpha/dt$. It may be useful to think of the interaction between individuals as having two components, say, a *transverse* and a *longitudinal* component. The transverse component is the interaction based on the current state of society and on the current state of its environment; the longitudinal component is the interaction based on the previous history of society. It is the component based on memory and on the transmission across generations; it is what is, to a large extent, transmitted in the early stage of life through education and upbringing. It could be tempting, then, to say that the longitudinal component promotes stability, and the transverse component promotes change and possible instability, but that is too simplistic a view, and the issue of stability will be taken up again in Chap. 11.

5.2.2 *Binary Interaction Strength and the Social Bond*

The model of the binary interaction presented in the previous subsection is very simple, and the calculations and their results may at first appear almost trivial. One reason for that is that much of the "social" aspect of the model is hidden in the parameter μ—the interaction strength—and through it in the concept of the social bond, and so in the current section we examine these two concepts in more detail.

The interaction strength, μ, as defined in Sect. 5.2.1, is a high-level parameter that hides within it a number of more detailed features of the interaction between two individuals. It expresses the total effect of the interaction, and as we now know, this effect is dependent on such factors as the content and format of the messages being interchanged, the relationship between the messages and the identities of the participants, the relationship (alignment and trust) between the two participants, the type of transmission channel, and the intensity of the interaction. But before considering each of these factors, we should first realise that the interactions can be divided into two groups according to their main purpose or intended effect: In the one group are the interactions that are aimed at making an individual change its identity; to accept new identity elements and, possibly, also abandon previously held elements. It is a matter of making the individual *believe* something.

In the other group are the interactions aimed at making an individual *do* something; belief without action will have no influence of the evolution of society. Our model of the binary interaction applies specifically to interactions in the first group; interactions in the second group have to be considered as part of a detailed treatment of adaptive action. Adaptive action is certainly an essential component of intelligence; it is the output of intelligence as a process. Our focus has been on the input side, on the binary interaction and the part of intelligence that results in a change to the identity, creating the precondition for adaptive action. Adaptive action is implied in various places in this book, but Chap. 7 is dedicated to a detailed discussion. However, with that in mind, the following discussion of the factors influencing interaction strength is applicable to interactions in both groups.

The concept of an information transmission *channel* was mentioned briefly in Sect. 4.1.4; we can now develop it in more detail within the framework of the binary interaction. That is, we shall always consider the channel to be between two persons; it is only the *situation* in which the channel is active that can display a range of participants. For example, in the situation where one individual is delivering a speech to a group of individuals, there are simply many identical channels being active simultaneously, between the speaker and each individual in the group. But each individual in the group may also receive multiple messages via multiple channels, in that it receives messages from some of the other group members in the form of body language, or shouts of "Yeah!" or "Shame!", etc.; this is what leads to group dynamics. A channel can be described from several different points of view, each one characterised by a parameter, say, a_i; not necessarily independent or orthogonal, and one possible set of parameters is the following:

$i = 1$: The *means* used to express the message, with three values: auditory, visual, and both. Auditory includes such means as speech, song, and music; and visual includes text, pictures, signs, and body language. There are other possible means, such as mechanical (touch, vibration) and olfactory (smell), but they would be of a secondary importance.

$i = 2$: The *flow* of the message, with two values: one-way and two-way. One way includes such activities as viewing, reading, and listening; two-way implies an exchange of information relating to the same subject, as in a conversation or an exchange of letters or emails. The significant difference is that the two-way channel supports a process involving the system consisting of the two participants in the exchange, and as such may result in the emergence of knowledge that was not present prior to the exchange. That is, while the exchange may only result in a consensus as a compromise based on the information exchanged, it may also result in genuinely new insight, new knowledge.

$i = 3$: The *permanency* provided by the channel, with two values: fleeting and stored. It is important to recognise that this refers to the channel, not necessarily to the message. For example, a conversation is fleeting, but the message may be stored in the memory of the participants. A stored channel implies that the message is contained in a storage medium, such as a letter, a book, a photo, a CD, or a computer memory (i.e. as a file).

$i = 4$: The *authenticity* of the channel, with two values: original and processed. Original means that the message is not altered by the channel and is delivered as produced by the acknowledged originator, whereas processed means that the channel has modified the message without acknowledging any such authorship. One example is where a reporter produces a highly biased and one-sided report of a news item; this is still an original message. An example of a processed channel is where a newspaper report quotes very selectively from a speech given by someone, thereby actually altering the message produced by the acknowledged originator.

Besides the channel, another factor in the interaction is, as already mentioned, the *situation* in which the interaction takes place. Here also there are number of aspects to consider, say, b_j, and two of these are:

$j = 1$: The *exclusivity* of the interaction, with the three values Yes, No Positive, and No Negative. If Yes, both participants in the interaction are occupied exclusively with this one interaction. If No, one or both of the participants are involved in other interactions at the same time, and then we have to consider if these other interactions are affirming or negating the message.

$j = 2$: The *environment* in which the interaction is taking place, with three values: indifferent, positive, and negative. The difference between the environment and exclusivity is that the latter relates to active interactions, i.e., involving other people, such as the interactions experienced in a crowd, whereas the environment is about passive interactions, such as an impressive building (e.g., a church), a show of power (e.g., a military parade or a nuclear-powered aircraft carrier), or a noisy environment.

The concept of a channel in the transmission of a message has been treated by various authors in different contexts; the most famous one probably being that provided by Claude E. Shannon in his seminal work, *A Mathematical Theory of Communication* (Shannon, 1948). Another treatment is given by Everett M. Rogers in his book *Diffusion of Innovation* (Rogers, 2003). He categorises channels as (1) interpersonal versus mass media and (2) localite versus cosmopolite. Interpersonal channels involve a face-to-face exchange between two or more individuals; mass media channels are means of transmitting messages that involve such media as radio, television, newspapers, and the Internet. Localite channels are those linking individuals inside a particular social system; cosmopolite channels are those linking an individual with sources outside its social system. Regarding our classification of interactions by purpose, in Sect. 4.2, Rogers states that mass media channels are best suited to create knowledge and spread information, whereas the formation and change of strongly held attitudes (our identity elements) is accomplished mainly by interpersonal channels. He also notes that it is often difficult for individuals to distinguish between the source of a message and the channel that carries the message; something that resonates with our parameter a_4 above. One issue with Rogers' categorisation is the restriction of "interpersonal" to a face-to-face interaction; this seems to exclude the use of such media as telephone, email, and Skype, and does also seem, at least on the face of it, to conflict with his statement that interpersonal channels may be either local or cosmopolite.

At this point we should recollect that what we have been considering here are the characteristics of the transmission process that enter into determining the interaction strength; what we identified by the parameter μ in the previous two sections. That is, these characteristics are concerned with the effect of a process that operates on the identities of the two participants in the interaction continuously during a certain time interval, with the *purpose* of achieving an increase in the alignment between the two identities in relation to a particular subject. But this transmission process is not the only factor determining the effectiveness of achieving this purpose. Two other factors are the *quality* of the message, represented by the overlap between the content of the message with the mental capability of the recipient of the message, as was discussed in the previous section, and the *reinforcement*, either in the form of repetition of the same message, or in the form of different, but supporting messages. Both of these factors are related to the concept of *cognitive advantage*; a concept that appears to have been introduced in the paper previously cited, *Idea Propagation in Social Networks: The Role of 'Cognitive Advantage'*, by Simpkins et al. (2010), where it was defined as "the acceptability of an idea to a particular (culturally-circumscribed) community. It determines the tendency of ideas to become established in a community, as well as the rate at which those ideas are transmitted from one individual to another." In our context of the interaction between two individuals this definition needs to be focused somewhat more sharply, and we shall refer to cognitive advantage as a characteristic of the relationship between a message and the identity of the recipient of the message.

If we now combine our view of identity in Sect. 3.3.2 with our model of interaction in Sect. 5.2.1 as applied to a single subject, the identity of individual *a* with respect to the subject S is a set of assertions, W_a, each one with an associated set of arguments, and the identity of individual *b* is a set of assertions, W_b, each one also associated with a set of arguments. The set W_0 consists of those assertions that are common to both W_a and W_b, and the arguments associated with an assertion in W_0 is just the union of the two sets of arguments. However, it is in the process of establishing W_0 that cognitive advantage arises, a process that is not explicitly included in the model, but is implied to be part of initiating the interaction process. In the context of an interaction between two individuals, cognitive advantage will be understood as the shared statements about the subject of the interaction, existing in the identities of the individuals prior to the interaction. The cognitive advantage is an alignment, as per Eq. 5.2, established through interaction; an alignment both participants are aware of.

In identifying W_0 as a measure of cognitive advantage, we need to keep in mind the simple nature of the model of the binary interaction presented in Sect. 5.2.1. It is the model of one, isolated interaction between two individuals regarding a single assertion. In most real situations, both of these individuals will have interactions with other individuals, or make observations, or even have internal cognitive processes with regard to this assertion, all of which will increase the number of relevant influences.

We can now briefly consider the concept of the *social bond*, which was the central concept of the book by that name, as it is very closely related to the concepts of interaction and alignment, which we introduced formally earlier, and to the concept of cognitive advantage.

> The social bond is the relationship between any two members of a society defined by the sum of cognitive advantage over all the subjects present in the identities of the two members, and it is the social bond, extended to all pairs of individuals in a set, that binds them together to form a society.

A few examples will make the nature of the social bond clearer:

Consider a group of individuals, each with an amount of factual knowledge in a different discipline or specialty. If their interaction consists solely of interchanging this knowledge, i.e., teaching each other facts, then this will not convert the group into a society, and the interactions do not qualify as establishing social bonds.

As a second example, consider a group of individuals whose identities show a high degree of alignment, but that have had no interaction with each other. This group does not constitute a society, and there are no social bonds between its members, as they are not aware of the alignment.

As a final example, consider your relationship to an element of the social structure, e.g., the nation. It is represented by such symbols as the flag and the national anthem, but if you were the only person in the world, the concept of a nation and these symbols would be meaningless. It is the fact that they imply a relationship to other individuals that gives them meaning. Such structural elements of society as associations, organisations, and nations are emergent properties of the system formed by the social bond.

The above examples demonstrate that the social bond is the result of our interaction with another individual that influences our behavior—our relationship to the rest of society—by changing our identity, and so it is dynamic or time-dependent. The interaction between two individuals may be intense for a period of time and then cease all together, but the effect, in terms of the increase in the social bond, stays on in the identities of the two individuals and continues to influence their behaviours toward other members of the society and their interaction with these members. This propagating effect of the social bond is one of the main determinants of the dynamics of the evolution of society, and we shall consider this aspect of it briefly in Sect. 5.2.4.

The social bond and the alignment may, at first glance, appear to be identical, but while that is true as expressed by the contents of the two identities involved alone, there is a very significant difference between them, as the second example above demonstrates. The social bond is an alignment of which both parties are aware, either because it has been developed jointly by them, or because they have acquired it through a common process, such as education or indoctrination. It has much in common with *trust*, which is a hallmark of a healthy society, and with *commitment*, in the sense of a willingness to reveal themselves and have the courage to stand up in public. This brings to mind Hannah Arendt's discussion of *vita activa* as political action (Arendt, 1958), where a central aspect of action is its revelatory character; the actor has to reveal himself, has to have the courage to stand up amidst his peers. And he must commit himself, so there is a heroic side to action. The importance of political (i.e., public) commitment has also been noted by other authors, e.g., Luria (1985), and its relevance to engineering was discussed in Aslaksen (2017a).

Thus, the use of "social" in the social bond has a dual significance. One is to indicate that the bond is an element of the complex system that is society; the other indicates that the bond is an enabler of a sincere engagement. This puts a requirement on any interaction for it to result in, or strengthen, a social bond: it must reflect the participants' identities. It is, of course, entirely possible to create statements that do not reside in an individual's identity; effectively creating a "virtual identity" for the purpose of promoting a particular outcome. This possibility, while always present, becomes easier to realise, and more difficult to detect, with increasing involvement of technology applications in the processing and delivery of messages. In particular, messages could be constructed by IT solely on the basis of information about the recipient and the intended change in the recipient's identity or behaviour, without being part of the identity of any person at all.

In the book (Richerson & Boyd, 2004), previously referenced in Sect. 1.3.4, the authors introduce two processes involved in the evolution of culture—imitation and adaption—and they illustrate them with numerous examples. How do these two processes relate to the concept of the social bond? To answer this question, we start by noting that imitation may result from observation of someone else's behaviour or product (e.g., as in copying the style of artists by studying their paintings). There is no direct involvement by the person being observed, and there is no element of persuasion, which is often so important in the interactions that establish a social bond. But imitation is not mindless acceptance; it is simply a case of where the cognitive advantage for the information contained in the observation is so overwhelming that the evaluation process appears to be automatic, and no modification of the identity is required. Imitation may also be the result of interaction with the individual whose behaviour is being imitated and may be the result of a lengthy process of persuasion. Just by seeing me imitating the behaviour of another person, you would not be able to know if this has come about from me observing that person or from having been persuaded by that person.

Imitation is something we can easily observe. It is still the main component of education, more so in the early years than in the later years, and it is something we all practice throughout life, to varying degrees, as we imitate successful people, people in positions of authority, and role models, and accept their judgements and views. But, as Richerson and Boyd point out (p. 99), if imitation was the only process, society would not be able to adapt to changes in the environment; everyone would just be imitating everyone else, and we end up with a society in which the alignment between any two individuals is 1. However, if the information received, either through observation or through interaction, does not immediately conform to the identity, the evaluation process will be more demanding and require more time and effort, and the outcome is either rejection or an adaptive action; the latter either as a physical action or as a modification of the identity, or a combination of both. This result will be transmitted to other individuals and evaluated by them, and it is this collective evaluation process, or *collective intelligence*, that results in the evolution of society, or, as Richerson and Boyd would say, in the adaptive capability of culture.

Thus, imitation and adaption are both accounted for in our information view; they are just separated by their position vis-à-vis the current state of the identity, as measured by cognitive advantage. This also shows that there is not a sharp division between the two processes, both can be active in the same circumstances; it is a question of relative importance. With regard to their relationship to the social bond, imitation is an expression of the existence of a social bond; adaption is an expression of the exercise of intelligence—individual and collective. The ability of society to adapt depends on its ability to exercise its collective intelligence, which we shall develop as a central measure of the state of society in Sect. 6.3.

We might note that there is something inherently *self-referential* in the concept of the social bond; it creates a society out of a collection of individuals, but the social bond exists only in the context of a society. As a brief side-step, we note that the notion of something being self-referential can be given a precise meaning in mathematics, to a large extent due to the work of Karl Gödel. But a more relevant application is the one made popular in the books by Douglas R. Hofstadter, *Gödel, Esher, Bach*, and *I am a Strange Loop*, (Hofstadter, 2007), in the exploration of consciousness and the meaning of "I". The same issues that are raised there could be raised with respect to the "consciousness" of society, in the sense that the collective intelligence of society is a product of the individual intelligence. Society's "consciousness" is an emergent property of the system formed by the consciousness of individuals interacting based on a social bond, and just as individual consciousness develops through childhood, social consciousness develops as society evolves.

Finally, it should be noted that there is a body of work under the title of Social Bond Theory, more generally known as Social Control Theory, based on the work by Hirschi (1969), which is concerned with the bonding of persons to social institutions and the effect of this bonding on their tendency to criminal behaviour. And the concept of a social bond relating individuals to society is used quite generally by many authors, which contrasts with the quite specific use in the present work.

5.2.3 Persuasion

Persuasion as a component of interaction was introduced in Sect. 4.3. The purpose of the current section is to consider some important aspects of this characteristic as it applies to the social bond; a generalisation to our information view is contained in Sect. 5.3.

Persuasion is not a very sharply defined concept, and is used with a wide range of meanings, including instruction, training, education, indoctrination, coercion, and brainwashing—any form of interpersonal interaction whatsoever. It is also a concept that has received considerable attention in academia, and in his doctoral thesis, Naas (1990) makes the interesting observation that the pre-Platonic Greeks, in particular as expressed by Homer in the *Illiad*, saw persuasion as a turning-away from oneself, threatening the identity of the individual, and therefore something that was to be avoided; a view that has quite some merit today. However, the Greeks and the Romans then developed models of persuasion that were aimed at *oratory*, which was seen as an art-form judged solely by its effectiveness, divorced from truth and any wider consideration of desirability, and this is the prevalent view today, in particular in the Anglo-Saxon justice system, which is essentially an oratorical joust between lawyers.

An excellent treatment of the subject matter is given in the doctoral thesis by Mark J. Stoda, *Symbolically seeking a change of mind: a model for comprehending, constructing, and critiquing persuasion* (Stoda, 2006), and the following observations are inspired by that work. After reviewing the various definitions and views of persuasion in the literature, he is careful to delimit his treatment to interpersonal persuasion, which is defined as being a deliberate activity accomplished symbolically and seeking a voluntary change of mind that favours the persuader's goal. This limitation fits very well with the use of persuasion within the social bond. He then goes on to review Greek and Roman developments of rhetoric, exemplified mainly by Aristotle and Cicero, respectively, and it is notable how Aristotle and, to a somewhat lesser extent, Cicero emphasize the importance of the moral quality of the persuader, something that would indicate that the persuadee is expected to have a direct personal knowledge of the persuader, but that is very unlikely in most of our interactions today. Moral integrity might indeed be a disadvantage in today's political arena, as was demonstrated by one of the most successful persuaders in the last century, Adolf Hitler.

Most of the other factors characterising successful persuasion identified by Aristotle and Cicero are equally valid today, such as knowledge of the persuadee's background and attitude, the ability to transmit a sense of passion, being attuned to the persuadee's emotions, and appearing concerned and personally committed. Progressing from this classical foundation, Stoda then analyses the more recent literature and integrates its various claims and viewpoints into a model of persuasion with a logical and useful structure, but before displaying this model, let me briefly highlight a few points that I found particularly relevant to the present work. The first is the important role of *uncertainty*; if there is no uncertainty, there is no need for persuasion. Not only does the persuader need to be cognisant of the full range of possibilities, but also understand the extent to which the competing possibilities are present in the persuadee's mind; this will determine whether to refute them or to just downplay or ignore them. The second point is the necessity for the persuader to expose his or her arguments to *public scrutiny*; to stand by them without making any concessions or compromises, no matter what the consequences. Even if this results in a long period of ridicule, ostracism, and persecution, people will start to think that if the argument is this important to the persuader, there must be something in it. We can see a connection with Hanna Arendt's *vita activa* (Arendt, 1958), as noted in the previous section.

The third point is the interpretation of knowledge as *believed information*, and that the argumentation that leads to it is a social process. It is not formal logic, but proactive empathy and practical reasoning within a public discourse. The existence and properties of the public discourse are central to the functionality of the collective intelligence and to our view of The Good Society, as will be further developed in Chap. 6.

Stoda's model of persuasion consists of two closely related parts: an entity-relationship diagram, and a process flow diagram. The entity-relationship diagram is constructed by heeding the imperative of focusing on the characteristics of the persuadee; this leads immediately to the recognition that the mental processing of a persuasive appeal consists of two distinct processes. One, the *peripheral* process, is a judgmental-heuristic process, a quick evaluation based on picking up *clues* in the persuasive appeal that activate pre-existing judgements regarding credibility or latent mental discords. Two, the *central* process, is a systematic and considered process involving analysis of the content of the appeal based on reason, conscience, and imagination. These two processes are basically the two processes defined in Kahneman (2002) and mentioned in Subsect. 4.1.3. Accordingly, the entities making up a persuasive appeal stand in a one-to-one relationship with the mental capabilities of the persuadee, resulting in the two-level diagram shown in Fig. 5.3.

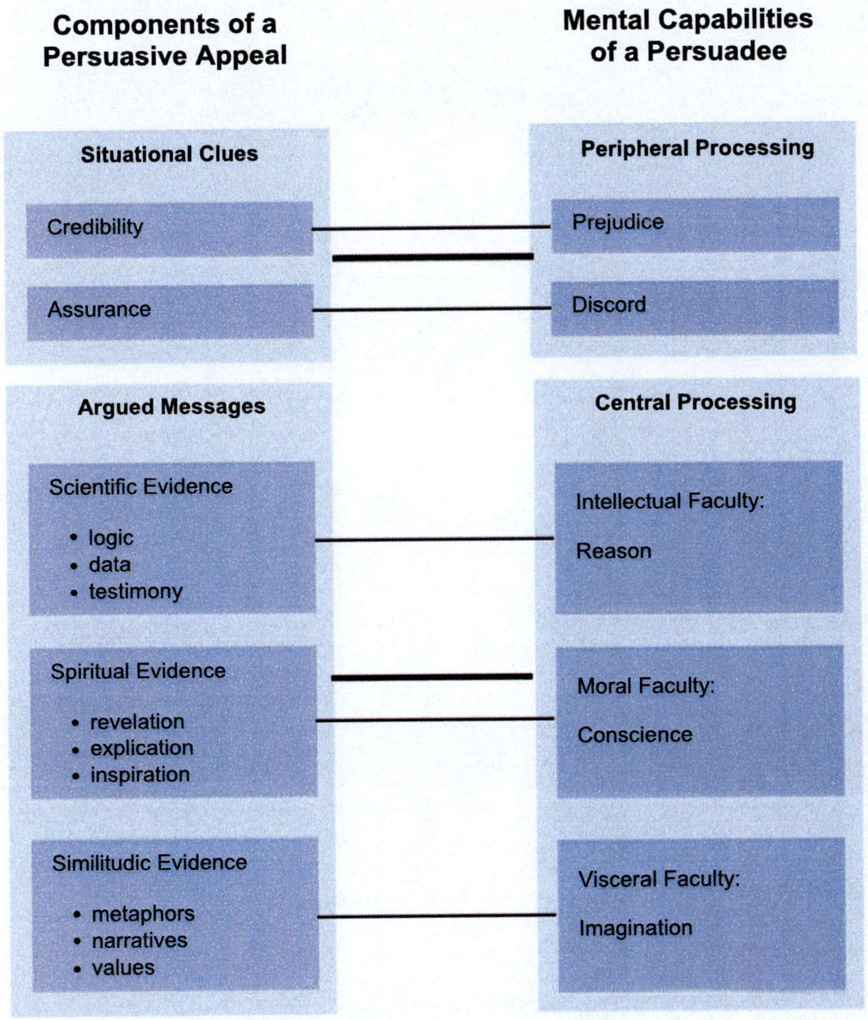

Fig. 5.3 The entities of a persuasive appeal and those of the persuadee's mental capabilities, and the relationships between them

The process flow diagram, shown in Fig. 5.4, illustrates how the persuadee's mental capabilities determine the outcome of the persuasion appeal, i.e., its success or otherwise. The initial stage of the process is an evaluation of the appeal based on the mental state of the persuadee relative to the content of the appeal, which can be characterised in terms of two parameters, aptitude and attitude, each of which has two main components:

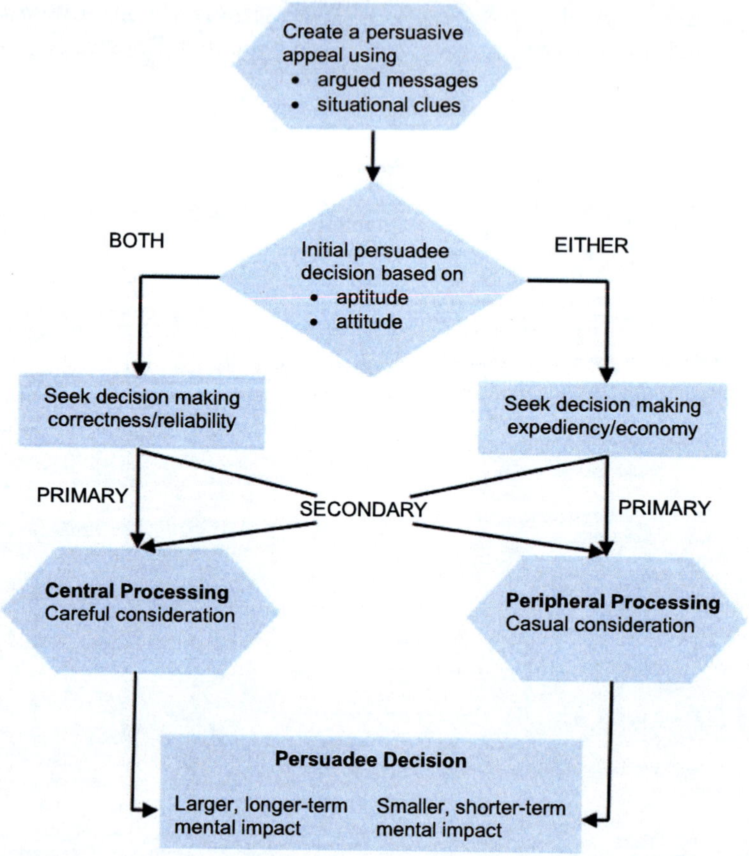

Fig. 5.4 Flow diagram of the persuadee's decision making process

Aptitude	Capability—physical and mental
	Cognizance—adequacy of education/experience
Attitude	Valence—evaluation of claim's agreeability
	Salience—commitment to claim evaluation
	—importance assigned to subject
	—affinity for critical evaluation

If both Aptitude and Attitude are positive, the persuadee will seek decision-making reliability, but if either Aptitude or Attitude are negative, the persuadee will seek decision-making expediency. However, it is not always a one-or-the-other situation; as indicated in the diagram, it is possible for both central and peripheral processing to be involved, and, as a result, the mental impact (i.e., the modification of the identity) to be a combination of longer-term and shorter-term elements. A detailed description and discussion of each of the entities in these two diagrams, together with numerous examples, are included in the dissertation (*op.cit.*), and we may refer to them at various places in the rest of the book.

While Stoda's model is very useful, it has, from our point of view, a few limitations that we need to be aware of. The first is that it does not consider the various channels involved in transmitting information between individuals, such as print (incl. email and social media), radio, and TV. The second is that it does not consider the frequency or repetition of persuasive appeals, and the third is that it does not consider the influence of technology and the opportunities technology offers for modifying the persuasion process. In particular, the opportunity for making it interactive. Also, the model specifically excludes any form of coercion; not only physical, but also mental. In this regard, a thorough and very readable discussion of both the psychological and physical aspects of the whole spectrum of persuasive techniques is contained in the book *The Science of Thought Control*, by Taylor (2004), with a focus on the more direct and oppressive forms, as applied by sects or oppressive regimes, and illustrated in the dystopian visions of Orwell and Huxley. But the book also touches on persuasion, as found in advertising, education and in the bias of the mass media.

This discussion of persuasion provides a suitable opportunity for commenting on the relationship between this form of interaction and the propagation of a meme. We introduced memes in Subsect. 3.3.2, as discrete, faithfully replicated units of knowledge. In the case of persuasion, the end-result is the establishment, in the mind of the persuadee, of a certain belief, in the form of a set of arguments supporting an assertion. But for this to happen, the interaction will generally involve other information; possibly praise for the persuadee or the promise of some sort of reward or advantage, and it will probably involve some information the persuadee rejects. Also, the thought processes of the persuadee may transform some of the information into different arguments supporting the assertion, reflecting his or her previous knowledge and experience. In short, the process of transmitting a belief is a much "fuzzier" process than transmitting a meme.

The readiness of society to accept and try out new ideas when they become available, as was discussed in Subsect. 3.4.3 with reference to Buchanan (1992), can be seen as a collective reflection of the persuadee's mental state in this regard. In the simple model of the binary interaction in the previous section, this is reflected in the quantity $(1 - W_0 - W_1)$. Returning to this model, persuasion is represented by the case where one of the individuals in the interaction, *a*, is intent on persuading the other individual, *b*, of *a*'s belief on a particular subject. As a boundary situation, we might assume that all of *a*'s identity is taken up by this one subject, so that $W_{a,1} = 1 - W_0$, and that *b* has no preconceived differing beliefs about the subject, i.e. $W_{b,1} = 0$. Then, for a value of the interaction strength, μ (and there is only one relevant component) equal to 0.1, we find the following dependence of the alignment rate on the initial level of alignment, W_0 (Table 5.2).

This result can be expressed by saying that the effort to erase the last bit of independence, or lingering doubt, in individual *b* takes a disproportional amount of effort.

Table 5.2 The alignment rate, α_0, as a function of the initial level of common belief, W_0, for the case where $W_{a,1} = 1 - W_0$, and $W_{b,1} = 0$

W_0	α_0
0.1	0.0603
0.3	0.0376
0.5	0.0198
0.7	0.0072
0.9	0.0007

Table 5.3 The alignment rate, α_0, as a function of the strength of the diverging belief in individual *b*, $W_{b,1}$, for three cases of the interaction strengths μ^+_{ab} and μ^-_{ab}

	μ^+_{ab}/μ^-_{ab}		
$W_{b,1}$	0.2/0.1	0.15/0.15	0.1/0.2
0.1	0.0648	0.0531	0.0384
0.3	0.0458	0.0434	0.0352
0.5	0.0357	0.0374	0.0326
0.7	0.0299	0.0333	0.0305
0.9	0.0263	0.0304	0.0288

A further case arises when individual *b* actually has some differing views on the subject, i.e., $W_{b,1} \neq 0$. In that case, both μ^+_{ab} and μ^-_{ab} are relevant, and Table 5.3 shows the alignment rate as a function of $W_{b,1}$ for three choices of μ^+_{ab} and μ^-_{ab}, again with $W_0 = 0.1$ and $W_{a,1} = 0.9$.

Comparing the two outside columns shows the increasing importance of persuading individual *b* to give up its differing views as $W_{b,1}$ increases. It is interesting to note that even at the highest value of $W_{b,1}$ it is better to put equal effort into positive and negative persuasion (the centre column) than to concentrate on getting *b* to give up its differing views (right-hand column)

5.2.4 Some Additional Results

In *The Social Bond*, the binary interaction is developed in much greater detail by means of a number of more complex models. In this subsection, I shall just mention some of the insights that are useful for our purpose in this book.

Group alignment. The model considers a group of *n* identical individuals, so that all the binary interactions have the same strength and there are only two interaction parameters, μ^+ and μ^-. The identity of an individual has three parts, as before: W_0, W_1, and W_2, but the part that was only a single W_2 now contains the identity items the individual has received from all the other $(n - 1)$ individuals in the group. The two parts W_0 and W_1 are as before, but the third part now contains $(n - 1)W_2$ parts. The alignment between any two individuals in the group is given by

$$\alpha = \frac{W_0 + nW_2}{W_0 + 2W_1 + (n-2)W_2} \tag{5.5}$$

and, except for a particular case, the end-state is complete alignment, $\alpha = 1$.

Dominance. A single individual within a group, say, individual number 1, is more convincing than the other individuals, which we express by assigning a greater interaction strength to this individual than to the other individuals of the group. That is, we have two sets (+ and −) of interaction parameters, $\mu_1 > \mu_i > 1 = \mu_2$. The effect of this is that the associated W_2, identified as W_2^a, will be greater than the others, identified as W_2^b, and that the identity of individual 1 becomes dominant. If we express this in terms of a parameter, the *dominance*, κ,

$$\kappa = \frac{W_2^a - W_2^b}{W_2^b} \tag{5.6}$$

the interesting result is that, for $\mu_1^- = \mu_2^- = 0$, the dominance initially rises rapidly, but then decays to a final value, as shown in Fig. 5.5. Both the maximum value and the steady-state value increase with the increase in the ratio μ_1^+/μ_2^+. The steady-state values are 0.638, 1.400, 1.793, and 2.103, and in the limit of a very large value of the ratio, it is straight forward to show that the steady-state value is 2.2.

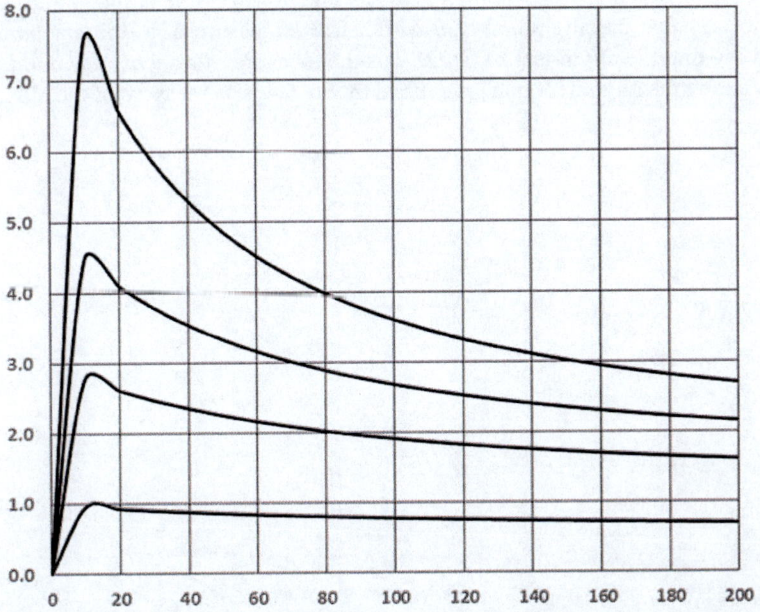

Fig. 5.5 The dominance, κ, of individual no.1 as a function of time (in units of time), for the case of $\mu_1^- = \mu_2^- = 0$, $W_0 = 0.1$, $W_1^a = W_2^a = 0.2$, $\mu_2^+ = 0.05$, and $\mu_1^+ = 0.5, 0.3, 0.2$, and 0.05, from the top to the bottom curve. The interaction strengths, μ, are in units of per unit time

What is happening in this case is that all the identity elements peculiar to individual no. 1, W_1^a, are transferred to each of the other individuals fairly quickly; that is, by a time close to the peak dominance, after which the elements peculiar to the other individuals, i.e. W_2^b, , are exchanged until their identities reach maximum capacity (normalised to unity).

However, if, for a fixed value of the ratio μ_1^+/μ_2^+, μ_1^- is increased from zero, the result is that the decrease is reduced and eventually eliminated, so that the initial large dominance remains. What is happening is that W_2^b is prevented from increasing, which leaves room for W_2^a. to develop further. Both of these cases illustrate situations well known in reality: getting in and making "the hard sell" early and then cutting off any further deliberations, and the importance of reducing the existing beliefs and structures before establishing a new belief. The first stage of revolution is destruction of the existing order and increased uncertainty, or even chaos.

Propagation. We can use an extension of the simple model, defined by Eq. 5.1, to get a first look at what is involved by having three people, individuals *a*, *b*, and *c*, interacting. They are assumed to be all from the same basic cultural background, so W_0 is common to all three, and, as before, individuals *b* and *c* have no preconceived opinions or ideas regarding the subject under consideration, i.e. $W_{b,1} = W_{c,1} = 0$, and $\mu_{ab}^- = \mu_{ac}^- = \mu_{bc}^- = 0$. Individual *a* starts to persuade *b*, and as *b* accepts elements of *a*'s belief, *b* starts to persuade *c*, so that we now have two measures of the intensity of the interaction: the rate of alignment of *b* with *a*, and the rate of alignment of *c* with *a*. An example of the two alignments as functions of time is shown in Fig. 5.6.

If we say that the alignment of *b* with *a* is the measure of a *primary bond*, we would say that the alignment of *c* with *a* is the measure of a *secondary bond*. And, again, taking the alignment rate, as defined by Eq. 5.4, as a measure of the alignment

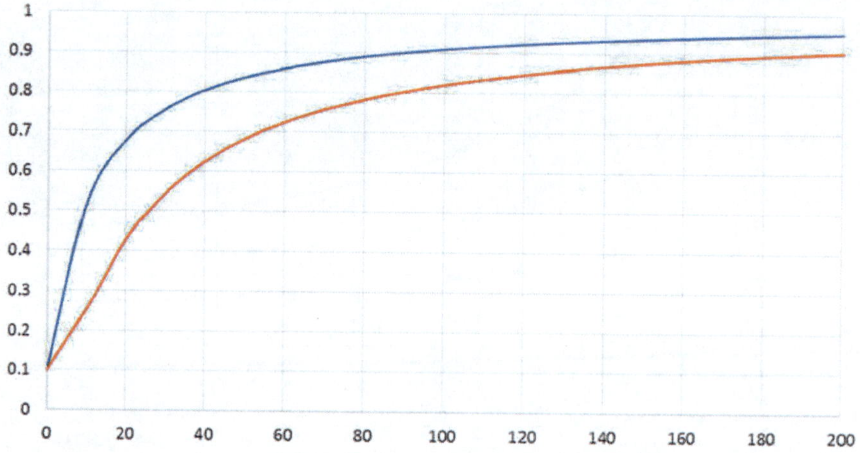

Fig. 5.6 The alignment of individual *b* with individual *a*, upper curve, and the alignment of individual *c* with *a*, lower curve, for the case where $\mu_{ab}^- = \mu_{ac}^- = \mu_{bc}^- = 0$, $\mu_{ab}^+ = \mu_{bc}^+ = 0.1$ per unit time, $W_0 = 0.1$, $W_{a,1} = 0.9$, and $W_{b,1} = W_{c,1} = 0$. The horizontal axis is units of time

process, the above example yields that the ratio of the primary to the secondary bond is 0.0404/0.01482 = 2.726. What this result shows, is that, if we think of individual *a* as the Master, individual *b* as a disciple, and individual *c* as a convert, it is inefficient for the Master to continue persuading the disciple past a certain point instead of starting with a new disciple.

5.3 The Individual's Interaction with Society

5.3.1 Definition of the Interaction

The previous section considered the interaction between two individuals, and while this is the basic interaction taking place in society, it does not actually mention or relate to society at all. To describe society starting from this basic interaction is the bottom-up approach, and it would be a very difficult and laborious task, analogous to describing the human being starting from the electromagnetic interaction between electrons and atomic nuclei. In this book we are mainly taking a top-down approach, starting with a high-level description of society as an unstructured system of identical individuals and with the interaction in the form of the individual's interaction with society as a whole, including the physical environment in which it is embedded. This description is illustrated in Fig. 5.7, and if we compare this with Fig. 3.2, the difference is that the interaction between individuals is now generalised to an interaction between the individual and the body of individuals, with that body being represented, in our information view, by what we shall denote as the *circulating information*.

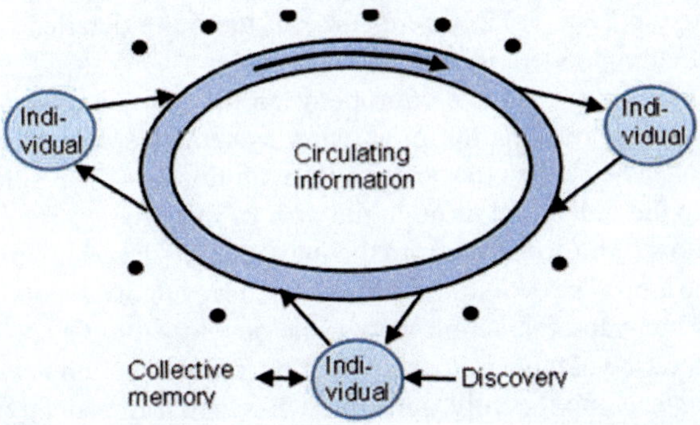

Fig. 5.7 A pictorial representation of our high-level description of society. The *circulating information* is accessed by and added to by the individuals; *discovery* is a means for the individual to obtain information from Nature through targeted actions (e.g. measurements), with the interaction with the *collective memory* sources (journals and data bases) shown separately

The individual in Fig. 5.7 is as described by the model developed in Chap. 4, with the three information flows shown in Fig. 4.4 corresponding to the flows shown in Fig. 5.7, with the understanding that the common information repository is part of society—it is what we in Subsect. 4.1.2 identified as society's *collective memory*. The *circulating information* is the large amount of information moving around in society, and each individual is exposed to, or accesses, a small part of it; that is the main part of the information flow labelled *b* in Fig. 4.4. Each individual also contributes information into the circulating information—this is how society evolves; it is the information flow labelled *c* in Fig. 4.4. And discovery is the information flow labelled *a* in Fig. 4.4.

If we focus on the class 3 of the circulating information, which is our primary interest, the processing of that information by the body of individuals is one part of what we already in Subsect. 3.3.1 identified as the *collective intelligence*; the other part is the reflexive and deliberative interaction—the participation in the *public discourse*. The operation of the collective intelligence is, essentially, what Habermas calls *deliberative democracy*, and the circulating information is his *public sphere* (Habermas, 2006); the only difference being that deliberative democracy is embedded in social philosophy, whereas the collective intelligence is defined in terms of our high-level model of society embedded in systems engineering. Implicit in the high-level model of society in Fig. 5.7 is the free flow of information between individuals; the flow is generated by the individual and is composed of the identities of every individual, as noted in Sect. 4.2.

In both Figs. 4.4 and 5.7, the arrows indicate information flows, and these flows are measured in information items per unit time. This is certainly an important measure of interaction, but it is only one of the many required to fully characterise interaction; it is what we identify as the *intensity* of the interaction, as was indicated at the end of Sect. 3.4.1. Section 5.2.1 defined the *strength* of the binary interaction as its ability to align the identities of the two participants in the interaction, and this was further elaborated in Sect. 5.2.2, where some of the more detailed factors entering the definition of strength were identified.

When we now turn to the interaction between the individual and society in our view of society as an information-processing system, we are faced with a quite different situation. The change to society—its evolution—is the result of the change society effects in the individual through the flow of information from society to the individual; the flow of information from the individual to the circulating information simply closes the loop. The evolution of society is the result of a second loop, between the information view and the action view to be developed in Chap. 7, and we will defer the consideration of what influences the flow of information between these two views to that chapter. Consequently, our approach is to first consider that the evolution of society is determined by the flow of information to the individual, and this flow of information—labelled as *b* in Fig. 4.4—has undergone a great transformation as society evolved, mainly under the influence of the development of technology, as mentioned briefly in Sect. 2.3, where we depicted that development in terms of three phases. In the first phase, the transmission of information was essentially direct from person to person, mostly verbal, but also by hand signs and body language. The only storage of information was in the human memory, and the only spatial propagation of information was in the movement of humans.

The second phase is characterised by the encoding of information in various forms, such as signs, letters, drawings, paintings, and sculptures. This introduced a dependence on technology, in the form of tools for producing the encoding, and in the form of the media to contain and store the encoded message. With the advent of printing and photography the cost of production was reduced by orders of magnitude, but at an increasing dependence on technology. Throughout this phase it was the case that while the transmission was not restricted to being in real time and spatially within the reach of the human voice, the spatial transmission of information depended on the medium being physically transported, as by the postal service. The storage capacity was greatly expanded, as was the temporal aspect of the storage (we have clay tablets from more than 4000 years ago). This phase also introduced a novel feature: The possibility of anonymity and/or false attribution of authorship. Without face-to-face interaction, "the written word" could be given an authority that was unverifiable, ranging from divine to scientific, and "fake news" took a big step forward.

The third phase, which started with the invention of the telegraph, is the current phase of human interactions dominated by telecommunications and information technology. This technology includes landline telephone, radio, television, digital data transmission, such as the Internet, and mobile telephone, and in addition to only transmission of information, the technology provides the means for processing the information. Compared with the two preceding phases, the two important differences are that most of the communication is not a two-way interaction, but a one-way provision of content, often for the purposes of advertising or entertainment, and that the transmission does not involve the transport of a physical medium as the carrier of the information, which means a reduction in cost by orders of magnitude. The result is that information, in its various forms, has become a very significant aspect of modern life, and that what is shown as the flow of information in Fig. 5.7 is very much more complex than just the direct interchange between two individuals described in Sect. 5.2. There is, of course, still significant person-to-person interaction, but as far as class 3 information is concerned, the overwhelming part is provided through some medium. Some of the characteristics of the binary interaction, such as persuasion, are reflected in today's interaction, but in a different form, as we discuss below.

5.3.2 Some General Features of the Interaction

Based on the foregoing, the interaction of the individual with the circulating information can be seen as being determined by two factors: the individual's ability to access and process information, and the influence of the *medium*, defined in Subsect. 4.1.4 as medium b, and which we there argued should be associated with the individual rather than with the circulating information. The ability to access and process information is composed of two components: the ability to accept signals from the various input channels, which is essentially the ability to see and to hear, and the ability to

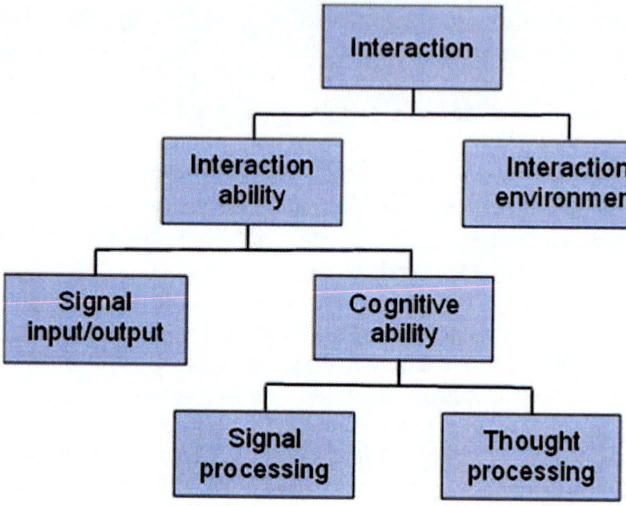

Fig. 5.8 A possible breakdown of the simplified concept of interaction into its main determining factors

process these signals into information, understanding, and knowledge, which is our cognitive ability. And, finally, our cognitive ability can again be decomposed into two components: a basic ability to process signals into information, and the ability to understand what this information means and turn it into knowledge. In Kant's schema (Kant, 1781), the first ability was that of representation, turning sensations (signals) into intuitions and concepts, as classes of intuitions; the second was the ability to think, to make judgements about representations (i.e., thoughts), determine their truth value, and turn them into knowledge. This breakdown of the interaction process is represented graphically in Fig. 5.8.

With our immersion in information technology, a more immediate analogy might be to think of the two components of cognitive ability as processor (hardware and operating system) and software, with the latter in the form of learning applications (e.g., neural networks), with the obvious difference that the networks can not only learn, but also create new networks. The newborn child has only the first component; all the "software" that enables cognition is acquired through the child's interaction with its environment as it grows up. Hence, interaction increases the cognitive ability, and the cognitive ability increases the strength of the interaction; it is a feed-back process. Thus, the "cogito, ergo sum" must be taken with a pinch of salt; it is not enough to have the ability to think, it is also necessary to have something to think *about*. The Thinker might seem to sit in splendid isolation, but any thinking going on is a result of previous interactions. The modern individual is a social creation.

The increase in cognitive ability through the lifetime of an individual is at first exponential, until at some point it is limited by a number of factors: The information content of the environment, the capacity of the processor, the rising demand for maintenance of the acquired knowledge, as well as the physical requirements of the supporting body (need for sleep, nourishment, exercise, etc.), together with the demands of routine work. Accordingly, the limit on the cognitive ability will vary considerably among individuals, but irrespective of the variability, the important fact is that the cognitive ability of each individual—the ability to process the incoming information (visual and auditory)—has a fixed value, and a consequence of this fact is demonstrated in the following.

All the items in the stream of information arriving at an individual are processed by the component of the cognitive ability shown as signal processing in Fig. 5.8, but only the items in classes 2 and 3 require subsequent thought processing. The classification of information input is highly subjective, and will also depend on the situation the individual finds itself at the time, but if we think of this classification in terms of the *importance*, *q*, of an information item (measured on some scale) *to the individual*, the information stream can be classified by a distribution, *r*(*q*), such that

$$\int_{q_1}^{q_2} r(q)dq$$

equals the number of items with importance in the range $q_1 \leq q \leq q_2$ impinging on an individual per unit time, and

$$\int_{0}^{\infty} r(q)dq = intensity.$$

The relationship between importance and information classes is illustrated in Fig. 5.9.

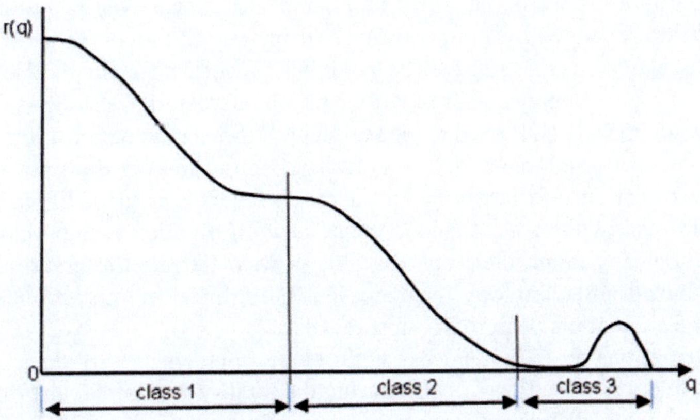

Fig. 5.9 The distribution of information items according to their importance to the individual, *q*, and the relationship to their classification

The individual's processing capacity is shared between signal and subsequent thought processing, such that the capacity for thought processing is what is not required for signal processing, and of this capacity, the processing of class 2 information takes precedence in almost all individuals. It follows from this highly simplified picture that the strength of interaction depends not only on the rate of arrival of class 3 information items, but on the availability of processing capacity for these items, which depends again on the rate of arrival of class 1 and class 2 items.

In Fig. 5.9, the value of q at the boundary between class 1 and class 2 is related to the concept of *attention*; any item with importance less than this value does not catch the individual's attention. One of the many ways in which to increase the importance of an item is to repeat it; with each repetition it moves to the right within class 1 until it finally crosses over the boundary value and comes to the attention of the individual. This effect is well known in advertising; the undesirable side-effect, from our point of view, is that it increases the load on the processing capacity, and thereby reduces the capacity for processing of class 3 information.

Another factor that enters the strength of the interaction is its *quality*. We might immediately understand this as applying to the information being received and relate it to such a concept as the truth of the information. But what is considered to be the truth is highly subjective; "the world was created about 6000 years ago" is the truth to some and nonsense to others. The evaluation of an information item will always depend on what the individual already knows, and so the quality of the interaction cannot relate to the content of the information alone. It is a characteristic of the transmission of the information between the source and the receiving individual, and it has two parts to it: The fidelity of the transmission, i.e., the degree to which the received information is identical to the information created by the source, and the correct identification of the source.

In the case of face-to-face interaction, the quality takes on its maximum value (which we might arbitrarily set equal to 1), and this is true even if the interacting individuals are separated by a large distance (using e.g., Zoom or Skype), although, with virtual reality, this may not even be true much longer. The quality is also maximal in the case of interaction by means of written or printed material, as long as the author is correctly identified and known to the recipient. However, an increasing proportion of the information available to an individual via online sources does not allow this degree of verification, and the digital format makes it very easy to modify information without approval by the source, and to create new information with no basis in fact and/or without any identifiable source. This is why, despite the great benefits of telecommunications technology, the increase in the strength of interaction is slowing down, as we shall come back to in Subsect. 10.1.2.

The interactions an individual has with its environment are so varied that it is difficult to find any useful way of structuring the totality of them. Trying to classify by the organ involved does not seem to be very definite; a conversation between two people involves not only the voice and ears, but body language and facial expressions. The hands can produce messages through such varied means as writing, typing, painting, drawing, carving, and the like. Another classification might be by how direct the interaction is, with face-to-face the most direct, then through voice by telephone, then through writing (letter, book, newspaper article, etc.), then through clothing and ornaments, and then through architecture and sculpture as the least direct; it is probably a classification very open to disagreement. Another classification might be in how effective it is in reaching its intended 'audience', and there are no doubt many others.

A different approach is to, at least initially, classify the interaction by its intended *purpose*, such as to transmit facts and ideas (teaching/learning), describe, amuse, threaten, praise, and many others, and as a point of departure, we can recognise four main groups of purposes: To convey factual information; to find a solution to a problem through dialogue and evaluation; to reach a mutual understanding or belief, as a combination of the understanding and beliefs present in the two participants in the interaction, i.e. a combination of elements of their identities; and to induce an action. The first two groups distinguish themselves from the third in that they can result in new identity elements; the interactions in the third and fourth (and to a lesser extent the second) groups have an element of *persuasion.* If we give some consideration to our own experiences, we can recognise that a not inconsiderable part of our interactions can be described as having, at least partly, the purpose of persuasion. If I tell someone about a good book I have just read, or a good movie I have just seen, I am not only, or even mainly, transmitting facts about it; my purpose is to persuade them that it is something they should read or see. In so much of our interactions there is a strong element of persuasion; we want the other person to believe what we are presenting. This probably has a deep psychological explanation; our fear of being different, of not fitting in to our society, as well as the converse, our fear of someone that is different, and therefore a potential disruption to our society. It correlates with our view of evolution that, as a species we, explicitly or implicitly, understand that cooperation—forming and maintaining a society—is key to our survival, and we fear and oppose anything that we perceive as threatening it.

In the context of this book and its explicit purpose of investigating the effect of the interaction between individuals on the evolution of society, as well as our view that changes to society are driven by changes to the identities of the members of the society, we can see that persuasion is the main component of the interactions we are interested in. If we recall the picture we formed, in Sect. 3.3, of an individual's identity, persuasion is the process of enlarging the subset of the identity containing the elements relating to the subject of the persuasion. The size of the subset is a measure of the strength of the individual's belief in the truth of the subject, and of the probability that that the individual will act in accordance with this belief, given the opportunity. But this probability depends not only on the strength of this one belief; the evaluation process will search for other beliefs in the identity that are relevant to a particular input or set of inputs, and so an important aspect of persuasion is a 'negative' one; persuading the individual to discard any conflicting beliefs, as was also the case with the binary interaction.

Section 5.2.3 considered persuasion in some detail, utilizing a particular scheme (Stoda, 2006), but this was done in the context of binary interactions, so that the persuasion involved two persons with their particular characteristics. Now we are looking at an individual—an average over all members of the society—subjected to

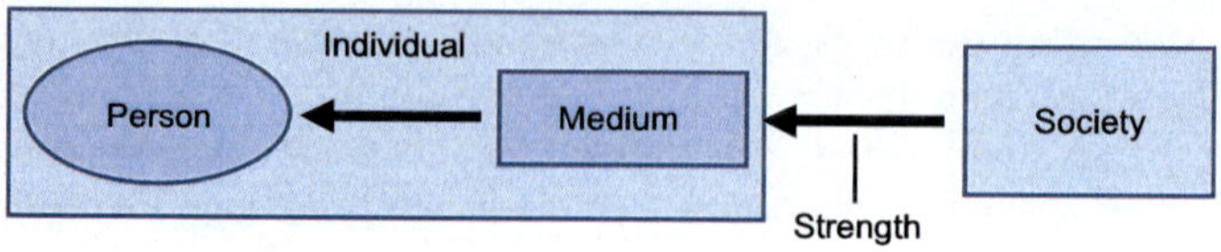

Fig. 5.10 The role of the medium in the top-level description; it modifies the way the individual sees society and the strength of the input side of the interaction

a stream of information generated by an impersonal body—society. The strength of the interaction therefore depends on the relationship of this stream to the individual. But consider, for a moment, a particular person in the society. This person receives a particular part of the circulating information, determined by the person's interests and education, interface to the information stream, and the time available for receiving and processing information. That is, each person "sees" and "feels" society differently, and the strength of the interaction—the degree to which the person's identity is influenced by this interaction—is also different for each person. Not only in its absolute value (however we measure that), but above all in regard to the myriad of channels and interfaces involved in the interaction. For some persons the most important interface may be the workplace, for others it may be religious communities, social groups, or numerous special interest institutions, including political parties, and for most it will be various media channels. In the information view, a person is inseparable from its information input channels, and together they form part of what we already in Sect. 1.3.2 called a hybrid (the other part is the technology involved in converting information into action).

Thus, in this general description of the interaction, the picture we should have is the one shown in Fig. 5.10, where the person is as before, and the medium is a composite of the channels available to the person "averaged" over all persons in the society. As the channels are largely defined by technology, the nature of that composite and of the averaging will be treated in Chap. 8. The block labelled "Society" contains all the information available in the society.

5.3.3 The Information View

Following the general considerations in the previous subsection, which included physiological aspects of information transmission and processing, we now return to our view of society as an information-processing system and to the model of the individual presented in Chap. 4. In that view, and given our focus on the operation of the collective intelligence, the strength of the interaction can be defined as follows:

> The *strength* of the interaction between the individual and society is the extent to which the information input to the individual from the circulating information changes the identity of the individual.

To relate this definition to our model, we first consider the flow of information items arriving at the individual at the rate of μ_a. This flow is composed of items in the three classes:

class 1: $\mu_a(1 - x)$
class 2: $x\mu_a$
class 3: $yx\mu_a \equiv \mu_1$.

The flow is first processed by Process A to identify those items worthy of attention—the items in class 2. Process A then evaluates the class 2 items and, as one result of the evaluation, some items are identified as being of class 3 and requiring more extensive processing, which is performed by Process B. It is the outcome of this latter process that changes the identity and may lead to changes to society, and we can restate the definition of the strength of the interaction as the rate at which items in the identity are replaced by items from the circulating information.

In our model of the information flows in society, this rate depends on two things: the rate at which medium b extracts items from the circulating information, and the modification of the flow by the medium. Of course, it also depends greatly on the composition and volume of the circulating information, and as we noted earlier, the 'richness' of this information has increased over time; it varies from nation to nation and is dependent on such factors as the state of technological development. The approach taken in the present development of a view of society is to accept whatever the circulating information is as a given, typically as present in the West. The factors and influences that steer and influence the evolution of society are all concentrated in the three media.

5.3.4 Dynamics of the Interaction

The picture in Fig. 5.10 is of a static system; our interest is primarily in the evolution of society. We then need to expand Fig. 5.10 by adding the individual's information output and the feed-back loop involving the physical domain, as shown in Fig. 5.11.

In Fig. 5.11, the individual's identity was introduced in Sect. 3.3.2, the collective intelligence was introduced in Sect. 3.3.1, and society's belief system was defined in Definition 3.7. Adaptive action—the realisation of changes to society's belief system as changes to the fabric of society—is the subject of Chap. 7. But how does the collective intelligence actually change the belief system? That is, how does the commonality, κ, of an identity item go from $1/n$ to n? It is clear that our model, as it has been presented so far, needs to be extended in order to be able to account for the dynamics of the belief system, as it does not allow for any other values of κ than $1/n$ and 1. In effect, our model presents a static picture of society, and an approximate one at that, as we have employed the rectangular approximation in order to give a simple meaning to the parameter α. This picture is useful for describing the state of a society, or for comparing societies, at any particular point in time; for a description of dynamics, the picture we need to have is something like what is shown in Fig. 5.12.

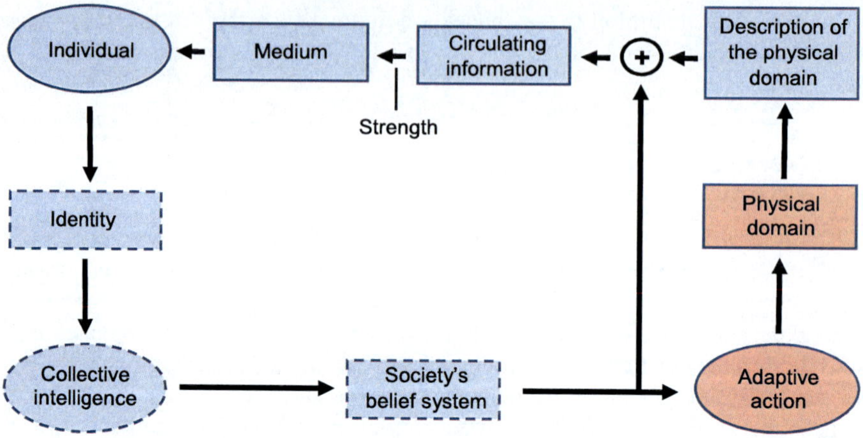

Fig. 5.11 The evolution of society through the transformation process, involving both the information view and the action view; the two are related through their participation in the feed-back process. The dashed outline indicates that the item contains identity elements only

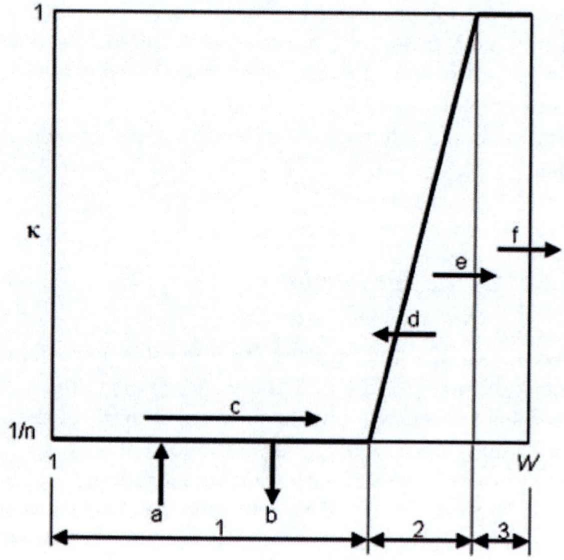

Fig. 5.12 The distribution of society's identity items according to their commonality, κ. The new group of items, 2, contains those items with a commonality less than 1, but that are, on the average, moving towards the right, i.e. increasing their commonality. Group 3 contains, as before, the $\alpha \cdot w$ items of society's belief system, and the items in group 1 all have $\kappa = 1/n$. The arrows labelled by lower-case letters are flows of identity items

The flows of identity items indicated in Fig. 5.12 are as follows:

a. New items created by individuals as a result of the operation of Process B, as discussed in Sect. 4.3.
b. Items discarded by individuals as a result of the operation of Process B.
c. The movement of items towards the boundary of groups 1 and 2, resulting from the interchange of information between individuals.
d. Items that stop progressing in the process of increasing their commonality (i.e. ultimately unsuccessful persuasion), and end up either with $\kappa << 1$ or being discarded altogether (b above).
e. Items, having reached $\kappa = 1$, becoming members of society's belief system.
f. Items discarded or replaced from society's belief system.

In Fig. 5.12, the different identity items that exist in the society are arranged along the horizontal axis, and the vertical axis shows how many there are of each type. In group 1, the items are unique, $\kappa = 1/n$; in group 2 there are $n \cdot \kappa$ items of each type, and in group 3 there are n items of each type, so that the total number of items circulating in the society is the sum of all the elements in these three groups, $n \cdot w$.

To develop an understanding of the process illustrated by Fig. 5.12, we start out, as usual, with a high-level picture of the process that changes the identity by accepting assertions from the stream of assertions circulating in society, for the case that all assertions have $\nu = 1$. The probability that a particular item of the set $n \cdot w \cdot (1 - \alpha)$ in the circulating stream will be accepted by a particular individual within one unit of time is given by

$$P = \frac{\mu_1 p}{w}; \tag{5.7}$$

where p is the *persuasiveness*, introduced in Sect. 5.2.3. The probability that it will be accepted into any one of the $n - 1$ identities is $(n - 1) \cdot P$; and the length of time until it is accepted, t_1, is equal to $1/(n - 1) \cdot P$, and for this item, κ is now equal to $2/n$. The probability that one of these two identical items will be accepted by one of the $(n - 2)$ identities that have not yet accepted this item is $2P \cdot (n - 2)$, and we see that $t_i = 1/(i \cdot P \cdot (n - i)$. Then, after $n - 1$ time intervals, the assertion will have achieved $\kappa = 1$, and the total time, $T(n)$, is given by

$$T(n) = \frac{w}{\mu_1 p} \sum_{i=1}^{n-1} \frac{n}{i(n-i)} \equiv \frac{w}{\mu_1 p} F(n) \tag{5.8}$$

where $F(n)$ is approximated to a high degree of accuracy by the function $1.039 + 4.633 \cdot \log(n)$, shown in Fig. 5.13.

Now, we do not have any data on the values of w and μ_1. A crude guess (and that is all it is) is that the number of statements of belief stored by an individual at any one time, i.e., the number of items, w, in the identity, is around 100, and that the number of information items relating to the identity arriving as input per day (this fixes the unit of time) is about 1. We are likely to reject most of these inputs, so p is

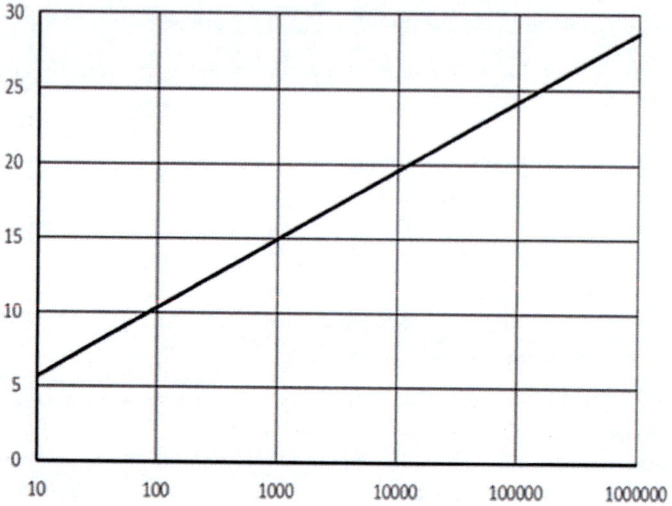

Fig. 5.13 The function F(n). where n is the number of individuals in the society

small number, say, 0.01, so that the time for an assertion to rise from $\kappa = 1/n$ to $\kappa = 1$, i.e., to be included in society's belief system is on the order of $30 \cdot F(n)$ years. Leaving the numerical value aside for the moment, what might at first be surprising is the modest increase in F(n) with increasing n. The reason for this is our assumption that each individual that accepts the information item s into its identity also makes this fact public; this is the presence of the i in the denominator of the sum in Eq. 5.8. This assumption is simply a restatement of the basic assumption of our model of the individual and its role in society, as it was stated in Chap. 4: The identity of the individual is reflected in society's belief system. We shall get back to this later; here we just note that if this public role is diminished, the value of F(n) increases; it takes longer to effect a change to society's belief system.

The model also shows the expected change in the rate of growth of κ, as illustrated in Fig. 5.14. Getting the first converts is always hard work; after that the growth is rapid, with an increasing number of believers in the public domain, but converting those last few unbelievers is not easy (A similar behaviour is displayed by the well-known Bass Diffusion Model (Bass, 1969).)

This picture of the progression of the commonality of an identity item is highly simplified; in effect, it is a picture of the items in the flow labelled e. in Fig. 5.12, neglecting the process that reduces the value of the commonality on its way upward (stream d in Fig. 5.12) and greatly increases the time to reach $\kappa = 1$. The reality is that in a society of identical individuals, and in which the level of persuasion is the same for all assertions, it is highly unlikely that society's belief system will ever change, or that, starting with $\alpha = 0$, that a belief system would ever develop. That a belief system does, in fact, develop in every society is due to the early years of the individual's life cycle, in which the exposure is not to a circulating information

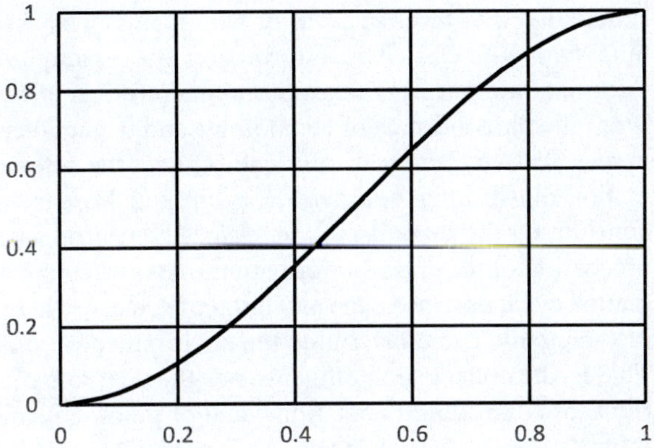

Fig. 5.14 The growth over time of the commonality, κ, for a single identity element. The time scale is in fractions of T(n) in Eq. (5.6)

of uniform persuasion, but to a highly selected set of information with a level of persuasion close to 1, in the form of upbringing and education. This emphasises the importance of education, which is the subject of Chap. 9.

The only significance of the process shown in Figs. 5.13 and 5.14 is as an indication of the growth of an assertion with a much higher level of persuasion than all others in the circulating information. A further complication of the process will result from increasing the value of ν above 1—increasing the strength of the belief system, this will reduce the rate of change, but the details of this effect are complex and well beyond the level of detail of this model. Replacing an existing identity item will be less likely; in Sect. 4.3 we had suggested a 1/ν dependence. Then, to replace one assertion, or belief, with another, there must be ν individual replacements of identity items, resulting in one new identity item which, although it has only a single argument, has an assigned value of ν equal to that of the replaced assertion, as described in Sect. 4.3. We might expect that each successive replacement becomes easier for two reasons: One, the belief to be replaced is successively weakened as its arguments are being deleted, and two, the new belief is becoming increasingly acceptable as arguments are being added. However, the latter effect depends on the rate at which the arguments are being presented; if the operation of Process B in accepting one argument for the new belief is still fresh in the memory, the acceptance of a second argument for the same belief is more likely—a fact well known in effective persuasion techniques. This process will presumably take more time the greater the value of ν, and there is likely an increase in the stress associated with it, as there is a prolonged period during which both conflicting beliefs are present.

At this point, the development of the interaction has been progressed about as far as intended for this high-level description, and before going on to consider some of the processes that constitute the environment in which the description is located, and then finally its relevance to the stability of society, it might be useful to briefly recapitulate its main features. The description represents society as an *information-processing system*, consisting of *individuals* as the distributed processors and the *interaction* between them in the form of information exchange. The information is subdivided into three *classes*, of which class 1 is essentially a background for the other two, with class 2 the information of everyday life, and class 3 our beliefs. The information processing takes place in a physical environment consisting of *belongings* reflecting the state of technology, and the actions resulting from the processing mirror the two classes of information, with the actions resulting from class 2 information constituting the *economic cycle*, and those resulting from class 3 information constituting the *transformation cycle*. The central assertion of the model is that the processing of the class 3 information—the *collective intelligence*—and the transformation cycle determine the evolution of society, whereas the class 2 information and the economic cycle determine the environment in which the evolution takes place. That is, the collective intelligence has the dual role of maintaining society's equilibrium (or controlling fluctuations) and of guiding its evolution—This subdivision of information and associated action is useful, but its limitations are obvious.

5.3.5 The Public Discourse

The processing of class 3 information—the collective intelligence—is performed by the system consisting of the individual processors (brains) interacting through the circulating information, an interaction often identified as the *public discourse*. Both the processing and the interaction are influenced by the physical conditions in which the individuals find themselves; these are the operating conditions of the collective intelligence considered in the next chapter. But there is another aspect to the operation of the collective intelligence, and that is the social framework in which the public discourse takes place. This framework has several components. One is a set of one or more competing total ideologies, as defined by Karl Mannheim and referred to in Sect. 11.4. This was also a topic in the Frankfurt School, where members such as Jürgen Habermas drew on the Marxist idea of ideology as a distortion of reality for its role in communication, wherein participants find that power relations prevent the open, uncoerced articulation of beliefs and values.

Partly through this association with power relations ideology is often given a negative interpretation, as narrowmindedness or inflexibility, but interpreted simply as any strongly held non-religious belief it is almost always present in the public discourse. Sometimes explicitly stated, but most often present as an underlying colouring or interpretation of the concepts involved in the interaction. As an ideology is generally characteristic of groups of people that interact more frequently, its effect on the interaction within such groups is a narrowing of the discourse, whereas in the interaction between people from different groups the effect is either unproductive disagreement or a speaking past each other. Ideology is taken up again in Sect. 11.4.

Another component is the level of participation. Accepting all the physical restrictions and existing power relations, to what extent are people using the available opportunities for engaging in a discourse with others about identity items or class 3 information items? Or, in other words, to what extent does the average member of society realise the importance of participating in the public discourse and thereby in the process of societal development? The importance of this participation was recognised by Mannheim, who saw the increasing involvement of the whole population in the political process as a key aspect of that process. He also thought that sociologists had a key role to play in facilitating the process, but that is not really the case today, with sociologists ensconced behind the walls of academia.

The level of participation is related to the intensity of the interaction and through it to the strength of the interaction, as we introduced it earlier. In the last several decades the level of participation in the interaction between individuals has increased significantly due to the presence of various social media, but how much of this involves class 3 information is difficult to say. For the development of our model, the crucial question is: Is the increasing use of electronic media increasing or decreasing the strength of the interaction? That question is given some consideration in Sect. 8.5.1, but it is something that requires a great deal more empirical research before we can have a clear understanding, even if only a qualitative one, of how the process of forming society's belief system is changing under the influence of the electronic media.

5.4 Interaction and Structure

5.4.1 Interaction Leads to Structure

The development of our view of society has, until now, been at the highest level of abstraction, with society as an unstructured body of individuals. That is, not only have we assumed that the individuals are all identical as information processors, but also that the interaction between the individuals and society is the same for all individuals. However, there is a very significant difference between these two assumptions. The distribution of individuals' characteristics is a feature of our species; it is something that has remained more or less constant over the last 10,000 years and, ignoring genetic engineering, will remain so for the foreseeable future. The degree to which the assumption is justified depends only on the sensitivity of whatever we are considering to the variations within the distribution. The distribution of the interactions between individuals is very much a feature of society and has changed dramatically over the same time-period. The assumption of equal interaction could only be a true representation of the most embryonic society, with a very low intensity of interaction.

There are a number of factors that lead to inequality in the interaction between individuals, and one approach to organising and understanding this body of factors is by means of a thought-experiment: Consider a collection of individuals, initially non-interacting and existing in a particular physical environment characterised by such factors as distance between individuals, or local density within the collection; barriers to communications, such as mountains or wide rivers; and available communications infrastructure, from roads and bridges to today's information technology. The individuals themselves are characterised by several factors, such as strength and

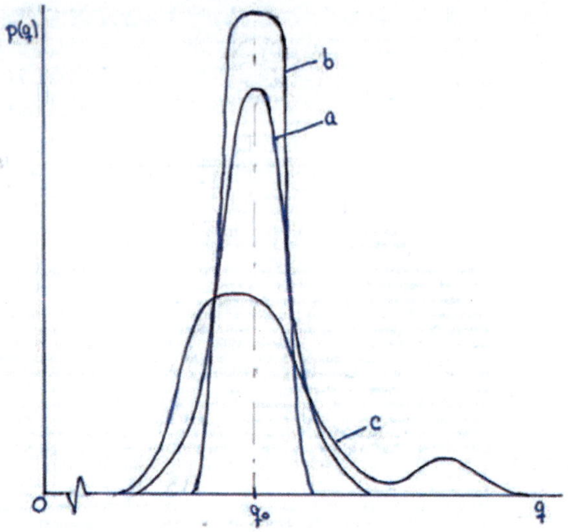

Fig. 5.15 Three distributions of the Quality of Life, q. The curve labelled *a* is that of a collection of individuals, without any interaction, illustrating the intrinsic variation in ability. The curve labelled *b* shows the effect of interaction under the assumption that the individuals show a high degree of altruism, and the curve labelled *c* shows the effect of interaction under the more realistic assumption that the more able exploit the less able and form a ruling class

information processing capability, each of which has a certain distribution over the collection, resulting in such a composite measure as the Quality of Life (QoL) to also be distributed over the collection, as is illustrated by the curve labelled *a* in Fig. 5.15. This measure will be further detailed in Chap. 6 and taken up again in Chap. 10; here we may rely on our intuitive understanding of it.

Now allow the interaction potential to increase slowly; the environmental factors will lead to a grouping of individuals according to the ease with which this potential can be realised as actual interactions. And within each group, the interaction will lead to a modification of the distribution of the QoL, and there are basically two outcomes. In a society with altruistic members, with the more able helping the less able, the effect is to make the society more egalitarian, as indicated by the distribution labelled *b* in Fig. 5.15. But in the more likely outcome, the difference in abilities will be amplified, with the more able taking advantage of the less able and embed this difference in a further structuring of the society, resulting in a distribution of the quality of life as indicated by the curve labelled *c* in Fig. 5.15.

So, the interaction between individuals, which turns a collection of individuals into a society, unavoidably results in a *structuring* of that society. And, in the past, that structuring was prominent in both the information view and in the action view. In the physical domain the structuring is still very real in various forms of *inequality*, which will be a central concept in following chapters, and which manifests itself in various ways, including unequal distribution of wealth, unequal level of control of the common infrastructure, and in a division of labour, but also in physical entities like classes and institutions. In the information domain the situation is somewhat different, in that the access to information has been made more readily available through information technology, from the printing press onward, thereby reducing the inequality in this regard, which is why it is valid, as a first approximation, to consider the individuals as identical information processors, as we are doing in most of this book. But the increase in the interaction, combined with the increasing amount of information available, is resulting in what we might classify as an informal structuring, in that the structural elements do not have any written or otherwise documented definition; they are collections of individuals that have developed a certain commonality in their beliefs and opinions, i.e., in their identities. In terms of our model, this means that the $(1 - \alpha)$-part of the individual identities are no longer all different; groups of individuals are defined by the commonality of one or more of their identity elements in the $(1 - \alpha)$-part of their identities. How such groups are formed and maintained through a process involving the mutual interaction between the individual and the communication media is the subject of Sect. 11.2.1.

The following subsection looks at a particular and well documented case of how the interaction between the members of a society influences its structure, in this case, the spatial separation into settlements. It also allows us to identify a parameter that was introduced in Sect. 4.3; that of the stress perceived by the individual arising from unresolved conflicting beliefs or opinions.

5.4.2 Structure Determined by Interaction-Induced Stress

If we look back in time, the only practical means of transmitting information was by word of mouth, and the determining factors were population density and mobility. Here we can draw on the seminal, and for us highly relevant, work of Roland Fletcher in studying numerous settlements, both ancient and more recent (Fletcher, 1995). In this work, Fletcher describes how the growth trajectories of settlements depended on the relationships between residential density, settlement size, and what he calls "the material assemblage" available to the occupants (which I identify as the extent of technology and its applications), and this growth was constrained by two limits: the upper limit on a tolerable residential density, determined by the stress experienced by an increasing level of interaction between the individuals within a settlement (I-limit), and the limit on the settlement's areal extension, set by the distance over which the prevailing communication system could operate adequately (C-limit). The interaction consists of all the sensory inputs experienced by a person due to the other people in the settlement, including through such channels as sound, body language (i.e., through vision), odours, and physical means (e.g., crowding, bumping into), as defined in (*ibid* 70). Communication is the exchange of messages between people, both face-to-face and mediated by various applications of technology (message tokens, clay tablets, etc.), and where the face-to-face exchange could be arranged through movement of the persons involved, so that means of person transport (roads, bridges, horses, carriages, boats, and more recently mechanised transport) became part of the communications technology. It is interesting to note that although goods transport uses much the same infrastructure as people transport, it is not included as a factor in determining the C-limit (*ibid* 82).

From Fletcher's analysis of the great amount of data collected by himself and other authors, two significant realisations emerge. The first is that, for a settlement to function as a cohesive entity, there needs to be a certain level of information exchange across the settlement, and therefore, for a given communications technology (in the above sense) there is a definite limit to the areal extension of a settlement. The minimum size at which the benefits of a sedentary versus a nomadic lifestyle could be demonstrated seems to have been on the order of 10^4 m^2. These agricultural settlements grew until they reached a limit of 10^6 m^2, with communications provided mainly by walking. This limit was only exceeded once the technology for building compact urban settlements, in the form of retaining walls, drainage, roads and bridges, was developed and applied, together with various forms of writing, and such settlements developed rapidly in size up to a limit of 10^8 m^2. This limit was maintained, with a few exceptions, until thermal power started its spectacular development at the beginning of the Industrial Revolution in the mid-eighteenth century, first the steam engine and then the internal combustion engine, supporting a revolution in personal transport and in printing. This was accompanied by such technology developments as electricity and telecommunications, and industrial cities grew in size to the next limit, 10^{10} m^2, which brings us to the present.

The second realisation is that there is an upper limit on the level of population density, and that this level has remained essentially constant throughout the evolution of settlements. Equating this limit to a limit on interaction (behavioural) stress, it says that despite all the changes to the environment in which interpersonal interaction has taken place (the built environment), the effect of this interaction (i.e., the stress and its limit) has remained unchanged. That is, this limit is a fundamental characteristic of the human psyche, of the human as a social being.

Fletcher describes the development of settlements in terms of three variables:

p population
a area of the settlement
d density of population within the settlement.

If we, at least to start with, take the density to be uniform within the settlement, we have the relation $a = p/d$, as shown in Fig. 5.16, and the growth of a settlement is represented by a trajectory in the p–d plane. The stress, s, experienced by a person due to the presence of another person, must decrease as a function of the distance, r, between the two persons, and let us assume a simple, exponential decrease,

$$s = s_0 e^{-\frac{r}{\sigma}}. \tag{5.3.15}$$

We have chosen r as the symbol for the distance because we now consider the situation where the person experiencing the stress is in the middle of a settlement within a circular area of radius r_0. In this case, the total stress experienced by this person, S, in units of s_0, is given by

$$S = 2px^{-2}\big(1 - e^{-x}(x + 1)\big); \tag{5.9}$$

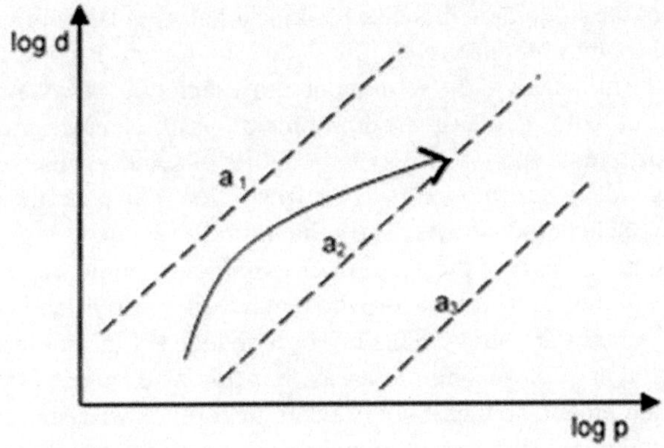

Fig. 5.16 Lines of constant area in the p–d plane, with $a_1 < a_2 < a_3$, and a possible trajectory of the development of a settlement

where $x = \left(\frac{p}{\pi\sigma^2 d}\right)^{\frac{1}{2}}$. The value of σ is determined by requiring that S is the same for the two endpoints of the I-limit in Fig. 4.5 of (Fletcher, 1995):

$$p = 10;\ d = 0.0642\ \text{persons/m}^2;$$
$$p = 10^8;\ d = 0.0294\ \text{persons/m}^2;$$

from which it follows that σ = 4.55 and the critical value of S (the I-limit) is S* = 3.83. The value of S/S* as a function of p and d is shown in Fig. 5.17.

This result shows what we would expect—behavioural stress is a local phenomenon; once the population reaches about 30 people, further growth at constant density does not increase the stress. However, with growth at constant density, the area

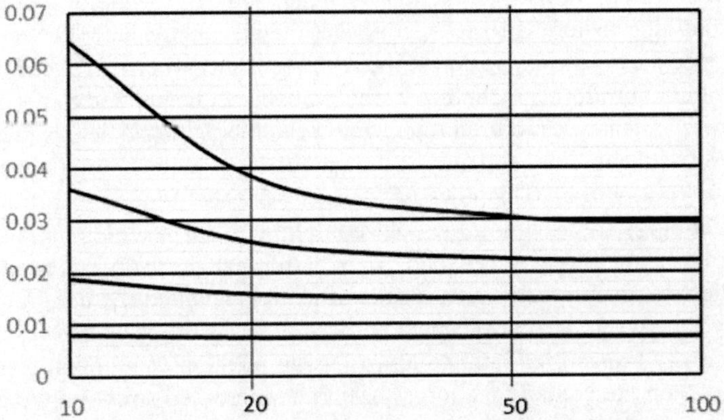

Fig. 5.17 Lines of constant behavioural stress experienced by persons in a settlement, as a function of population, p (horizontal axis), and density, d (vertical axis, in persons/m^2), as a fraction of the critical stress value, S*—the I-limit in (Fletcher, 1995)—with, from upper to lower curve, S/S* = 1.00, 0.75, 0.50, and 0.25

of the settlement increases proportionally, and the communication between members becomes increasingly difficult and inefficient. But it is this communication that turns the members of the settlement into a society; it is what underpins the fabric of society, and so this decrease in communications efficiency represents an increasing stress of a different type—a stress on the fabric of society. The two types of stress are not independent, or orthogonal, as the frustrations of a poor and overcrowded public transport system demonstrates. The behavioural stress experienced by a person depends on the homogeneity of the person's environment; people who are visibly different (hair style, clothing, skin colour) or who behave differently (language, gestures, body language) increase the stress. This effect is reduced if the settlement is segmented into racial and/or socio-economic areas, but this also reduces the communication across the settlement, so that there is a certain conflict between reducing the behavioural stress and increasing the social cohesion. A similar effect may take place once the density level is reached. The settlement needs to restructure itself and expand into a larger area, thereby lowering the density, before starting on another period of increase, resulting in a saw-tooth history of population density. With it, there is likely to be a certain structuring into local communities (boroughs, arrondissements, Bezirke), so that our assumption of a uniform density is only an approximation, and the sense of belonging to a local community reduces the behavioural stress, while probably also reducing the cohesion within the settlement as a whole.

Nevertheless, it is both significant and useful to distinguish the two types of stress, and we shall relate them to our high-level view, which is concerned with society as a whole. However, before doing that, there are a couple of issues we have to consider. Most of the settlements on which Fletcher's analysis is based existed prior to the Industrial Revolution, which gave these settlements two features. One, the developments in communications technology that permitted a transition across a C-limit were preceded by developments in the activities that sustained the settlements, first agriculture, and then, forms of extraction and manufacturing technology, in particular tools and building materials (bricks, concrete, metals, chemicals). The technology was developed during the expansion of the settlement and, when this expansion came up against the C-limit, this technology was available to be applied to the improvement in communications required to expand the settlement size across the C-limit. This, as well as the interdependence of the interaction and the communication, was discussed in (*ibid* Ch. 6). This does not detract from the importance of communications technology in providing the cohesion required for a settlement, but adds a certain variability to timing, speed, and extent of the changes in different settlements.

Two, the relationship between the area of settlements and the communications facilities was between two entities parametrised in terms of distance. Messages had to be physically transported, and so an extension of the area automatically meant a further load on the communications system and increased stress on the ability to maintain social cohesion. With the advent of telecommunications, this relationship changed, and the importance of distance in communications declined sharply.

A further issue is that, in the settlements studied by Fletcher, there was a clear distinction between the interaction, resulting from the behaviour of people, and the communication, as a means of maintaining coherence of activities across the settlement. The latter was largely an administrative function, and while the population would be very aware of its effects, as e.g., in taxation, and of the stress resulting if it was inadequate, as in increased lawlessness, it was something that the great majority of the population would simply accept as given. In a modern society, that is quite different, with an increasing proportion of the stress experienced by individuals arising from concerns about the state of the society and from responding to the demands of government at all levels, and an increasing proportion of the interpersonal, or behavioural, stress occurring via the communications infrastructure. The distinction between the two types of stress—on individuals and on society—is becoming less clear.

A final issue is that of structure; both of the settlement and of the society. As already mentioned, as the complexity of the society increased, there could be a segmentation of the physical space into racial and/or socio-economic groupings, and also in that people with the same interests and beliefs congregate within subareas of the settlement. A further source of spatial segmentation is the structuring associated with the division of labour, both hierarchically (levels of administration) and by trade (e.g., "tinsmith alley"). As a settlement grew in size and complexity, this spatial segmentation reached a point where the society could be considered as consisting of several settlements. This issue of spatial structure is not one that is relevant to our high-level model, but it is one that will play a significant role in the future development of settlements, as outlined in the final chapter of Fletcher (1995). However, in the present age—the information age—there is a different type of structuring that is important, as discussed below.

To see the significance of Fletcher's work in the context of our model, we need to relate his concepts of "interaction" and "communication" to the features of the model, i.e., Processes A and B, and the knowledge base Θ, as shown in Fig. 3.2. The "interaction" does not feature explicitly in the model, but we now identify the stress resulting from the interaction as resulting from an unacceptable degree of inconsistency within the knowledge base, Θ, and it is also experienced as a need for certainty or for closure. This is a well-known and -documented feature of the human psyche; Googling "need for certainty" gives more than a million results. The effect of inconsistency among the elements of Θ is the subject of a special field within cognitive science called cognitive dissonance theory, based on the idea that persons seek to minimise the internal psychological inconsistency of their cognitions (which equate to our identity items—the elements of Θ). Cognitive inconsistency is taken to be an uncomfortable state, and hence people are seen as striving to avoid or minimise it. A good description of cognitive dissonance theory is given in Chap. 4 of Daniel O'Keefe's book *Persuasion* (O'Keefe, 2002).

In our model the stress arises when the time available for executing Process B is inadequate in relation to the flow of new information being placed in Θ by Process A. This implies that there is an upper limit on the information processing capacity available to, and shared by, the two evaluation processes, and that, if the flow of information is great enough, Process A will tend to use up the available time, crowding out Process B.

Fletcher's "communications" consists of information that is in classes 2 and 3. In class 2 we find the information of an administrative nature, associated with the various processes that keep a modern society functioning, such as restrictions (laws) of various sorts, taxation, licensing, and reporting ("red tape"). In class 3 we find the information that maintains a certain minimum value of the alignment α throughout the society—the characteristics of the society, its culture. Therefore, if the communications infrastructure is stretched beyond its capacity, there are two separate consequences. One, the services that provide the benefits that constitute the rational for a society in the first place start to deteriorate. Two, the common belief system that supports cooperation disintegrates. Both can be viewed as exerting a stress on the fabric of society that either limits further expansion or, possibly, continues expansion as two or more societies.

Chapter 6
Operating Conditions of the Collective Intelligence

6.1 Some Basic Considerations

In the development of our understanding of society in terms of two views we have, so far, focused mainly on the information view. But this information relies on the physical domain for its existence, be it in the form of neuron patterns in a human brain, the state of memory cells in a computer, written or printed symbols, art objects, and all instances of the performing arts, so there is clearly a close relationship between the two views. And in the course of the development of this understanding of society we have also introduced a number of concepts, such as the collective intelligence, society's belief system, the common infrastructure, collective memory, and more. These are all related by virtue of being part of the description of the same entity—society, and we must now try to develop a clear picture of these relationships and their role in relating the two views, in that many of them have a dual definition, depending on in which view we are operating.

Starting with our definition of a *society* in terms of its main components, we expand the initial definition given in Sect. 1.3.1 as follows: A society consists of

- a set of *individuals*;
- a set of *interactions* between them; and
- the *belongings*.

In the action view, the individuals are described in terms of their behaviour and physical abilities; the interactions are described in terms of actions that establish some form of relationship with at least one other individual; and the belongings are as listed in Subsect. 1.3.1. In the information view, the individual is an information-processor, defined in terms of its information processing and storage capabilities, without any concern for the physical realisation of these capabilities; the interactions are exchanges of information, and the belongings are the various collections of common information that we identified as the *collective memory*.

As discussed in Sect. 3.3.1, the *state* of a society is defined by the values of a set of state variables, **S**; these variables describe the components of the society

E. W. Aslaksen, *The Evolution of Society*,
https://doi.org/10.1007/978-3-031-23054-7_6

defined above. The set can be decomposed into two parts: the variables describing the components in the action view, S_a, and the variables describing the components in the information view, S_i, and in both cases the extent of this set depends on the level of the description, in the sense of a top-down structured description (see Sect. 3.2). Finding measures of such variables is the subject of Chap. 10.

As society is a living organism the state is continually changing, so the variables are functions of time. That is, the variables are the variables of a process, defined in terms of a number of partial differential equations. We might now expect that this set of equations would also separate into two sets of equations, each set containing only variables of that view, but that is not the case. For while, at any point in time, there are two separate descriptions of the state of the society – one in the action view and one in the information view, the equations for the variables in one view involve also the variables in the other view:

$$\frac{d\boldsymbol{S}_a}{dt} = \boldsymbol{f}_a(\boldsymbol{S});$$
$$\frac{d\boldsymbol{S}_i}{dt} = \boldsymbol{f}_i(\boldsymbol{S}). \tag{6.1}$$

That is, variables of one view become process parameters in the other view, and in the context of the development of our understanding of the evolution of society documented in this book, with its emphasis on the information view, we can consider the variables of the action view to define the operating conditions of the process driving the evolution of society, and in Subsect. 3.3.1 we identified that process as the operation of the transformation process.

6.2 Operating Conditions of the Collective Intelligence

As illustrated already in Fig. 5.11, the transformation process involves three linked processes:

- the information-processing within the individuals (Chap. 4);
- the interactions between the individual and society (Chap. 5); and
- the realisation of the belief system in the state of the society (Chap. 7).

Of these, the first two processes constitute what we have defined as the collective intelligence; it is concerned with the processing of class 3 information and the evolution and maintenance of society's belief system. The third process is concerned with realising the belief system as actions that modify the structure of society; together they constitute the transformation process. So far, we have been concerned with the collective intelligence only, and mainly within the information view. What we now want to consider are the conditions under which the collective intelligence operates.

As with the information it operates on, a measure of the operation of the collective intelligence will have two dimensions; a *quality* dimension and a *strength* dimension, the latter introduced in Sect. 5.3.2. In the description of the collective intelligence, which has been within the information view, the operation was described in terms of quality only; it did not involve the strength, as in the intensity of the interaction, its responsiveness to changes in the economic cycle, its effectiveness in influencing the economic cycle, or any other characteristic. As such, the concept of the ideal operation, as defined in Sect. 3.3.1, is applicable to any society, ranging from an Aboriginal tribe to a modern nation-state. However, when we now want to describe the deviation of the actual operation away from the ideal, the description of the operating conditions must be in terms of characteristics of the specific society under consideration, which ultimately is the world society. In that society, a large number of characteristics are provided by various international organisations, such as the World Bank, International Monetary Fund, Transparency International, and Reporters Without Borders, and in an earlier publication, *The Social Bond*, the deviation of the operating condition away from its ideal state—which was there identified with The Good Society—was described as due to a number of *restraints*. These restraints were defined in terms of several macroeconomic variables, and a single, composite measure of the deviation from the ideal state was proposed. The approach there was to consider the restraints and the composite measure on the basis of individual nations, and then to form a population-weighted average over all nations to create a single measure of the operation of the collective intelligence for the world society, and this approach is outlined in the next section.

6.3 A Composite Measure of the Good Society

6.3.1 The Good Society

The measure of The Good Society is essentially one of information-processing, and until recently the processing power was provided by the collective brains of the members of society. But now information technology provides an additional means of processing, and control of this process is effectively the same as control of peoples' brains and of the evolution of society. This clearly raises some significant novel philosophical questions of an ethical nature—the beginning of an *information ethics*—and so one purpose of the measure proposed in this section can be seen as indicating one possibility of relating this ethical domain to measurable characteristics of our society. A focal point for activities related to information ethics is provided by the International Centre for Information Ethics (http://icie.zkm.de/, accessed 20.06.2022), which also maintains an exhaustive catalogue of published material on this subject.

The evolution of society is driven by the collective intelligence, which is formed by the interaction of individual information processors. Thus, there has been, from the emergence of the first societies, a dichotomy, or tension, between the individual and society, in that society provides the individual with the infrastructure that makes a good life possible, but it can do so only by modifying what it means to be an individual; this is inherent in any interaction. A high-level view of The Good Society, then, comes down to the relative importance of individual and of society, as raised briefly in Sect. 1.2, and such views are presented in various terms that can be visualised as forming a matrix, with the one dimension including such terms as freedom, ownership, equality, justice, and rights, and the other dimension including economic, political, and moral aspects of these terms.

As a starting point we can take what might be called the "Bible" on the Good Society, the book *The Good Society*, Bellah et al. (1991). It gives a thorough and perceptive account of the development of US society, and in doing so, identifies and discusses many of the issues and problems confronting modern society in general. But, for all its good qualities, it is a US-centric view of the world, and reflects a characteristic of the presentations that also becomes evident even in our very cursory review of the literature more generally: how the political landscape in the US differs from that in the rest of the world, in that it is much less ideological than we find e.g. in Europe, and more in the form of discussions about variations on a business plan for the maintenance of a superpower. One example of this is the article (Champlin & Knoedler, 2005), which gives a good overview of the state of the US version of the Good Society; a view which involves a bewildering use of variations of "liberal", but in which such a central concept as socialism is completely absent. However, the authors bring out the importance of the concept of freedom (and its interpretation), to which we shall return in Sect. 6.6, and, in quoting (Lakoff, 2003), of the political discourse and its rhetorical dimension, which we discussed in relation to persuasion, in Sect. 4.4.

An article that focuses on the political aspects of justice is *Building the Good Society*, by Cruddas and Nahles (2009). In it, the authors describe a new model of social democracy based on ideas of the British Third Way and the German Neue Mitte, championed by Tony Blair and Gerhard Schröder in 1999, and traces the limited success of this initiative to an uncritical embrace of the new global capitalism. The two Prime Ministers underestimated the destructive potential of under-regulated markets and believed that a class-based society had given way to a more individualised, meritocratic culture. The authors advocate a good society based on solidarity and social justice, which requires that each citizen has the resources—money, time, relationships, and political recognition—to make a good life for themselves. They state that for this to happen, the main task in the years ahead will be to create and consolidate political trust in public life.

Michael Walzer, in *What is "The Good Society"?* (Walzer, 2009), argues that there is not only one good society, but several; each generated by the constant flux of movements, with their associations, communities, and states. Goodness requires plenitude, with people participating in many different ways in many different groups. Amitai Etzioni also considers society as a collection of communities; each a combination of a web of affect-laden relationships among a group of individuals, and a measure of commitment to a particular culture (Etzioni, 2002). The article examines the value of bonds and their limits, and the dialogue that results in the establishment of moral values, including the limitation of such dialogues. The author states that for a society to be good, much of the social conduct must be regulated by reliance on the moral voice, rather than on the law, and the scope of the law itself must be limited largely to what is supported by the moral voice. The good society is defined as one that balances two values—social order and autonomy—rather than maximising one.

In an article published in the Australian forum *The Conversation*, Denis Mueller addressed a number of issues associated with a free press as an essential element of the Good Society (Mueller, 2016). He starts out by saying "a free press—now including all news media—is essential to the functioning of a capitalist democracy. Without it, citizens have no means of exercising the powers of what John Locke called the sovereign people. They would be robbed of the information necessary to making political and economic decisions, and of knowing what others in society, outside their immediate circle, were doing." He then goes on to demonstrate how, for all its importance, free press and speech is not an absolute value; we have laws concerning national security, defamation, and contempt of court—to name the most obvious—all of which constrain free speech, and even the "free" press will try to suppress information that goes against its own ideology, as he illustrates with a recent example.

In an interview published in 2001, John Kenneth Galbraith addresses the dangers to society of inequality and the concentration of wealth, and the political power that goes with it (Galbraith, 2001). In this regard, he points to the danger of excessive privatisation. He also raises (but does not pursue) the issue that, with an increasing corporatisation of our society, a new class is being created—the managerial class—that represents a further challenge to democracy. In Australia, an example of this are the managers of the superannuation industry.

Johannes J. Britz consider the ethical dimensions of the Global Information Society from a social perspective and as an application of the moral notion of justice (Britz, 2008). He emphasizes that a precondition for the Good Society is a moral consensus, and in citing (Lötter, 2000) he recognises that there may be features in the existing social structure that make the introduction of a just Information Society difficult. An example is ownership; ownership of information will be different to that of physical property.

In her book *Access to Information as a Human Right*, Cheryl Ann Bishop uses the process of conceptualisation (i.e., developing the implications inherent in a concept) of four basic rights—freedom of expression, information privacy, a healthy environment, and the truth about human rights abuses—to develop a justification for the right to information, and supports this by reference to a number of authors (Bishop, 2012, see also 2009). As a further conceptualisation, she quotes (Weeranmanty, 1994), who perceived that the right to information is a prerequisite for achieving world peace.

Focusing on economic aspects of the Good Society, Robert J. Shiller discusses the role of financial capitalism in society (Shiller, 2012 and 2013). His view is that, if developed on the basis of democracy and humanism, the financial market could mitigate many of life's risks, and that, to a large extent, the bad repute the financial market, in the form of the collateralised mortgage obligation, received after the GFC, was as much a result of public ignorance and greed as of any defect in the basic financial structure. Regulation of financial markets must take account of human nature, and to become an integral part of the Good Society, it must also conform to accepted society standards (rational or not). In addition to being both informative and innovative, the book is relevant to our understanding of the operating conditions of the collective intelligence, in that Shiller acknowledges and discusses two of our main themes: the central function of information in all aspects of modern society, including the operation of financial institutions and markets (e.g. pp. 9–12 and 160), and the fact that the control of capital equals power, "the almost unlimited power achieved by the very successful in business (and their heirs) is almost unknown in other walks of life" (p. 165). The "pros" and "cons" of financial capitalism are clearly presented and, given that financial institutions and markets are a fact of modern society, the issue is only how the "cons" can be reduced while enhancing the "pros", and Shiller gives numerous thought-provoking suggestions. His enthusiasm for financial innovation did briefly bring to mind the engineer that was working for peace by developing a more efficient weapon, but really the only question that remained at the end was what appeared to be a contradiction between his intention to democratise financial operations and the statement that new and improved financial instruments will be more complex and only understandable to a small elite of specialists. This is an issue that is found also in both law (increasingly complex terms and conditions for everyday transactions) and in politics (the citizens are a hindrance to effective execution).

There is a journal dedicated to the topic, *The Good Society*; a journal of the Committee on the Political Economy of The Good Society, published by Penn State University Press. PEGS is a nonpartisan, ideologically diverse, non-profit organization whose goal is to promote serious and sustained inquiry into innovative institutional designs for a good society.

The main point of this brief excursion into the literature about the Good Society is that all the contributions acknowledge the central importance of information—of the quality of the information and of its free interchange. Whether in finance, in politics, or in belief systems, the manner in which information is treated is a determining factor in the success of that aspect of society. The concept of the interaction between individuals in the form of information exchange is intended to express that in a model that is independent of the particular aspect or issue of society it is applied to. As the interaction is one of the two main components of intelligence (the other is the individual's inherent processing power, which it is currently less in our power to do anything about), and if we see the extent to which individuals are able to exercise their intelligence as members of society—which is what we have identified as the collective intelligence—as a measure of the Good Society, then that becomes, in effect, a measure of the effectiveness of the collective intelligence.

6.3.2 Definition of the Measure

The fundamental premise of the view of society presented in this book is that, in the long run, it is the exercise of the collective intelligence, as constituted by the interaction of the individual intelligence and its exercise through the application of will of all the members of the society, that determines the evolution of society, and that the extent to which that process is able to operate determines the extent to which the evolution of society proceeds along a path providing the greatest probability of survival. Consequently, the measure we are seeking should be a measure of the extent to which society supports and promotes the exercise of our intelligence. This is not the same as the state of society, **S**. But, as the discussion of The Good Society showed, the measure will involve some of the variables in **S**. For the moment, let us label the measure by ω, with $\omega = 1$ signifying perfect operating conditions. The variables chosen to represent ω in the particular model developed in *The Social Bond* can be considered to be of two types: one a characteristic of the individual—an aspect of the individual's moral character—and the other the restraining circumstances in which the individual finds itself. The reason for differentiating these two variables is that they relate to the individual in different ways. The moral aspect, which we shall identify with the rejection of corruption, while related to corruption as a measure of society, is something the individual can decide to participate in or not; it is, to a certain extent, a measure of a person's willpower. Whereas the restraining circumstances are beyond the individual's control.

Before continuing with the detailed development of the measure ω, it is appropriate to acknowledge that there are several other measures related to The Good Society contained in a significant body of work, often under the heading of Social Quality, and we will consider that briefly. They fall broadly into two groups: those that focus on "hard" data, such as data on income, wealth, education, and health, and those that focus on "soft" data, based on people's perception of wellbeing and happiness, as obtained from surveys. Some measures also combine the two types of data.

We can approach a definition of such a measure by recognising that, as the evolution of society is driven by the individual, we can consider the purpose of society to be to realise the potential of the individual and look for a variable that measures the extent to which society fulfils that purpose, recognising that while 'potential' covers a multitude of capabilities, we might expect it to also include the ability to process information of class 3. In Maslow's hierarchy of needs (Maslow, 1943) this is the top (fifth) level: the need for self actualisation. Another definition is that the purpose is to provide its members with the greatest *quality of life* (QoL), a concept that has received considerable attention, and that was briefly introduced in Sect. 5.4.1. It expresses the general well-being of individuals in terms of their expectations for a good life, relative to the economic and socio-cultural context in which an individual lives. There is an international society for quality-of-life studies, with its own journal, *Applied Research in Quality of Life*, and a great number of articles and books on this subject have appeared over the last fifty years or so.

There are basically two approaches to constructing a QoL measure: One is to define macroeconomic measures that are deemed to be major determinants of the wellbeing of the nation's members, such as *per capita* GDP, life expectancy, and level of education, as well as inequality in the distribution of these measures within nations. An example of this approach is the Inequality-adjusted Human Development Index

(IHDI) produced by the United Nations (https://hdr.undp.org/data-center/human-development-index#/indicies/HDI, accessed 20.06.2022).

The other approach is to create a subjective measure of wellbeing by conducting a survey on a sample of the members of a nation regarding how they consider they are located on a multi-dimensional scale of significant factors influencing a person's happiness. An example of this approach is the World Happiness Report (https://worldhappiness.report/, accessed 20.06.2022) a report that is issued annually by the United Nations Sustainable Development Solutions Network, and based mainly on the Gallup World Poll (https://www.gallup.com/analytics/232838/world-poll.aspx, accessed 20.06.2022). Another initiative in this group is the one by the World Health Organisation that defines QoL as an individual's perception of their position in life in the context of the culture and value systems in which they live and in relation to their goals, expectations, standards, and concerns. It is a broad ranging concept affected in a complex way by the person's physical health, psychological state, personal beliefs, social relationships and their relationship to salient features of their environment (https://www.who.int/healthinfo/survey/whoqol-qualityoflife/en/, accessed 20.06.2022). A further measure is provided by the OECD Better Life Initiative (http://www.oecdbetterlifeindex.org/#/11111111111, accessed 20.06.2022).

The measure proposed below falls squarely into the first, or "hard data" group, and was chosen for its easily demonstrated relevance to the free exercise of intelligence, as we have defined it, and, above all, for the availability of data for a significant part of the world from organisations that can be expected to continue to produce this data on a regular basis.

If we denote the moral aspect by *m*, and the extent of restraint by *r*, the dependence of ω on these two parameters can be expressed by the equation

$$\omega = 1 - (\mathrm{r} + \mathrm{m})/2, \tag{6.2}$$

but no significance should be given to this particular expression at this point in our development, as we have not defined how *r* and *w* are measured.

Our aim is to determine the current value of ω for each group in the society of interest, and then form the average. For this to be possible, the data entering into the definition of ω must itself be both available and well defined, and it must be possible to assign it to the chosen grouping. In practice, this means choosing an *evaluation model* of ω in terms of parameters that are both significant and for which data is available. Starting with the restraint, there are many possibilities, both as regards the number of restraints and the definition of each restraint. The following proposal is just one such possibility:

a. A significant factor is clearly the financial situation in which the individual finds itself, as measured by income per person per year, *q*. If the *subsistence level* at the location where the individual lives is denoted by *s* ($/year), the *financial restraint*, r_1, is defined to be of the form

$$r_1 = e^{-\frac{q-s}{q_a}} \tag{6.3}$$

where $q \geq s$, $r_1(q < s) = 1$, and q_a is the income averaged over all the members of the society. The interpretation of this restraint is that at (or below) the subsistence level the individual has essentially no time or energy to exercise intelligence; all effort is absorbed by mere existence. Nor is there any surplus income for acquiring information. As the income rises above the subsistence level, the restraint is reduced, but the reduction depends not directly on the value of this increase above the subsistence level, but on the relationship of this value to the average income in the society. If this is the most appropriate measure of the effect of increasing the income above the subsistence level is open to discussion; the choice in Eq. 6.3 is reasonable, but somewhat arbitrary.

b. The access to and level of education is a prerequisite for obtaining and evaluating information, and thereby generating the knowledge which forms such an important part of exercising intelligence. As the exercising of intelligence we are considering relates to the evaluation of the impact of proposed changes, often involving the application of new technology, on the evolution of society rather than details of the change itself, the knowledge required is of a general rather than of a detailed nature (such as is required for creating technology). Consequently, the lower levels of education are more important; literacy is more important than having a Ph.D. We might therefore propose to evaluate the *attained educational level* as 0.6 for primary education, 0.9 for secondary education, and 1.0 for tertiary education, and define the *educational restraint*, r_2, as

$$r_2 = 1 - \text{attained educational level} \tag{6.4}$$

c. Having both the time and the education to process information into knowledge still leaves the issue of being able to obtain the information. That ability can be considered to consist of two separate factors: One, *access* to the media carrying information (printed matter, radio, television, internet, social media) via such channels as newsagents, public libraries, broadband, and mobile phone. Two, the *quality* of the media, reflected in the level of censorship, diversity, ownership and control, accuracy and completeness of the information, and so on. Access is evaluated on a scale of 0 1, with the value 1 signifying full access to all existing information channels, as would be the case in many parts of the developed world. Similarly, quality is evaluated on a scale of 0–1, the value 1 signifies the absence of censorship, a vigorous multiplicity of sources, and a high degree of independence. We therefore define an *information restraint*, r_3, as

$$r_3 = 1 - \text{access} \times \text{quality}. \tag{6.5}$$

d. Being able to formulate an adaptive action is one thing; being able to action it is a different matter, and so lack of access to participation in a democratic and political decision process becomes a further restraint. Democratic, in the sense that all involved have the same rights, is not enough; for the outcome of the process to have any effect, it needs to be a political process. That is, a process that is part of the public affairs of a society and of its government. Consequently, if we denote the level of political rights by *u*, of civil liberties by *v*, and of democracy by *w*, all measured on a scale of 0–1, we define the *process restraint*, r_4, as follows:

$$r_4 = 1 - (u + v + w)/3 \tag{6.6}$$

Having chosen and defined four components of the restraint, *r*, there remains to decide how these four components are to be combined to form *r*. There are numerous ways to do this, depending partly on what importance one assigns to each component. For example, one might consider that the financial restraint, r_1, is so important that if it goes to zero, the other components are insignificant, and so decide on a combination of the form $r_1(r_2 + r_3 + r_4)$. The combination proposed does not make such a radical distinction, but allows for a certain weighting of the components:

$$r = \frac{c_1r_1 + c_2r_2 + c_3r_3 + c_4r_4}{c_1 + c_2 + c_3 + c_4}. \tag{6.7}$$

A particular choice of the constants c_i is used in the sample evaluation presented at the end of this subsection, ($c_1 = 2$, $c_2 = c_3 = c_4 = 1$) but a standardised set of values would have to emerge as a matter of consensus among users of the model.

As was discussed earlier, the extent to which persons exercise their intelligence in the form of taking goal-oriented adaptive action depends on both the moral fortitude (or willpower) of the person and the restraints presented by the society in which the person is embedded. So, what determines a person's will to overcome or reduce the restraints? We might think of personality traits, such as strength of character, motivation, and the like, but how would we measure them? Interrogating or testing enough individuals in whatever society we are considering in order to get a significant measure of the average would seem a hopeless task. Rather than starting with the individual, a different approach would be to ask: What characteristics of a society reflect the strength of will of its members? However, a difficulty with this approach is that restraints are characteristics of the society, whereas will is a characteristic of individuals. Is there a characteristic of society that does not represent a significant restraint, as compared to the four we have identified, but that is a good indication of the average strength of individual will? One such characteristic might be the level of corruption; we would expect that the will to oppose corruption is in some way related to the will to overcome restraints, but corruption is not in itself a significant restraint on exercising intelligence.

This issue also points to a characteristic of the measure of the Good Society proposed here: It is *indirect*, in the sense that it infers the Good Society from *objective*, measurable parameters of society, such as the attained level of education. A different approach would be to define the Good Society *directly* in terms of *subjective* features, and then conduct surveys to let people themselves express to what extent they consider

Table 6.1 Global values of the restraints and the resulting value of the global measure of the operating conditions of the collective intelligence

Restraint	Sample size	Value
Financial	0.94	0.729
Education	0.69	0.427
Information	0.21	0.491
Process	0.88	0.412
Moral fortitude	0.18	0.662
ω	0.18	0.404

society to be good. This is the approach taken by the European Social Survey (www.europeansocialsurvey.org/, accessed 20.06.2022), which conducts a survey by means of face-to-face interviews in the majority of European countries every two years. The data is made freely available for non-commercial purposes.

The *moral fortitude*, *m*, shall be expressed by the *level of corruption*, *c*, measured on a scale of 0 to 100 with 100 being very clean and 0 being very corrupt. The relationship between *w* and *c* will, without any further justification, be taken to be as follows:

$$m = 1 - e^{-\frac{100-c}{\sigma}}, \tag{6.8}$$

where σ is a scale factor that can be adjusted as a better understanding of the problem complex emerges through use of the evaluation model. In determining the result shown below, the value of σ was chosen as 50.

With the data available in 2016, and detailed in Appendix A of *The Social Bond*, the values obtained for the various parameters are listed in Table 6.1. Thc sample size is the proportion of the world's population for which data was available.

6.3.3 *Economic Inequality*

As is demonstrated by the values in Table 6.1, data for the complete set of macroeconomic variables was only available for some nations, making up about 18% of the world's population. However, all of these restraints show a significant dependence on economic factors, and a problem with our current politico-economic system is the increasing economic inequality. Despite all sorts of tinkering, from tax initiatives to international assistance and assertions that the trickle-down effect of the free economy will have a levelling effect and eventually eliminate inequality, as hypothesized by Kuznets (1955), the inequality is increasing. A recent and thorough examination of the historical development of the factors involved in economic inequality, in particular, the accumulation of capital, is provided by Piketty (2014), although some of the conclusions he draws from the data have been the subject of debate, such as in

Summers (2014). But before considering some of the implications of this increase in inequality we need to be more precise about the definition of 'inequality'. There is wealth inequality and income inequality, with the latter defining the position of the individual within the economic environment in which it finds itself—its ability to sustain itself and, above all, to participate in the public discourse. This measure is related to both the level of average income and to the distribution of income within the society; that is, to income inequality.

Wealth inequality has arisen through the conjunction of several factors; factors that are closely related through their common genesis—the industrial revolution. First of all, the role of capital changed dramatically, from being tied to land and to whatever improvements could be achieved in cultivating it, to a flexible instrument that could be invested in the provision of an ever-expanding array of products and associated services. The production process involved two classes of people—those that provided the labour and those that provided the capital, and its expansionary nature created two pressures—the pressure to provide new opportunities for investing the accumulating capital, and the pressure to consume the flow of products and services. The result is an ideological concept—the so-called "free" market; and a concentration of capital in a small part of the population—the Transnational Capitalist class (TNC) (Robinson, 2014), with a large part of the population focused on consumption and distracted from the public discourse.

Secondly, the industrial revolution ushered in an explosive development of information technology—printing, and then telegraph, telephone, TV, and the internet. This has had a positive effect on the operation of the collective intelligence, in that it has made information more readily available and at very low cost, and the interaction between individuals independent of distance. But it has also had a number of negative effects, principally due to the ease with which it allows information to be manipulated, so that it is increasingly difficult to ascertain its truth, and through the sheer volume of information, which makes it difficult to select what is important.

This leads to a third factor, which arises as a result of the pressure to consume and the facility provided by information technology, and it is the power of advertising. Or, perhaps more generally, the ability to persuade. Through every possible medium, the individual is subjected to an all-pervasive stream of information intended to persuade the individual to act, to buy, to believe, or to reject, and the result is that the individual loses some of its individuality and becomes, to some extent, a creation of the advertiser. And it is not only products and services that are being advertised, but also the ideology that underpins the capitalist economic model and the consumer society. And while the political systems differ, the economic model is much the same in all parts of the world, be it China or the US, and this is what makes the concentration of wealth such a universal measure of suppression of the collective intelligence.

Together, the two measures of economic inequality form the single most important indicator of the operating conditions of the collective intelligence—we might think of it as a sort of principal component—and in the next two sections we develop these two measures in more detail.

6.4 Income Inequality

In accessing data for any measure of income inequality, we have to be aware that such measures depend on whether we consider gross income, which is the sum of all sources of personal income before tax and other redistribution measures, or disposable household income. The latter is the more appropriate measure in a social context, as the household is (still) the most universal social structural component. The most recent OECD Rapport for its 27 reporting countries, (https://www.oecd.org/social/soc/49499779.pdf, accessed 20.06.2022), shows that while the income for the lower decile increased by 1.3% per year from the mid-1980s to the upper 2000s, the income for the upper decile increased by 1.9% per year, and the average Gini coefficient increased from 0.29 to 0.316. If gross income is used as a measure, the Gini coefficients are higher, because taxes and other redistribution measures have the effect of lowering the inequality.

Also, the influence of income inequality on the operating conditions of the collective intelligence is a highly non-linear one, as is reflected in the models presented in the following subsections and is readily demonstrated by the following example: The ability of the individual within a nation to participate in the social discourse is sometimes presented as being strongly related to the most-often presented measure of the economy, the *per capita* Gross Domestic Product. Strongly, but not exclusively: If one person is able to have the best education, access to all possible sources of information, and spend all its time thinking, while everyone else is kept in total ignorance and just slaves away, the average information-processing capability will be very low, but if that one person is rich enough and owns all the production capacity, the average *per capita* GDP could be reasonably high. This example is, of course, a utopian extreme, but it illustrates that inequality must be an important economic parameter when it comes to the influence of the economy on the collective intelligence and the ability of the individual to participate in the public discourse.

6.4.1 A First Model

This model considers that at or below a certain income level, the individual has essentially no time or energy to exercise intelligence; all effort is absorbed by mere existence. Nor is there any surplus income for acquiring information (buying newspaper, subscribing to mobile phone service, etc.). This level is often thought of as the *social poverty line*, or SPL. As the income rises above the poverty line, the restraint is reduced, but the reduction depends not directly on the value of this increase above the SPL, but on the relationship of this value to the average income of society. A possible realisation of this relationship is given by the following expression:

$$x = e^{-\frac{u-s}{u_0}} \tag{6.9}$$

where x is the measure of the restraint presented by the individual's economic situation; u is the personal income, with $x(u < s) = 1$; s is the SPL at the location (nation) of the individual; and u_0 is the world mean income—all in current PPP dollars per year. In this expression the world mean income was chosen, rather than the national mean income, in order to account for the international character of the information industry, but the model is otherwise the same as expressed in Eq. 6.3.

The SPL, s, even when measured in PPP dollars, will vary between locations; an income above the SPL in Somalia will not sustain a person in Singapore. And it will also increase over time in a given location as the mean income increases. A recent World Bank paper (Joliffe & Prydz, 2017) demonstrates that the daily SPL, s^*, can be related to the median daily income, u^*, through a linear relation:

$$s^* = 1.15 + 0.5u^*, \tag{6.10}$$

where 1.15 is a daily minimum (intercept) in 2017 PPP $. For any particular year, t, the intercept, q(t) is equal to 1.15·d(t), where d(i) is the deflation relative to the year 2017 (or inflation for i > 2017). The yearly SPL income, s, in current PPP $, is given by

$$s = 365 \cdot q(\mathrm{t}) + 0.5 \cdot u_{\mathrm{m}}, \tag{6.11}$$

where u_{m} is the median yearly income.

With this definition of the SPL, the smallest value of x in Eq. (6.9), corresponding to a personal income far above the SPL, is obviously 0, and the greatest value, corresponding to a personal income below the SPL, is obviously 1. But in the utopian case of absolute equality in the world and with personal income far above the intercept, we obtain the paradoxical result that the value of x is $e^{-0.5} = 0.607$. The issue here is that the SPL, as defined in Eq. 6.10, was developed in (Jolliffe & Prydz, 2017) based on the actual situation in the world today, and therefore its use in Eq. 6.9 is appropriate for determining the restraint on the operation of the collective intelligence in today's world.

Data regarding the distribution of income within nations is provided by several organisations and in various formats. One is the World Bank, which provides the data in the form of the shares of the five quintiles plus the shares of the lowest and highest deciles of the GNI (World Bank GNI shares). Let these percentages be denoted by v(i), i = 1 to 7. The World Bank also provides the *per capita* GNI (PPP current international dollars) for each nation on a yearly basis (World Bank GNI), denoted by v(0), as well as for the world, denoted by u_0. The yearly deflation factor, d(i), is provided in (World Bank deflation).

The median income, u_m, is given by the expression

$$u_{\mathrm{m}} = 0.05 \cdot v(3) \cdot v(0), \tag{6.12}$$

and the *relative poverty line, s' = s/u*$_0$ is given by the expression

$$s' = (365 \cdot q(\mathrm{t}) + 0.025 \cdot v(3) \cdot v(0))/u_0. \tag{6.13}$$

The average of x over the population, $\overline{x}$, is given by the sum of 7 quantities:

$$\overline{x} = 0.1 \sum_{i=1,2,6,7} e^{-w(i)} + 0.2 \sum_{1=3,4,5} e^{-w(i)} \tag{6.14}$$

with $w(\mathrm{i})$ as follows ($w(\mathrm{i}) \geq 0$):

$$\begin{aligned}
w(1) &= 0.1 \cdot v(6) \cdot v(0)/u_0 - s'; \\
w(2) &= 0.1 \cdot (v(1) - v(6)) \cdot v(0)/u_0 - s'; \\
w(3) &= 0.05 \cdot v(2) \cdot v(0)/u_0 - s'; \\
w(4) &= 0.05 \cdot v(3) \cdot v(0)/u_0 s'; \\
w(5) &= 0.05 \cdot v(4) \cdot v(0)/u_0 - s'; \\
w(6) &= 0.1 \cdot (v(5) - v(7)) \cdot v(0)/u_0 - s'; \\
w(7) &= 10 \cdot v(7) \cdot v(0)/u_0 - s';
\end{aligned} \tag{6.15}$$

The national income inequality measure is then this average value of x.

The numerical calculations are based on the World Bank data, which is quite complete for the time period 1990 to 2019, with some values for 2020. The few missing values in the income distribution data were completed as follows:

i. Where the first value appeared after 1990, the missing values were set equal to the first value.
ii. Where gaps appeared in the sequence of values, the missing values were determined by linear interpolation.
iii. Where the last value appeared prior to 2020, the missing values were set equal to the last value.
iv. Where no yearly values were available, the entity (nation, principality, etc.) was deleted from the data base. However, the population covered was not less than 97% of the total world population in any one year.

Individual nations show a considerable variation over time, as well as differences between nations; the values of $\overline{x}$ for a few nations in the years 1990, 2000, 2010, and 2020 are shown in Table 6.2.

6.4.2 A Second Model

Consider a society with n_0 members (to be defined) and form them into a set ordered by increasing income, $\mathrm{y}(n)$, $n = 1, \ldots, n_0$, and $\mathrm{y}(n+1) \geq \mathrm{y}(n)$. For large values of n_0 we can treat the variable $x = n/n_0$, $0 \leq x \leq 1$, as a continuous variable, and the income distribution function $\mathrm{y}(x)$ as a continuous, monotonically increasing function. The

Table 6.2 Values of $\overline{x}$ for a selection of nations

Country	1990	2000	2010	2020
Australia	0.377	0.370	0.395	0.392
Chile	0.728	0.654	0.617	0.603
China	0.920	0.836	0.723	0.669
India	0.902	0.886	0.857	0.828
Indonesia	0.775	0.772	0.743	0.731
Nigeria	0.833	0.876	0.842	0.868
Sweden	0.292	0.306	0.327	0.330
Great Britain	0.390	0.390	0.413	0.426
United States	0.351	0.347	0.380	0.388

cumulative income distribution, Y(x), is given by

$$Y(x) = \int_0^x y(x')dx', \tag{6.16}$$

and we shall denote the cumulative total, Y(1), by β. The Gini coefficient, G, is given by

$$G = 1 - \frac{2}{\beta}\int_0^1 Y(x)dx \tag{6.17}$$

Let us now, for the purpose of this illustration, choose the simplest form of y(x),

$$\mathrm{y(x)} = y_0 x^\alpha, \quad \alpha > 0,$$

where, as will become obvious, we exclude the case α = 0, which corresponds to complete income equality. We then have

$$Y(1) = \beta = \frac{y_0}{1+\alpha}, \ or \ y_0 = \beta(1+\alpha); \tag{6.18}$$

and

$$G = \frac{\alpha}{2+\alpha}. \tag{6.19}$$

To make the connection with our model of society, we assume that the ability of the member to participate in the collective intelligence is directly proportional to the income, up to an income y^*; for incomes above y^* the ability to participate does not increase. The income y^* may be identified with what is called the *living wage*. There is a corresponding value of x,

$$x^* = \left(\frac{y^*}{y_0}\right)^{\frac{1}{\alpha}}$$

which shows why $\alpha > 0$. The operation of the collective intelligence, CI, as a fraction of its maximal value y^*, is the cumulative value of this participation,

$$CI = y_0 \int_0^{x^*} x^\alpha dx + (1 - x^*)y^* = \beta(x^*)^{\alpha+1} + (1 - x^*)y^*. \tag{6.20}$$

If we now make y^* the unit of monetary value, so that CI, y_0, and β are measured in units of y^*, then

$$\mathrm{CI} = \beta(\mathrm{x}^*)^{\alpha+1} + 1 - \mathrm{x}^*; \tag{6.21}$$

with

$$x^* = \left(\frac{1}{\beta(1+\alpha)}\right)^{\frac{1}{\alpha}}. \tag{6.22}$$

The function CI(G, β)is shown in Fig. 6.1. In this figure, the dash curve indicates the limit of the model, arising from the condition $x^* \leq 1$, and some allowance must be made for the coarseness of the calculation grid. But the basic message is unmistakeable (and not surprising): At low values of β, the better strategy for increasing the collective intelligence is to increase β, whereas for β values above the nominal income, y^*, it is much more important to decrease the inequality.

Now, to make this model more specific, and, as an example, relate it to an actual society—Australia, we need to define the nature of the "member" and that of β. In the context of economic wellbeing, the Australian Bureau of Statistics (ABS) considers that the individual is best accounted for as a member of a *household*, and the disposable income required for a given level of wellbeing is dependent on the size of the household. But not proportional to the size; the *equivalent size* is given as 1 for the first adult (i.e. 15 years of age or over), then 0.5 for each additional adult, and 0.3 for each child. Then, for a lone person household, the *equivalised* income is equal to actual income; for households comprising more than one person, it is the estimated income that a lone person household would need to enjoy the same standard of living as the household in question, which is the income of the family divided by the equivalent size. Thus, in our little model, the "member" is the household, the income distribution is the distribution of the equivalised incomes of these households, and

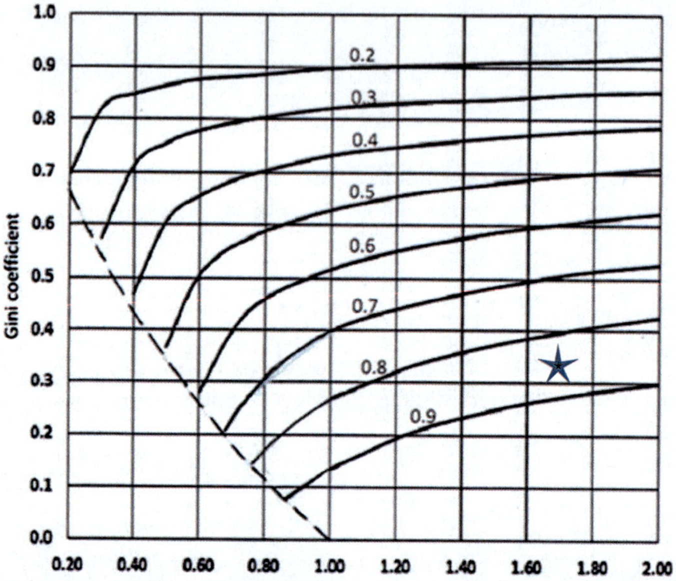

Fig. 6.1 Curves of constant value of the collective intelligence, as a fraction of its maximum value, in the plane spanned by the Gini coefficient and β, interpreted as the mean household disposable income relative to the living wage, y^*. The star indicates Australia's position in this evaluation

β is the ratio of the average disposable income to the living wage. From the ABS (6523.0—*Household Income and Wealth, Australia,* 2015–2016), we can extract the following data:

Mean household disposable income: $52,468
Median household disposable income: $44,356
Gini factor: 0.325.

Views on how a living wage should be related to such income data varies greatly; one calculation, adopted by the Greater London Authority and reported in (https://www.economist.com/the-economist-explains/2015/05/20/how-a-living-wage-is-calculated, accessed 20.06.2022), is that it should be 60% of the median disposable income, plus 15% for contingencies, which in the case of Australia would make it $30,606 and $\beta \approx 1.7$, which is shown as a star in Fig. 6.1.

The importance of the *distribution* of the aggregate parameters used in the neoliberal economic model, described in Sect. 12.3.2, has become an important issue in the last two decades or so and has led to a significant body of work based on empirical studies, such as the seminal work of David Card and Alan B. Krueger, *Myth and Measurement: The New Economics of the Minimum Wage* (Card & Krueger, 1997). An overview of this development, which constitutes a criticism of the neoliberal model, is provided by an article by Boushay (2019).

6.4.3 A Third Model

A further measure of the influence of income inequality on the operation of the collective intelligence is provided by the collective intelligence *participation rate.* It is defined as the fraction of the population with an income above 0.69 times the median income, where the individual income is the household disposable income divided by the number of persons in the household.

The income distribution is mostly presented as the total income within certain segments of the population. In particular, the OECD database (https://www.oecd.org/social/income-distribution-database.htm; IDD-Key-Indicators.xlsx, accessed 20.06.2022) gives the total income for the following segments:

Bottom 10% $= a_1$
Bottom 20% $= a_2$
Bottom 40% $= a_3$
Top 40% $= a_4$
Top 20% $= a_5$
Top 10% $= a_6$.

Let the income distribution be simplified as constant within each decile of the population, as shown in Fig. 6.2.

The total income within each decile, Y(i), as a fraction of the total income, is then given by the following expressions:

$$\begin{aligned} &Y(1) = a_1 && Y(7) = Y(8) = (a_4 - a_5)/2 \\ &Y(2) = a_2 - a_1 && Y(9) = a_5 - a_6 \\ &Y(3) = Y(4) = (a_3 - a_2)/2 && Y(10) = a_6 \\ &Y(5) = Y(6) = (1 - a_3 - a_4)/2 && \end{aligned}$$

and

$$\sum_{i=1}^{10} Y(i) = 1.$$

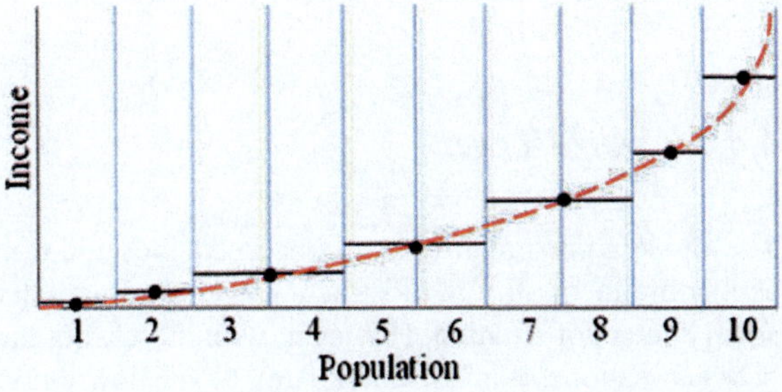

Fig. 6.2 The total income within each population decile, reflecting the data provided by OECD

If the individuals within the population are numbered by *n* from lowest to highest income, let *x* be the normalised variable *n*/N, where N is the total number of people in the society, and *y*(*x*) be the income distribution function, indicated by the dashed curve in Fig. 6.2. If y(x) is also approximated by a similar step function, then within each of the seven the steps, $y(j) = 10 \cdot Y(j)$, $j = 1\text{–}7$, and the median value of income, M, as a fraction of the mean, or *per capita*, income, is given by the expression

$$M = 5 \cdot (1 - a_3 - a_4) = 10 \cdot Y(5).$$

As a check on this expression, the OECD data for Australia in 2018 is:

Median income: $51,935
Mean income: $61,171
$a_3 = 19.6\%$; $a_4 = 63.4\%$

so that Median/Mean = 0.849, and M = 0.85; a very close correspondence.

To determine the proportion of the population that has an income above $y_0 = 0.69$ times the median income, the dotted curve in Fig. 6.2 is approximated by a quadratic function,

$$y(x) = u + vx + wx^2,$$

with the constants u, v, and w determined by fitting to the three points at $x = 0.15$, 0.3, and 0.5:

$u = 2.784 \cdot Y(2) - 2.373 \cdot Y(3) + 0.588 \cdot Y(5)$
$v = -7580 \cdot Y(2) + 10.769 \cdot Y(3) - 3.187 \cdot Y(5)$
$w = 4.900 \cdot Y(2) - 8.577 \cdot Y(3) + 3.677 \cdot Y(5).$

The value of x corresponding to y_0, x_0, is then given by the expression

$$x_0 = \frac{-v + \sqrt{v^2 - 4(u - y_0)w}}{2w};$$

and the proportion of the population that has an income above y_0—the *participation rate*—is then equal to $1 - x_0$. Results for a few countries, as copied from the calculation spreadsheet, are as shown in Table 6.3.

6.4.4 World Level

The models presented above concerned individual nations, which is appropriate, as income inequality is most acutely felt and perceived within the closer interpersonal contact of a nation. However, with increasing interaction and interdependence between nations, it is also interesting to see how income inequality is developing in the world as a single society, and using the data from Model 1, i.e., the World Bank

Table 6.3 Calculation of the participation rate for selected countries

Parameter	Country					
	Australia	Chile	UK	US	Sweden	India	China
OECD a1	0.027	0.019	0.024	0.016	0.036	0.014	0.009
OECD a2	0.072	0.051	0.069	0.053	0.087	0.041	0.019
OECD a3	0.196	0.144	0.189	0.166	0.226	0.123	0.094
OECD a4	0.634	0.721	0.648	0.676	0.594	0.749	0.765
y(0.15)=10*Y(2)	0.450	0.320	0.450	0.370	0.510	0.270	0.10
y(0.3)=10*Y(3)	0.620	0.465	0.600	0.565	0.695	0.410	0.375
y(0.5)=10*Y(5)	0.850	0.675	0.815	0.790	0.900	0.640	0.705
u	0.281	0.184	0.308	0.154	0.300	0.155	-0.197
v	1.114	0.862	0.906	1.525	1.501	0.658	2.067
w	0.049	0.245	0.220	-0.514	-0.613	0.637	-0.538
Participation	**0.729**	**0.699**	**0.736**	**0.716**	**0.763**	**0.670**	**0.635**

data, there are (at least) two approaches. The simplest is to calculate the weighted (by population) average of the national inequality measures; a more elaborate approach is to convert the national data to an income distribution for the world population as a whole. The interesting point is that, in both cases, the measure has remained practically unchanged over the 31-year period 1990–2020.

In the first case, the weighted average of $\overline{x}$ over all nations (accounting for $> 97\%$ of the world's population) can be approximated by the constant value 0.75016, with an rms error of 0.00623 over the 31-year period.

In the second case, the income shares, by quintiles, show small fluctuations around the following values:

First quintile: 0.020
Second quintile: 0.027
Third quintile: 0.064
Fourth quintile: 0.164
Fifth quintile: 0.725.

We can, in principle, use the same measure as defined in Sect. 6.4.1, with the individual's economic situation expressed by Eq. 6.9, but the interpretation and the resulting value will now be different, as the poverty line is the same for all individuals, irrespective of their locations. We would expect the value to be greater that the average over national values shown above, due to the effect of wealthy people driving up the poverty line, and this is indeed the case:

Poverty measure: 0.881; rms error: 0.00183

Again, the small rms error over the 31-year period demonstrates how little income inequality for the world as a whole has changed over this period.

The measure defined in Model 3—the participation rate—can also be determined for the world's population by using the INTERCEPT function in Excel and the first three quintile values above; the result is 0.644.

6.4.5 A Comment

This discussion of income inequality should not be interpreted as implying that there is anything wrong or undesirable about income inequality per se, it is only meant to show how and to what extent income inequality can have a negative effect on the operation of the collective intelligence, as reflected in the measures introduced. Given the great range of capabilities, energy, and desires of people, a considerable variation in income is to be expected. As far as the operation of the collective intelligence is concerned, there is no negative effect of high individual income as long as this income is spent on consumption, as in a lavish lifestyle. (The impact on the environment is a different matter.) But there are two negative effects: One is that income can be used to pervert the operation of the collective intelligence; that is, in some form of corruption. The other is that surplus income is invested; with investment comes ownership, and with ownership comes control. The result of both effects is to give the individual power, power not only over private affairs, but also, in the real world, over the operation of the collective intelligence and over the resulting adaptive actions. However, both the nature and the effects of wealth inequality are different to those of income inequality, and this is the subject of the next section.

6.5 Wealth Inequality

Wealth inequality is measured in much the same ways as income inequality; by examining the wealth distribution, or by a Gini coefficient, but its influence on the operating conditions of the collective intelligence is quite different to that of income. Individual wealth is not in itself a significant determinant in the information processing within individuals; as long as the income is sufficient, wealth would be at least as much a negative as a positive influence. The influence of wealth comes through the ability to control the interaction between the individual and society, and while this was always the case to some extent, it has now taken on a critical importance through the rise of information technology, as outlined in Sect. 2.3.4. A further influence is through education in those countries where the public education system is deficient, with private education providing a better start in life for those who can pay for it.

However it is measured, wealth inequality is increasing, as is demonstrated by the data in Piketty (2014) and in the Oxfam Briefing Paper referenced earlier, a document that includes an extensive list of references to relevant data. The data shows an overall increase in wealth and that the number of people living in extreme poverty has been halved in the period from 1990 to 2010. But it also shows how the increase is skewed towards the wealthier half of the world's population, although the exact extent of this imbalance depends on the definition of personal wealth, and its interpretation is the subject of some debate (https://www.oxfam.org.au/wp-content/uploads/2016/06/OXF003-Tax-Havens-Report-FA2-WEB.pdf, accessed 20.06.2022). Oxfam is primarily concerned with alleviating poverty and with the plight of the poorest half of the population; in Chap. 11 we shall consider the relationship of wealth inequality to power. In the context of this chapter, we are interested in the influence of wealth inequality on the operation of the collective intelligence, and although the link may be somewhat indirect, there is no doubt that wealth inequality has a negative effect on the operation of the collective intelligence. But the nature of this effect—the form of the relationship between inequality and the exchange of information—has changed in several ways over time. Taking the West as an example, at the height of the power of the Catholic Church and prior to the advent of printing with moveable type—say, around 1200 AD—there was great wealth inequality, and its effect on the exchange of information was very direct, in that only the wealthy and powerful could afford the reproduction of manuscripts by hand. Then, as printing technology developed and information reproduction became less and less expensive, this dependence on wealth decreased, and there was a blossoming of the creation and exchange of information and ideas—first the Renaissance and then the Enlightenment. But in the last hundred years or so, two developments have resurrected the negative effect of wealth and power on the exchange of information. One is the emergence of autocratic/dictatorial/oligarchic governments—communist, fascist, nationalist, or simply dictatorships, where the media are severely controlled. The other, in the liberal democracies and the one of concern here, is the domination of the media by a small class of extremely wealthy persons—part of what has been called the *Transnational Capitalist class*, TNC (Robinson, 2014).

The exchange of information in the circulating information, which was originally a face-to-face interaction, is today mainly an interaction mediated by information technology, and thus easily influenced by the owners of the technology, as well as by the owners of the companies that provide the major source of income for the media companies through advertising. This feature of the interaction is not new, and has been the subject of numerous publications, including a section in (Aslaksen, 2020); another recent publication is a UNESCO discussion paper (Mendel et al., 2017). The concentration of ownership of the media in a relatively small number of individuals and families is a peculiarity of this industry, as was documented in the paper (Djankov et al., 2003). And the reason was given as the high amenity of ownership, in the form of prestige and social visibility and influence, as compared with other industries, where the benefit is more focused on the financial performance. The concentration may take several forms, such as horizontal concentration of many titles in the hand of one company, vertical concentration of media outlets, printing houses and distribution channels, and cross-ownership of non-media related companies in media houses. And there are such mergers and acquisitions going on all the time; just recently

AT&T acquired Time Warner in a landmark $85 billion deal and The Walt Disney Co. acquired up 20th Century Fox for $71 billion.

It is important to emphasize that this dominance is not realised by limiting the freedom of publication; on the contrary, it has never been easier to make one's opinions public. The Internet and social media platforms, such as Facebook and Twitter, are available at very moderate cost or even for free. The dominance arises out of volume and with it the ability to catch people's attention. If in earlier times the individual's attempt to gain attention was like shouting into a vacuum, it is today like shouting into an extremely noisy environment. There is no need to worry (or at least not too much) about opposition to the established neoconservative order; nobody can hear anything but the loudest voices.

The transnational nature of both the media and the capitalists controlling them makes wealth inequality on a national basis less significant as an economic measure in the present context; we need to consider wealth inequality on a world basis. However, inequality, as it is usually expressed, e.g., by the Gini coefficient introduced earlier, is not really the most appropriate measure here; what is important is the *concentration* of wealth in the TNC. As an illustration of this, consider a society with such an extreme concentration of wealth that half its wealth is owned by a single person. Depending on how the remaining half is distributed, the Gini coefficient can vary within the range 0.5–1.0, without changing the fact that the wealth and its associated power are basically concentrated in a single person. What we would like is a measure that reflects the distribution of wealth in the uppermost part of the "wealth pyramid", a concept propagated by Credit-Suisse and documented in their yearly Global Wealth Report (https://www.credit-suisse.com/about-us/en/reports-research/global-wealth-report.html, accessed 20.06.2022). Let the adults in the world be numbered by $x = 1, 2, 3, \ldots$, in order of decreasing wealth, $y(x)$, so that $y(1)$ is the wealth of the world's wealthiest adult. The Global Wealth Report, published yearly since 2010 by the Research Institute of Credit Suisse, provides the values of x for a number of values of y, of which the upper four are 10, 50, 100, and 500 million $US. It now turns out, based on the data for the years 2010–2019, that a function of the form

$$y(x) = \frac{a}{x^b}$$

can be fitted to these four values with an RMS error of less than 3%, yielding a set of values (a, b) for each year, as listed in the following Table 6.4.

The combined wealth of the n wealthiest adults, $Y(n)$, is given by the integral

$$Y(n) = \int_1^n y(x)dx = \frac{a}{1-b}\left(n^{1-b} - 1\right), \tag{6.17}$$

and the values for $n = 10$, 100, and 1000 are shown in Table 6.5.

The above calculation demonstrates how data from those reports can be used to determine the combined wealth of the world's n most wealthy adults, or conversely,

Table 6.4 Curve-fitting parameters a and b for each year in the period 2010–2019

Year	a ($US million)	b
2010	114.9	0.68484
2011	89.7	0.65919
2012	90.4	0.65984
2013	97.2	0.65748
2014	129.5	0.66658
2015	144.0	0.67648
2016	165.6	0.68131
2017	187.2	0.68827
2018	116.5	0.65083
2019	121.5	0.64831

Table 6.5 The combined wealth of the n wealthiest adults, in $US billions

Year	$n = 10$	$n = 100$	$n = 1000$
2010	389	1192	2851
2011	314	1001	2508
2012	316	1007	2520
2013	341	1090	2740
2014	449	1415	3498
2015	492	1530	3714
2016	563	1735	4177
2017	630	1923	4572
2018	412	1332	3389
2019	431	1400	3576

how many adults make up the top of the pyramid above a given combined wealth value. The question is then: What combined wealth value would include the majority of the adults who own and control the greater part of the industry that produces the majority of the circulating information? It would appear practically impossible to provide an answer to this question that would be authoritative and reasonably accurate, but considering that the capitalisation of the major media corporations is somewhere around 2 trillion $US in 2019, and given the previous comment about the prestige and influence on opinion provided by ownership, we might, in a somewhat arbitrary manner, set this level of combined wealth at 5 trillion $US, which is about 1.4% of the global wealth (Credit-Suisse, 2019 report), and consider this group of people to constitute the TNC. The results for the years 2010–2019 are shown in Table 6.6.

Using this measure of wealth concentration, the WCI, as a measure of the domination of the circulating information by the TNC needs to be accepted with a grain of salt, because, while the members of the TNC have the interest in generating and preserving wealth in common, the context in which this interest is promoted varies

Table 6.6 The number of adults constituting the TNC, defined as the group owning 1.4% of the global wealth, shown as both as an absolute value and as a percentage of the world's population

Year	TNC size adults	% of all adults	Wealth con-centration index
2010	950	0.021	65
2011	1979	0.044	32
2012	1777	0.039	36
2013	1743	0.037	38
2014	1149	0.024	57
2015	849	0.018	79
2016	655	0.014	103
2017	649	0.013	107
2018	2037	0.040	35
2019	2482	0.049	29

The *wealth concentration index*, WCI, is the ratio of the group wealth (1.4%) and group size, so that a value of 1 would signify a uniform distribution

considerably between different regions and countries, such as the US, China, and Saudi Arabia. As a consequence, the form and content of the input to the circulating information varies considerably along such dimensions as political ideology and religion.

The crucial feature of the wealth distribution data in the current context is the accelerating trend towards a concentration of wealth at the very upper end of the distribution. In the years since 2010, the share of the world's wealth owned by the richest 1% of the world's population has been steadily increasing, to the extent that it now is approximately equal to 50%. That is, this 1% of the population has as much wealth as the remaining 99%. But even more significant is the concentration within this 1%: About 4500 individuals, out of an adult population of 4772 million, each have wealth exceeding 500 million USD, and we note that most of these ultra-wealthy individuals reside in the Western part of the world, i.e., North America and Europe (Credit Suisse, 2015). These individuals constitute what William Robinson identifies as the Transnational Capitalist Class, or TNC (Robinson, 2014), and to the extent that they are able to coordinate their actions, this TNC is a force that can challenge most national governments and play a significant political role on the world stage. Hence, not only is capitalism, in its current form, resulting in increasing economic inequality, but it is creating a new entity, a new structural element in the world society as a system. And it is a very powerful element; wealth is power, not least in its ability to influence government through lobbying, as discussed by Hanna (2013). In the US, by 2012, direct lobbying had reached $3.28 billion, up from $1.44 billion in 1998—more than double. And in the previously referenced book by Shiller (Shiller, 2012), a chapter is devoted to the issue of lobbying in the US.

6.6 Freedom

6.6.1 Freedom in the Context of This Book

Freedom is a concept we meet frequently in many contexts, such as in philosophy, politics, and social science, and also more specifically as freedom of the press, freedom of speech, freedom of religion, freedom of navigation, and so on. The concept of freedom has been at the core of various activities over the last few centuries, ranging from very practical ones, such as the opposition to slavery, to highly esoteric speculations in philosophy. This wide scope can be narrowed considerably by two means: Firstly, by realising that the concept can be split into freedom *to*, in the sense of choice of action, and freedom *from*, in the sense of the absence of constraint. Secondly, by specifying the context in which the concept is applied; there is a difference between the freedom of a football player to play within the rules of the game and the freedom of an artist in creating a painting or a sculpture. In the context of developing an understanding of the evolution of society, as it is presented in this book, freedom is important because the call for freedom has been used as the motivator for many of the changes, big and small, that have made up the evolution of society.

Intuitively we probably all have a general understanding of what is meant when the word 'freedom' is used, without analysing too closely the implications raised by the context in which it is used. We tend to think of freedom as something that is inherently good and as a characteristic of the ideal human being; reality is a devaluation of this ideal state. This is reflected in such a definition of freedom as we find in the Oxford English Dictionary; "the power or right to act, speak, or think as one wants". This simplistic meaning is related to the existence of early man, the individual hunter roaming through the land with no obligations and with interactions limited to the occasional coupling with an available female or the elimination of an intrusive male. Free, but what he could do with that freedom was very limited, both by his natural environment, over which he had no control, and by the very limited repertoire of activities available to him. Speaking; well, he could talk to himself, I suppose.

A similar case is the concept of the value of a human being as something absolute and intrinsic. But human beings that have no interaction with other humans, no-one knows that they exist, and even though they compose the most beautiful music, write the most gripping and profound prose and poetry, and have discovered untold scientific truths, leave no trace of this, such "virtual" human beings have no value. The value of a human being arises solely through his or her interaction with other humans, and so it is with freedom—it has no meaning outside of its social context. And in society, the freedom of the individual is always subject to a compromise between the benefits of restricting individual choice in the interest of efficient cooperation and the potential cost of foregone individual contributions.

A measure of the influence of society on the individual is the strength of the interaction between individuals, and as an illustration of the relationship between interaction and freedom recall the analogy introduced in Sect. 2.1. Just as we use different concepts to describe the movement and energy of molecules in the three states of a collection of water molecules, our description of freedom must reflect the society to which it refers.

6.6.2 Freedom in the Context of Society

As society evolves along the lines set out in Sect. 2.3, the progression can be thought of in terms of a loss of *personal freedom*. Being part of a society means accepting certain rules and laws, and they represent a restriction on the behaviour of the individual. However, to use freedom as a measure, we need to provide a definition that is suitable for its use in this context, as freedom is a word that is used so frequently and with such a broad spectrum of meanings. First of all, and drawing upon the major work of Patterson (1991), we need to realise that the concept of freedom is not an inherent feature of the human character; it is a product of Western civilisation with its roots in antique Greece and with a development that has continued into the present. But even the concept of a distinct human character as something fixed, something we can strive towards fulfilling, as when Maslow says, "Man demonstrates in his own nature a pressure toward fuller and fuller Being, more and more perfect actualization of his humanness." (Maslow, 1968), is deceptive. Humanness, unless we mean having one head, two arms, and two legs, is not something fixed we can strive towards; it is something we are continually developing through our membership of society. It is impossible to know what "humanness" will be, say, a thousand years from now. Consequently, when we apply freedom as a measure over the evolution of society, it is our current understanding of the concept we are applying.

As an aside, the tendency to consider ourselves to be *it*; the final goal of evolution, and for our current concepts, belief, and understanding to have some form of eternal validity, is evident in other contexts also, and one closely related to freedom is ethics. In his book, *The Ethics Project*, Kitcher demonstrates how our concepts of ethics and of what is ethical must, if they are to be relevant and effective, relate to our society, the environment in which our actions take place, and as this is a continually developing environment, the same must be the case for ethics. The issue is not primarily about ethical truth, but about ethical progress. Ethical truth is not preordained and something towards which we are striving; instead, we should be striving for each progressive change to be truthful (Kitcher, 2011).

Secondly, and again following Patterson, freedom can be viewed as a composite concept with three components:

- Personal freedom: the absence of restraints on the one hand, and the ability to realise one's desires on the other. These two aspects of personal freedom—negative and positive—are just two sides of the same coin.
- Sovereignal freedom: the power to act as one pleases; regardless of the wishes of others.

- Civic freedom: the capacity of members of society to participate in its life and governance.

The distinction between personal and civic freedom reflects the view of the individual as something that has an existence separated from society. As we do not adopt that view, but see the individual as intrinsically an element of society, civic freedom is absorbed into individual freedom. Sovereignal freedom, as the desire to have unrestricted power over others, may seem at odds with our intuitive understanding of freedom, but that view is a result of Enlightenment rationalism; historically persons, such as the pharaohs, have had that power—the power of a god—and others have clearly wished they had it. In that absolute sense it has no relevance in the present context, but in a relative sense, within the limits experienced by the slave-master and the slave, it is a useful concept, although, as is shown below, it can also be absorbed into personal freedom.

Thus, we have arrived at a point where there is no need to distinguish between the three components of freedom for the purpose of defining a measure of freedom in the context of society; we should perhaps just call it human freedom, recognising that the human is inseparable from society—the human is an element of society. However, before we can develop this measure further, we need to distinguish another meaning of the term 'personal freedom'—as the ability of a person to realise its own humanness. A discussion of this aspect of freedom, as well as a comprehensive literature survey, is contained in the doctoral dissertation by Grodi (1995), where it is identified as consisting of Acceptance of Self, Acceptance of Circumstances, Choice, Non-conformity, Authenticity, Responsibility, Spirituality, Humility, and Letting Go. It is something a person must work out on its own, it is a personal experience, more or less independent of society, and is therefore not really a component of the freedom of the individual as an element of society. It would be tempting to say that the freedom we are looking at is a framework within which this intensely personal freedom can be realised, but that is not correct, as the latter can be realised even in circumstances of external coercion and restrictions.

Freedom, with its two aspects—the freedom *from* something, and the freedom *to* something—can be thought of as the totality of actions made possible by the state of development of society, minus the restrictions society places on our actions. Of course, that totality is only virtual, as without restrictions there would be no society, but this picture is nevertheless a useful conceptualisation. It does require us to admit two types of restrictions: those that restrict us from doing something and those that restrict us from not doing something, i.e., that restrict our freedom by compelling us to do something, but this just reflects two types of laws. In this manner we are replacing the simplistic definition of freedom with what we might think of as the *richness* of society; the number of actions or opportunities society offers its members.

For our definition to be relevant, it must at least account for four different aspects:

a. The restrictions, as set out in the legal framework of the society, that apply only to a subset of the members;

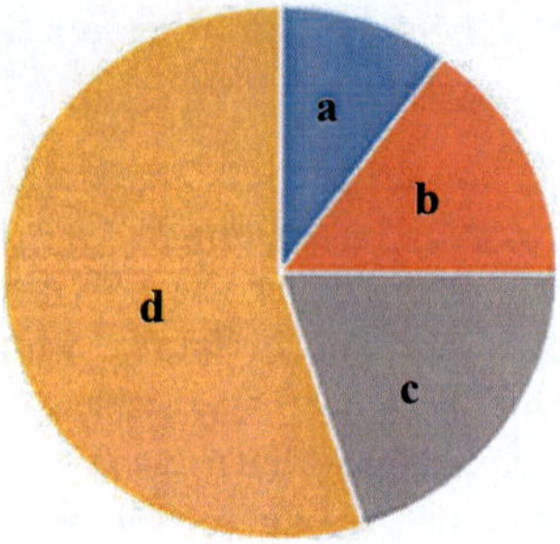

Fig. 6.3 Illustrating the totality of actions possible within a society as a circular disk, and the three types of restrictions as the sectors **a**, **b**, and **c**, leaving the sector **d** to represent the actions available to all members of the society

b. de facto restrictions on the rights of any particular member of society relative to those on the rights of any other member, arising from the structure of the society (economy, culture/customs, class structure, etc.);
c. the restrictions placed on society as a whole (through legislation, religion, etc.); and
d. the number of activities available to members of the society, what we termed the *richness* of the society.

We can regard these four aspects as leading to a partitioning of the total of possible actions, as illustrated in Fig. 6.3. That is, the whole circular disk represents the total of possible actions available to members of a society, and the sectors a, b, and c represent the actions that are proscribed by the corresponding restrictions, leaving the sector d as the actions available to all members of the society, i.e., the richness of the society.

The first aspect, and the associated restrictions, **a**, in Fig. 6.3, can, on the one hand, reflect sovereign power, insofar as this power is legitimised within the social framework, but it can also reflect a cultural norm, such as restrictions that apply to minorities, or restrictions that apply only to women, and in an ideal, democratic society there would be no such restrictions. The restrictions associated with the second aspect, **b**, in Fig. 6.3, are ones we encounter in our daily routines, often of a hierarchical nature, and these restrictions are the most insidious ones and the ones that concern us most, because they are 'hidden'. A recent treatment of such restrictions, albeit with a focus on women, is contained in the doctoral dissertation by Welch (2010). Her central concern for social freedom is the obstacles created by hierarchical relations to interaction and to the ability to choose as expressions of self-determination, and she constructs a liberatory theory of social freedom by addressing problems of choice for daily life engendered by social forms of systemic oppression and present normative conditions needed to overcome hierarchical relations in the social sphere. One of the issues she raises in this regard is that of "the invisible foot"—the extent to which most rational choices for oppressed persons are coerced via institutional inequities or inequalities, making the choices seem voluntary and thereby reinforcing the inequities and the identification of the persons as belonging to a separate social group. The effect is, in many ways, very similar to that arising from the concept of 'enframing' that was introduced and discussed by Hill (1988):

the effect of interpreting our experience in accordance with an entrenched world view, belief, or philosophy, and suppressing anything that cannot be made to fit into it, although enframing includes aspects of the legal framework also. We shall encounter an example of this in the form of a paradigm, in Chap. 11.

The restriction in group **c** are the ones imposed by the legal framework on every member of society. This is the main group, and contains almost all our laws and regulations, dealing with everything from warfare to burning rubbish in the back yard. These restrictions have arisen in response to changes in society, i.e., changes to the interactions between individuals, and the driver behind these changes has generally been new knowledge leading to new technology resulting in new applications of technology. This sequence is represented in Fig. 6.4, and it illustrates that the restrictions are most likely to be created towards the end of the sequence, when the effects of the changes on society have become very noticeable. The sequence begins with science, with mankind's insatiable curiosity and quest for knowledge and understanding. As the consequences of the new knowledge are generally poorly understood at this point, there are very few restrictions; exceptions are e.g., stem-cell research and research on humans. The next process in the sequence uses the new scientific knowledge to develop technology, the knowledge of how to create solutions to practical problems. As part of this development, unwanted effects become obvious, and restrictions are put in place to prevent these. The third process in the sequence utilises technology to create applications that meet needs expressed by society. This process, which relies heavily on engineering, takes place in the element of the functional structure of society identified as industry, and here numerous restrictions are required to ensure that the wider requirements of society are met, e.g., with regards to safety and disclosure of product characteristics, but also on the structure of society itself, such as inequalities. The final process is the acceptance and use of the applications by society, and through this experience many effects become visible that require further restrictions.

The meaning of freedom represented by Fig. 6.3 is not one of absence of restrictions, but as the richness of actions available to the individual, involving both the individual alone (discovery and reflection, as identified by μ_b and μ_c in our model) and the individual's interaction with the circulating information (identified in our

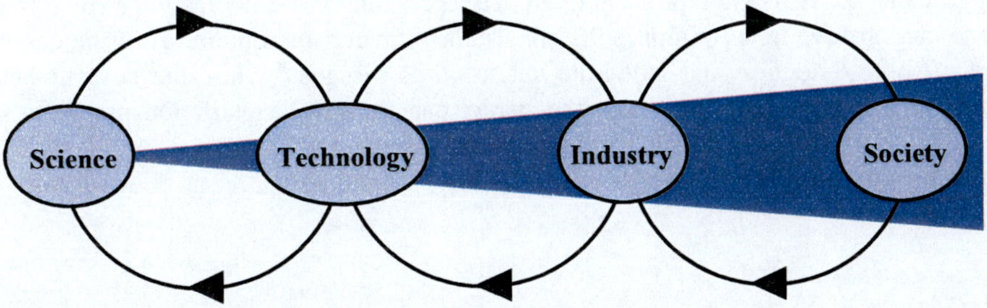

Fig. 6.4 A simplified sequence of processes leading from scientific knowledge to meeting the needs of society, and the accompanying introduction of restrictions in order to eliminate unwanted effects

model by μ_a). As society evolved over the last 10,000 years, the number of restrictions increased exponentially from the Code of Hammurabi (1760 BC), which fitted on a single stele, to the laws and regulations of a nation like Australia today, which number in the tens of thousands. The actions available to the individual have increased to a similar extent, although an exact numerical comparison is not possible. As an example, consider the actions made available to each of us by the complex social subsystem consisting of cars, roads, and support functions such as petrol stations, repair services, etc. We enjoy the benefit of reducing the impact of distance on many of our activities—work, visiting friends, going on holidays, and so on. There are numerous restrictions attached to owning and driving a car, including traffic rules, requirements for passing driving tests, requirements on the condition of the car, on associated inspections, and on insurances, as well as on the condition of the driver (vision, hearing, drugs, and alcohol), but most of us accept these restrictions as appropriate and the associated loss of individual freedom as well worth the benefit.

And this leads us to what intuitively must be a central characteristic of society: the balance between what society offers the individual and the demands placed on the individual by society. And an important aspect of this is that it is not only about the actual affordances and demands, but about how these are *perceived* by the individual. This balance is part of what I call the essence of society; a concept that will be developed further in Sect. 10.1.1. We have already encountered various reflections of it, as in the balance between free and bound energy, and in the extent of the common infrastructure (exemplified as the balance between government expenditure and GDP), but making it more concrete and perhaps even measurable is not an easy task.

The above view of freedom is applicable to both parts of the transformation process—the collective intelligence and adaptive action, but with a different emphasis in each part. In the case of the collective intelligence, the focus is on the interaction of the individual with society, as illustrated in Fig. 5.7, and on the participation in the public discourse, where freedom is commonly thought of as freedom of expression. This is the subject of the next section.

In the case of adaptive action, the focus is on the ability to convert belief into concrete changes to society. One might think that, having reached a consensus of opinion on a matter, the realisation through government action would be almost automatic, but that is not the case. Even in a representative democracy like Australia one can observe how opinion polls consistently favour one option, as in the cases of offshore detention and indefinite detention of refugees, while the government pursues the opposite option. This effect, which can be seen as a restriction on society's freedom of action, is a feature of the political system, and an analysis of this effect occupies most of Chap. 7.

6.6.3 Freedom of Expression

Our high-level model of society as an information-processing system assumes that the identity of every individual is present in the circulating information, so that whenever an individual updates its identity as a result of the processing of new information, this is immediately available to all other individuals. In practical terms, this means that the individual enjoys *freedom of expression* and all that this entails—freedom of speech and unrestricted access to the circulating information through the various channels provided by the existing level of information technology. This is clearly an ideal situation; in reality few persons come close to it, and most of us are far removed from it. For example, a journalist and a newspaper editor have an ability to put their thoughts into the circulating information that few of us can equal. Consequently, in a developed society any measure of the freedom of expression of the individual, as an average over all adults in the society, would take on a value quite far removed from the ideal. That, in itself, does not disqualify freedom of expression as one measure of the operating condition of the collective intelligence; in particular, as it could still indicate if the condition is improving or deteriorating, but it would be very difficult to define as a quantitative, measurable variable.

It is also worthwhile to realise that the value of this variable, whatever the detailed definition, has probably not changed monotonically. A simple, small, tribal society, where the public discourse was conducted around the fire in the evening might have had a value of this variable higher than what it has in our society.

6.6.4 A Concluding Remark

This chapter was concerned with identifying and discussing the conditions that affect the operation of the collective intelligence—the process that establishes and maintains society's belief system. One main part of this process is the information exchange between individuals that takes place via the circulating information; a process that is often identified as the public discourse. The other main part is the information processing performed by each individual, an evaluation of the received information of the basis of stored information, with the result of the evaluation being added to the circulating information as well as updating the individual's stored information. This is all described within the information view of society, but as both information exchange and processing are physical processes, the operating conditions are located in the action view, i.e., in the physical domain.

The ideal operating conditions are those that provide identical opportunities for every individual to access the circulating information, and this was reflected in the negative effect of *inequality* and its expression in economic terms. However, in the context of our aim to understand the forces that drive the evolution of society—the transformation process—the collective intelligence is only one half of the process; the other half is the process that converts changes in society's belief system into changes in society—what we in Chap. 3 called adaptive action—and that is the subject of the next chapter.

Chapter 7
Adaptive Action

7.1 The Concept

Already in Chap. 3 we saw how information and action can be viewed as two sides of the same coin—society, and in our description of the relationship between them we have chosen information to be the primary side—information determines action. This is true for the actions of daily life, resulting from information in class 2, but it is also true for the actions that lead to a change of society, resulting from information in class 3, and these are what we call *adaptive actions*. So far, we have only been concerned with the exchange and processing of information, and we have defined intelligence in terms of information-processing ability. In Sect. 3.2 we pointed out that a change to society—a step in its evolution—requires not only a change in belief as a result of the processing of class 3 information, but action to reflect the change. Together, information processing and action constitute the transformation cycle.

A more common and holistic definition of intelligence is as the ability to effect *goal-oriented adaptive behaviour*. A small sample of articles in three handbooks (Sternberg, 1982; Wilhelm & Engle, 2005; Wolman, 1985), as well as an overview by Sternberg and Salter (1982), demonstrate that the focus of research on intelligence is on understanding, quantifying, and measuring that behaviour, either in a classical "external" fashion through stimulus–response descriptions, which form the foundation of intelligence tests, or in a more recent "internal" fashion that sees the behaviour as the result of information-processing processes. As a result of this latter approach, various process models emerge as models of intelligence. The interesting feature of all these articles is that, while the "adaptive behaviour" part of the definition receives in-depth attention, there is almost no discussion of what the goal of the "goal-oriented" part of the definition is.

We now need to address the concept of adaptive action in some detail. And the starting point is to recognise that the concept operates on two levels—the level of the individual and the level of society. This reflects the two levels of information processing—in the individual and in society, as individual intelligence and collective

E. W. Aslaksen, *The Evolution of Society*,
https://doi.org/10.1007/978-3-031-23054-7_7

intelligence, resulting in changes to individual identity and in changes to society's belief system, as was indicated in Fig. 5.11.

7.1.1 The Individual Level

When we, as individuals, detect a change in our environment that we perceive as potentially having a significant influence on our lives, we may respond by taking an action to adjust our current situation accordingly. That is, if the change is perceived as threatening, we try to minimise or eliminate the threat; if it is perceived as beneficial, we try to optimise the benefit. This can be viewed as a two-step process; in the first step recognising the change, processing the information, as described in Chap. 4, and modifying our stored information accordingly. In the case of class 3 information, this means modifying our belief—our *identity*. The second step is to reflect this changed belief in adapting our behaviour accordingly. However, the second step—the *adaptive action*—is not an automatic consequence of the first; the decision to act is itself the result of a complex process, in which *will* plays a main role—the subject of Sect. 7.2.2.

To develop an understanding of adaptive action suitable for our purpose, we start with a very simple model of the individual, in which the information-processing model developed in Chap. 4 is augmented by input and output interfaces and by the knowledge of the capabilities of these interfaces, as shown in Fig. 7.1. The individual responds to an input in the form of the external stimuli representing a new situation by performing an adaptive action. The action is performed by the actuators (feet, vocal cords, etc.) available to the individual, which are controlled by the processes taking place in response to the signals provided by the sensors. The processes are dependent on the knowledge accumulated by the individual through experience and education, as detailed in Chap. 4, and the capabilities of the interfaces are described by such parameters as visual acuity, frequency range, smell, dexterity, strength, reach, etc.

If we refer to the model developed in Chap. 4, the change in identity that corresponds to a particular adaptive action—that is, the information that has the potential to result in the action—is a function of two variables: the received information, x, and the belief system used to evaluate the information—the identity; the state of which we might denote by y. The identity is a dynamic variable, formed by all the information previously received and therefore continually changing, and we can express this as

$$Adaptive\ action = A(x(t_0), y(t_0)); \tag{7.1}$$

With

$$y(t_0) = \int_0^{t_0} B(x(t))dt; \tag{7.2}$$

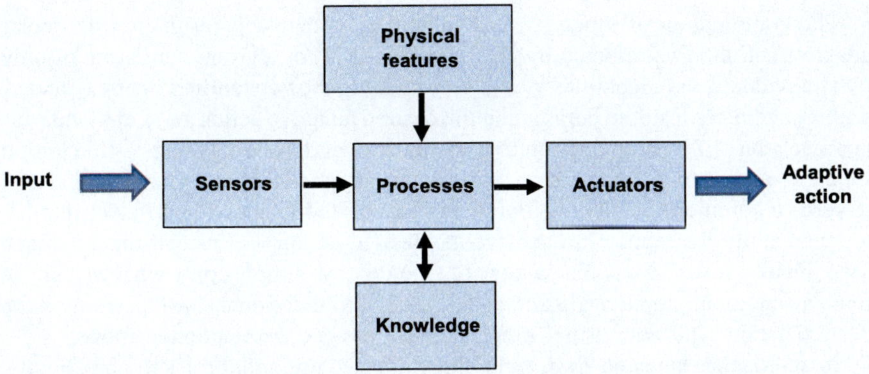

Fig. 7.1 A model of the individual incorporating both the information and action views

and where $t = 0$ indicates the time of birth of the individual, and t_0 is the time at which the adaptive action is to take place. Of course, A and B are generally complex functions with very many parameters and somewhat different for every individual, but the basic process expressed by these two equations is well known in politics, advertising and, above all, in religion—repeat something often enough, and it will be accepted as a fact and result in a corresponding adaptive action (vote for a candidate, go to war, buy a product, enforce conformance).

Now, because every individual's identity is different, the time it takes to modify the identities of the members of a society, or at least a significant portion of them, to the extent that they will exhibit the same adaptive action given the same input, can be quite long, even many years, depending on the shift in identities required. This problem can be overcome by tailoring the input to the existing identity. Ideally this would immediately result in the desired action, but at least it would reduce the indoctrination period considerably. This is illustrated in Fig. 7.2; again, in a highly simplified manner.

Fig. 7.2 The combinations of inputs, x, and identities, y, that will result in a particular adaptive action, A_0

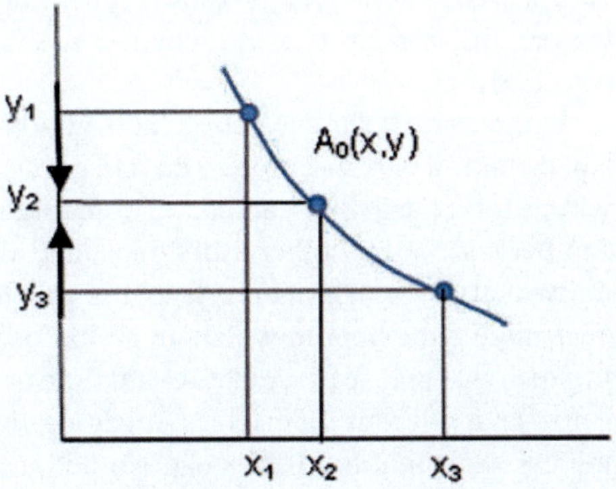

The combination of inputs, x, and identities, y, that will result in a particular adaptive action is represented by the curve $A_0(x,y)$. For a group consisting of only two individuals, with identities y_1 and y_3, providing the two tailored inputs x_1 and x_3 will immediately result in both taking the desired adaptive action A_0. Let us call this approach One. If it is too difficult to find inputs x_1 and x_3, it may be possible to find inputs x_1^* and x_3^* that, while not resulting in any adaptive action, will move y_1 and y_3 towards a common identity, y_2. Following this period of individual indoctrination, we then apply the input x_2 to achieve the desired action. Let us call this approach Two. Finally, if we have no knowledge of y_1 and y_3, we simply apply what we believe to be a convincing input again and again, until both individuals have performed the desired action. This approach, Three, is the simple one we mentioned above.

In approaches One and Two, technology has two important roles to play. Firstly, through data harvesting, mining, and analysis it becomes possible to determine the "value' of y for every individual. As individuals, our interactions with the outside world are conducted more and more through electronic data channels that are being monitored, both by service providers, and by the Government. Secondly, data processing power is also increasing exponentially, so that maintaining, updating, and extracting information from such big data sets is no longer limited by processing capacity. These two processes constitute the core of what Alex Pentland calls Social Physics (Pentland, 2014). So, as far as data storage and processing is concerned, technology has advanced to the point where the question is no longer "can we do it", but "should we do it", and we shall return to this issue regarding the application of technology in general in Chap. 8.

However, what is described above is a special version of the transformation process, one could say a perversion of the process, as it is applied in advertising and indoctrination It is a one-way promotion of a product or belief; it is not what we understand by the individual's participation in the transformation process based on the collective intelligence. It does not want a discourse; it wants an unquestioning acceptance. It does not use (or tries to suppress) the evaluation part of the information processing, blocking the responses (questions) arising from the evaluation so that they become fewer and fewer, slowly "freezing' the identity so that, in the end, acceptance is automatic, and the individual has become a consumer automation. Of course, this also results in a change to society, but it is unlikely to be in the right direction.

In the term "adaptive action", the action resulting from new information (belief) can be taking part in a protest march, going on a vegan diet, joining a political party, voting for a particular candidate, donating money to a cause, and so on; action that can be observed by other individuals and so provides an information input to these individuals—an interaction. But it is practically impossible for these observers to determine the extent to which an action resulted from the individual's participation in a dialogue and an associated evaluation or simply from the acceptance of a received command, without themselves engaging in a dialogue with the individual. That is, the essential feature of an adaptive action on the individual level is that it invites a dialogue; one-way interaction does not contribute to the operation of the collective intelligence. The extent to which the transformation cycle conforms to its ideal cannot be determined by examining adaptive actions on the individual level, only by examining the conditions under which it operates which, on the individual level, are essentially the operating conditions of the collective intelligence, as considered in Chap. 6.

7.1.2 The Society Level

On the level of society, adaptive action is the realisation of a change in society's belief system as a change in what we already in Sect. 1.2 introduced as the common infrastructure. The means of this realisation is itself a part of the common infrastructure, and for our purposes it is useful to consider it as composed of three components. The first accesses society's belief and expresses the changes to society for it to reflect this belief in some concrete form. The second component specifies the corresponding changes to the common infrastructure, and the third component realises the specified changes. The first two components comprise the *political framework*, the last is the *government*; the exact nature of all the components of this process depend on the particular society. For a typical democracy, the process is illustrated in Fig. 7.3, where the "Description of required changes to society" would be in the form of party platforms or manifestos, and "Specification of changes to the common infrastructure" would be in the form of acts of parliament or bills and amendments.

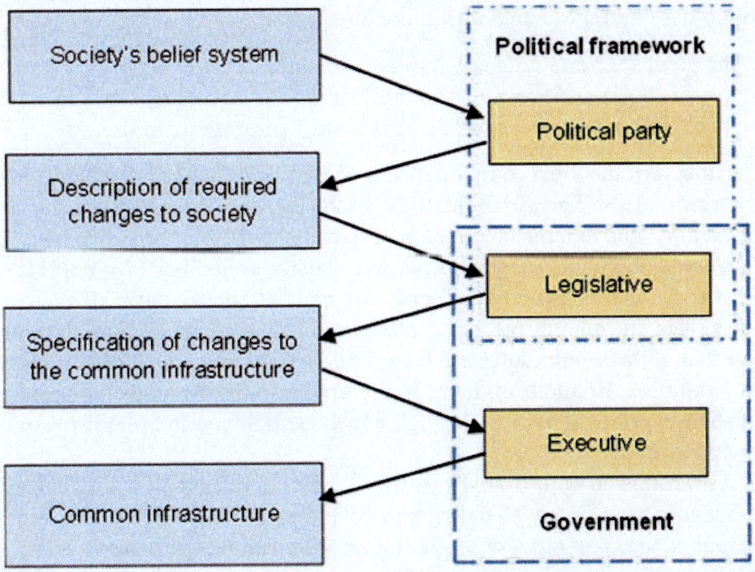

Fig. 7.3 The three components involved in adaptive action on the societal level and the entities through which they interact in order to realise changes to society's belief system as changes to the common infrastructure

If we think of the components as processes transforming input requirements into outputs meeting these requirements, and define deviations between ideal (i.e., meeting all requirements) and actual outputs as *errors*, then all three are subject to both systemic and random errors. Such errors can play a significant part in the evolution of society.

The political framework is the subject of Sect. 7.3, but first we briefly consider two factors that apply to both levels of adaptive action: freedom and will.

7.2 Freedom and Will

7.2.1 Freedom of Action

In Subsect. 6.6.2 we considered, albeit quite briefly, the concept of freedom in the context of society and noted that it must be seen as a balance, or compromise, rather than as something absolute. As members of society, we are never absolutely free, and would not want to be, we cherish our relationships to other members, and any relationship involves rules and restrictions. In our high-level model of society, the issue of freedom arises in two places: with regard to the collective intelligence, as freedom of expression (Subsect. 6.6.3), and now in connection with adaptive action, as the freedom to act. In this case, freedom is expressed in terms of the ability to change the common infrastructure, and seen from the perspective of the individual, that ability has several components, including:

- that the *means* for effecting change are available;
- that there is nothing *restricting* the individual from accessing those means; and
- that there is nothing *perverting* the functioning of those means.

In a democracy the most common means of effecting change is voting in elections for the various levels of government. In Australia there are three levels: local council, state or territory, and federal, elections are typically held every three years, and voting is obligatory. The vote is cast for a person, usually a member of a political party, to represent the individual citizen, hence the name "representative democracy". In this sense, it is an indirect means of effecting change. A less common, but direct means is that of the referendum; used regularly in Switzerland but only infrequently in other countries. In addition, there is the opportunity of membership and active participation in political parties, through which the policies and platforms of parties can be influenced.

The restrictions on voting can be overt, as in requirements on voting age, gender, criminal record, and financial status, and covert, as in making physical attendance difficult and a complex process of getting on the electoral roll, and the functioning of the voting process can be perverted by corruption and a lack of accountability. In Australia the situation is the opposite to restrictions: voting is compulsory and well organised, which is to be applauded.

The most damaging component, and the one posing the most acute problem for so-called "liberal representative democracies" today is the perversion of the process converting the vote—the "will of the people"—into adaptive action and appropriate changes to the common infrastructure. Of the promises made at election time many are never actioned, and of the actions taken, a significant proportion is not in accordance with the views of the majority of the population, as is sometimes demonstrated by opinion polls. There are several reasons for this distortion of the operation of the political process, but at the root of them all lies the fact that the process has changed its basic nature from one of serving and representing the people to that of a business—a business of buying and selling. Votes are bought by handouts (pork-barrelling) and by mass advertising campaigns, and the means are acquired by selling favours to special interest groups.

7.2.2 *Will*

From our identification of human intelligence as the driving force behind the evolution of society, we are led to accept the extent to which individuals are able to exercise their intelligence as an important measure of the process of evolution. However, referring to the model of intelligence developed in Chap. 4, and the definition of intelligence as goal-oriented adaptive behaviour, this model illustrates *how* intelligence operates, but *why* does it operate? What is it that *drives* the process? The finest intelligence in the world has no effect on the evolution of society if it is not used to generate adaptive actions. The force that drives the process is *will*, as noted already in Sect. 1.3.3. We could incorporate this in our simple model by inserting the will as a gate between the processing facility and the actuators, as shown in Fig. 7.4. Will is the force driving evolution; intelligence is the process steering evolution. They are two sides of the same coin; what we called "life". Or, in a similar vein, as Hannah Arendt said: "Will is our mental organ for the future, just as memory is our mental organ for the past" (Arendt, 1978).

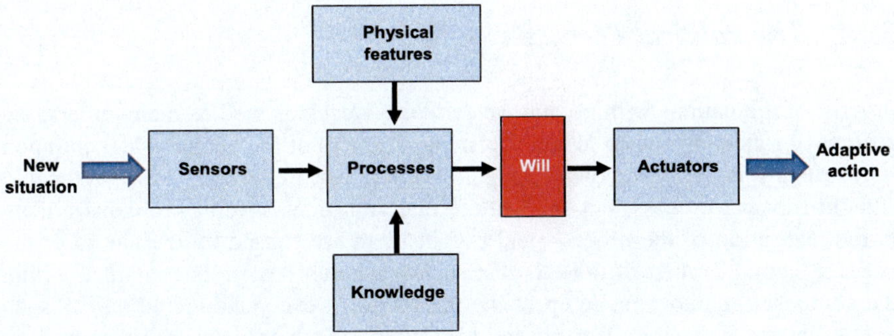

Fig. 7.4 High-level model combining intelligence and will

The concept of will, and its central role in human existence is, of course, not new, and was put forward most forcefully by Arthur Schopenhauer in his work *The World as Will and Representation* (Schopenhauer, 1969). In Book ii he describes the will to live as the blind impulse towards existence that is present in all things, and that uses knowledge for its purpose, as indicated in Fig. 7.4. But, maybe due to his pessimistic and melancholic disposition, he saw this will as condemning man to an unhappy, restless striving to fulfil base desires, and thought that escape from it was only possible by withdrawing from life into asceticism and meditation. That is not the view put forward here; as already noted, the development over the last several centuries has certainly been positive with regard to allowing humans to realise their potential.

In earlier times, the will to exercise intelligence (or capability), such as it were, was very directly challenged by conditions of existence, such as the need for finding food, providing shelter, engaging in person-to-person competition, and so on. Today, the challenges are much subtler and more concealed; the necessity to exercise intelligence (i.e., use willpower) is no less urgent, but it is not only about food and shelter, but about preserving and evolving of our society. Under this perspective, will becomes almost synonymous with virtue, as pointed out by Ellul (1992), a philosopher who otherwise has been highly critical of the expression of will in the form of technology.

A corollary to this is that the goal of the "goal-seeking adaptive behaviour" can be formulated as allowing unrestricted exercise of our intelligence, reflecting the realisation that it is our intelligence that determines the path of evolution. That is, the "free" in "free will" refers to the absence of external constraining conditions, and the adaptive action is such as to reduce these constraints. It is futile to ask what the ultimate destination of evolution is; we are not evolving towards something, some ideal state here or in Heaven. And it does not matter; what is important is that we know what the direction of evolution should be right now. In this sense, it is a differential definition of The Good Life.

7.3 The Political Framework

7.3.1 The Political Process

In most of the nations within what we call "the West", as well as many others, the process of reflecting society's belief system—the will of the people—in a common infrastructure is the one shown in Fig. 7.3. On paper, this looks like a reasonably straight-forward process, but this simple diagram hides several problems arising in the realisation of the process, and two of them are caused by the nature of the belief system. The first of these is that society's belief system is contained within the composite memory made up of the memories of every individual and as such is not directly accessible. It is not defined in any document or in any database, so the problem is how to create a representation of the belief system that can form the point of departure for the process of creating the common infrastructure; a set of requirements on the services to be provided by this infrastructure. And the same goes for changes to the belief system.

The political process attempts to solve this problem by at first reversing the information flow. Through its immersion in society (the political process is a component of the structure of the action view of society) the process generates a description of a common infrastructure that responds to what it believes the belief system to be, and then asks society to judge if it got it right or not. This is obviously a process that raises a number of questions, but before we consider this further, we look at the second problem caused by the nature of the belief system.

In our high-level model of society as an information-processing system, the system is uniform, or unstructured, and the belief system is a reflection of the common part of every individual's identity, i.e., complete consensus. In reality, society's belief system is a complex system, with a detailed structure and with dynamic interactions between the structural components. The political process attempts to handle this complexity in what is, greatly simplified, two different ways. One is to implement the proposed description of the belief system (or changes to the belief system) and observe how the performance of the common infrastructure is received. The complexity of society and its belief system is then reflected in the complexity of the response; some sections of society will accept and support the infrastructure, other sections will object to aspects of it in various ways, ranging from cheating and noncompliance to armed resistance. In this case, the successful operation of the political process, i.e., if it gets it right or not, depends on its ability to recognise what has to change in order to achieve a high approval rating and to implement these changes in a timely manner. And, of course, as the belief system changes, due to such factors a increasing wealth and level of education, increasing inequality, new technology, etc., the political process needs to be responsive; it must be a dynamic process.

The other way is to recognise a first level of structure in the belief system in the form of a number of major concerns, such as poverty, inequality, health, education, defence, etc. This allows the political process to also develop a structure, with each component of this structure presenting society with an understanding of the belief system based on a particular weighting of the concerns, and society's approval is provided by an additional process, that of voting. Each eligible member of society casts a vote for its preferred interpretation of the belief system or, in other words, for the interpretation closest to the member's own identity. The interpretations are defined in terms of a collection of *policies*, and each collection is produced by a *political party*. Political parties are the subject of the next section, and again in Sect. 12.2; here we need to complete the description of this form of the political process by addressing one further step: How to transform the result of the voting into a single set of requirements on changes to the common infrastructure? This is the purpose of the block labelled "Legislature" in Fig. 7.3, and while it consists of a group of people, often divided into two entities with slightly different remits, the process contained within it has two sub-processes. The first is for each party to be represented within the Legislature by a number of politicians determined by the number of votes the party received, according to a rule that varies somewhat between nations. The second sub-process is for the politicians to agree on a formulation of the policies as requirements on the services to be provided by the common infrastructure—requirements in the form of laws or Acts of Parliament. The process is an inefficient and Machiavellian combination of statesmanship, persuasion, arm-twisting, threats, and even occasionally an up-market form of bribery, but on the whole, although not in every instance, the outcome displays a recognisable resemblance to the will of the people. The democratic nature of the legislature can also be perverted by the presence of minor parties exploiting a balance-of-power situation and may also fail to correctly anticipate the effects of a particular formulation of a law; after all, there is no formal qualification required to become a member of the legislature.

The block labelled "Executive" in Fig. 7.3 is not simply an isomorphic transformation of laws into practice; the implementation in the form of institutions and/or procedures may result in effects that vary from both what society expected and what the legislature had intended. The degree to which the legislature delegates policy decisions to the executive, and the consequent degree of independence of the bureaucracy—what might be seen as a trade-off between politics and expertise—has been the subject of numerous studies within political science and public administration, such as Epstein and O'Halloran (1999), Huber and Shipan (2002), and Farazmand (2010), and a particular aspect of this issue will be an important factor further on. We shall not consider this whole complex of the functioning of Government and public administration; we simply accept that it is subject to numerous non-democratic forces. Our interest in the current context is with the effectiveness of the collective intelligence as both a driving force and a corrective force, and this effectiveness is being determined by two interfaces between the process shown in Fig. 7.3 and society: the extent to which it recognises and accepts the will of the people, and by the extent to which the people are able to know the outcome of the Government's management actions, allowing them to take corrective action. It is in the first of these interfaces that the political party plays a dominant role, which we consider in the next subsection. But we should recognise that there is a contradiction involved in the relationship between society and the Government.

The contradiction is one that has always been present and may be seen as an intrinsic feature of the concept of a society—providing the greatest freedom for individual expression while at the same time maximising the benefits of close interactions. The contradiction arises from the demand on Government to enforce the laws required to implement the decisions made by society on the one hand, and for the members of society to be unrestrained in the execution of their intelligence on the other hand. We object to being forced to do something against our will, but we have agreed to let Government use force if required, within prescribed limits. And we want that enforcement to be as effective as possible, so, as an example, we allow police to use the best technology available, such as automatic weapons, water cannon, tear gas, surveillance cameras, command-and-control systems, and so on. But if we refer back to the discussion, in Sect. 7.1.1, of using the manipulation of information as a means of influencing behaviour, we can understand how tempting it is for Government to restrict or modify the information provided to society.

7.3.2 The Political Party

The schematic description of adaptive action on the social level in Fig. 7.3 shows society's belief system, defined in Sect. 3.3.3, as the interface between the individual and the societal levels, accessed on the societal side of the interface by the *political party*. And in that figure the role of the political party is shown as that of a conversion or transforming process, presenting the content of society's belief system—the *will of the people*—as requirements on the actions of the government, as discussed in the previous section.

A possible definition of the concept of a political party is the following:

> A political party requires 1) continuity in organisation – that is, an organisation whose expected life span is not dependent on the lifespan of current leaders, 2) manifest and presumably permanent organisation at the local level, with regularised communications and other relationships between local and national units, 3) self-conscious determination of leaders at both national and local levels to capture and to hold decision-making power alone or in coalition with others, not simply to influence the exercise of power, and 4) a concern on the part of the organisation for seeking followers at the polls or in some manner striving for popular support. (LaPalombara & Weiner, 1966, p. 6)

This definition, and most others that can be found by searching the Internet, are general enough to allow the political party to appear in a variety of political structures, both competitive, multi-party structures and non-competitive, single-party structures. And there is nothing in these definitions that restrict the appearance of political parties to democracies. However, we shall focus on political parties as they appear in a multi-party democracy.

In order to manage the interface to government, a party is structured accordingly, i.e., into national, state, and local (council or county) levels. At each level it is represented in thc legislative function of government by its elected representatives, or Members. In performing their duties, these representatives have a dual allegiance—to the people who elected them, and to their party—and as a rough rule, the allegiance to the people relative to that to the party is strongest at the local level and weakest at the national level, as one would expect from the size of the electorates alone.

In one form or another, the concept of the political party is central to the operation of most nation-states. Whether a one-party state or a representative democracy, the political power is wielded by a relatively small group of people who manage the affairs of government in much the same way as the directors and senior executives manage a corporation—the *elite*. In both cases the members of the society constitute the market; in both cases the business strategy is a combination of persuasion and an adaption to market demands; and in both cases the objective is survival against internal and external competitors. And in both cases the "will of the people", which in our model is represented by the adaptive action resulting from the operation of the collective intelligence, is expressed in a passive fashion, by either accepting or rejecting what is offered.

It is here that there is supposed to be a significant difference between a one-party state and a representative democracy; the latter offers a choice between several alternatives, as well as the ability to propose new alternatives, whereas in a one-party state there are no alternatives, and the choice is, at best, between acceptance and passive rejection. In practice, there is actually not that great a difference between the two; in both cases, the governing party, or coalition of parties, prosecute the daily business of managing the state in accordance with a pragmatic evaluation of the existing circumstances based on their own interests, with public opinion just one of many factors—and the one most easily manipulated. And in both cases there is the same undercurrent in the population of frustration with the lack of a more direct and effective participation in the decision-making process; people feel they are not being heard. The level of frustration will, of course, depend on the extent to which the regime meets the basic needs of society, i.e., on the standard of living, but once these needs are met, and the level of education and leisure time increases further, the level of frustration rises again. In single-party and authoritarian states it is simply suppressed; in the liberal democracies it is exploited in support of minority grievances related to such issues as nationality, race, gender, and religion. To the extent that such actions are successful—possibly even resulting in a change of government—this display of people-power may appear as a win for democracy, but in reality it is circumventing the collective intelligence and represents a fluctuation away from the optimal path.

The multi-party and, in particular, the two-party system are also inefficient, at least in practise, if not in their ideal form, as the competitiveness and antagonism of the election campaign is carried over into the operation of the legislature between elections. As an example, in Australia much of the time during the sitting of Parliament is spent on childish finger-pointing (literally) and point-scoring, and the time between sittings is spent on building the base in the constituency for the next election. The productivity, in terms of developing well-considered responses to the many issues and problems arising, is much lower than it could and should be; the only country that approaches the ideal is Switzerland.

A single-party system should, again in principle, be able to avoid the inefficiencies of the multi-party system. And if we define a democratic system as one that realises the will of the people, then a one-party system could, in theory, be just as democratic as a multi-party system, with the party organisation providing the structure for the upward movement of that will from the grassroots to the government. However, in practice the single-party system is more susceptible to the establishment of a ruling class that squanders much of what should be a gain in efficiency in realising the will of the people on maintaining its own privileged position.

The current turmoil on the political front is the subject of daily reports and comments in newspapers, newsletters, and blogs, with the rise of populist, right-wing political groupings and rhetoric as a common theme. And while the easily discernible causes include the effects of economic inequality and mass migration, there is an underlying dissatisfaction with the political process in general; people feel that they are not being listened to and their concerns not being recognised. Nowhere was this more starkly demonstrated than in the 2016 US presidential election, where the most outrageous statements were accepted as a welcome respite from the usual bland drivel; where the candidate who spoke most clearly to the underlying concerns, Bernie Sanders, was sidelined by the political machinery; and where something like

40% of the eligible voters were so disillusioned that they could not even be bothered to vote. And if the breakdown of the political system in the West's leader needed a further demonstration, it was provided by the storming of the Capitol on 6 January 2021.

In Europe the situation is complicated by the European Union creating an additional structural component in the political system, in which the balance between the legislature—the European Parliament—and the bureaucracy—the European Commission—is skewed strongly in favour of the bureaucracy. This appears to be an additional significant factor in the present unrest within the EU, and one cause of the emergence of populist right-wing or nationalist parties within the member states (and, of course, Brexit). The issue of right-wing parties was analysed already in Rooduijn (2015), and some articles on the subject of the EU's problems are Social Europe (2017), Archick (2017), and more recently Karacsony (2021). The future of social democracy in Europe is discussed in Meyer and Rutherford (2012), and in Chap. 8 of that book, *The Greatest Happiness Principle: An Imperative for Social Democracy*, Christian Kroll makes the observation that "People no longer vote for the same party over and over again, out of loyalty, but have become consumers of political ideas who feel free to decide anew, before each election, who deserves their vote". This change in the behaviour and expectation of voters should be reflected in a corresponding change in the political system and in the function of the political party.

There are many signs of dissatisfaction with the political system here in Australia also. Among these are the numerous complaints about the performance of our political leaders to be found in the Letters section of the Sydney Morning Herald, the rise of minor parties and independents, and the latest Scanlon survey, which found that support for the statement "the system of government we have in Australia…needs major change or replacement" was 39% (Scanlon, 2021).

The interesting point of all the discussion and analyses of this current situation is that there is practically no consideration of the possibility that there might be something wrong with the role of the party within the political process; it is so firmly entrenched in the common understanding of the process that questioning it is unthinkable; a classic case of enframing. Taking a leaf from Thomas Kuhn's description of scientific progress, in which an existing belief system, or *paradigm*, is further developed and defended in the face of mounting evidence of its inadequacy, until a new idea is put forward that solves the problems and leads, in what is termed a scientific revolution, to the acceptance of a new paradigm, we might ask if the time is not ripe for a revolutionary change to the political paradigm—a re-examination our political system, in particular, the role of political parties, in the light of what technology has to offer.

Almost every other social activity has been significantly influenced by technology: Shopping, via the Internet and home delivery services; entertainment, via download on demand; social interaction, via social media; common information, such as phone numbers and time-tables, via the Internet; financial transactions, via the Internet; and so on. In short, the ability and convenience of the individual to pursue all these activities have been greatly enhanced; only the ability of the individual to participate effectively in the political process has not improved and has avoided taking advantage of technology. In *The Social Bond* a proposal for abolishing the party system and strengthening the individual's direct involvement in the political process was put forward, based on current advances in IT. There is no real technical or cost reason why such a scheme could not be implemented; the reasons are, firstly, that the party system has developed into quite a substantial business, and secondly, that the competition between parties deflects attention and effective action away from privileged positions in society. The reality is that there are significant and powerful sections of society that have no interest in promoting a more inclusive political system, and that see "the will of the people" as an obstacle in pursuing their interests. And in most cases, this is not a matter of reasoning based on a sound understanding of the nature of society as a complex system, but simply an ideology. A further discussion of this problem and the above-mentioned proposal for a solution are contained in Sect. 12.2.

7.4 A Final Note

To round off this first main part of the book, which has developed a view and model of society, it might be helpful to display its basic features in the form of the diagram shown in Fig. 7.5, and to see to what extent it can be considered to be a contribution to social theory.

According to Popper, in *Logic of Scientific Discovery* (Popper, 1959), a scientific theory should meet the following criteria:

1. Be internally consistent.
2. Distinguish between empirical and logical elements, making the logical form of the theory explicit.
3. Add something to existing theories and not conflict with them.
4. Conclusions can be shown to conform to reality.

With regard to the first criterion, the description provided by the model is internally consistent if we allow for the interpretation the parameter α given in Sect. 5.3. As stated there, the model presents a static picture of society, and an approximate one at that, as it employs the rectangular approximation in order to give a simple meaning to the parameter α. This picture is useful for describing the state of a society, or for comparing societies, at any particular point in time; but to account for the dynamics of even this simplified view of society the model would have to be expanded considerably. As it stands, it is intended as a supporting tool in other, high-level views of society, exploring such aspects as the role of education, the influence of private vs. public education, the effect of economic inequality, the concentration of control of information technology, and the impact of alliances and polarisation in the world society, just to mention a few.

If an "empirical element" is understood as a quantity that is defined in terms of how it is measured or observed, the view and its model do not contain any empirical

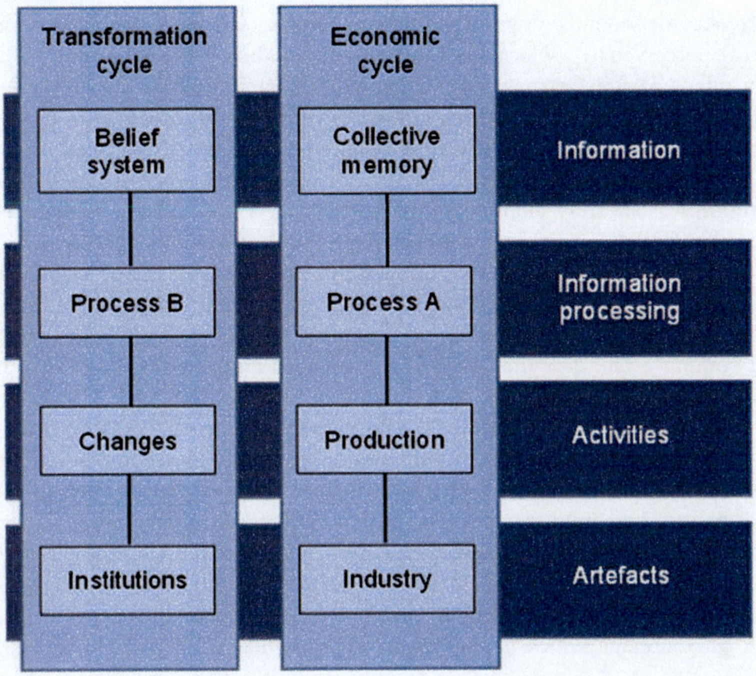

Fig. 7.5 A graphic summary of the view presented in the preceding sections. The activities in each of the two cycles are the result of the processing of the society's information content, with *Production* encompassing all those activities that sustain and grow the current "version" of society, whereas *Changes* represents those activities that change one "version" into the next; i.e., those that determine the evolution of society. The artefacts encompass the physical belongings of society, grouped into those related to Production, ranging from tools and facilities to such infrastructure as arable land and transport infrastructure, and collectively identified as *Industry*, and those related to the *Institutions* involved in governing the society, such as parliament, departments, courts, and law enforcement

elements. The introduction of, and relation to, measurable variables will only arise through application of the model.

The view and model are not intended to expose any conflict with existing theories, but to provide an overarching view—a common point of departure for exploring and developing existing theories further. If this sort of top-down approach adds anything is a subjective matter; I consider it an invaluable tool in handling complexity.

The main conclusion, or assertion, is that fluctuations in the evolution of society are accompanied by restrictions and manipulations in the flow of information, and to test this we can look at two types of fluctuations: wars and financial crises. The years leading up to the First World War—a small fluctuation of 1% in world population (21 million out of 1.8 billion)—were characterised by two features restricting the flow of information; the first being the formation of two opposing alliances, each one limiting and modifying information in order to justify their existence and increasing belligerence; the second being the German emperor's need to see his own resentment at what he perceived to be Germany's inadequate standing in the community of world powers reflected in the German people. A somewhat larger fluctuation was the Second World War—about 3% in world population (75 million out of 2.3 billion)—which was preceded by exposing the German people to an almost hysterically distorted version of reality. The Vietnam and Iraq wars are more recent examples of where information was severely manipulated; neither of these would have happened if the truth had been made available to the American public.

Both the stock-market crash in 1929, with its ensuing Great Depression, and the Global Financial Crisis in 2008/9 are examples of fluctuations that would not have occurred had the flow of information been unrestricted; in hindsight "one could have seen it coming". The data that would have allowed an accurate assessment of the risks involved and the associated probability of a collapse of financial markets were, in principle, present, but were either not provided to the institutions that could have taken corrective action or simply suppressed or ignored in the hope that it would all sort itself out. There are many mechanisms in our society that allow the suppression of information, and there are many motivations for taking advantage of them—economic, political, religious. But even if information is not suppressed, any new information has to contend with what we already know; in the case of our belief this is expressed by Eq. 7.2. And in that knowledge base, the more recent information tends to be dominant, which is why failure is often preceded by great success.

Chapter 8
Technology

8.1 What is "Technology"?

8.1.1 The Meaning of "Technology"

The meaning of the word "technology" in daily speech is broad and highly context dependent, as can be seen by looking up the word in Wikipedia. The word relates to the field of human activity that may be described as the modification of elements of the natural surroundings in order to meet a need; what we might call a *purposeful* modification. It started when humans developed the mental ability to recognise the possibility of such a modification and the physical dexterity to realise it, and the purpose included giving visual pleasure or increasing one's self esteem (painting, ornaments, sculptures), worshipping a deity (monuments, temples), providing shelter (dwellings), increasing mobility (roads, bridges, boats), providing food (traps, weapons, agriculture), preparing, serving, and storing food (bowls, pots, plates), and so on. This is roughly what the ancient Greeks identified as *techné*, any creative manual activity and the products that arose from it, and, except for the most elementary stone tools, its development has occurred within the period of interest in the context of this book—the last 10,000 years.

Philosophers and most sociologists have a broad concept of technology and would accept such a definition as "artefacts and their development, production and use". Philosophers have produced a substantial body of work under the heading of "Philosophy of Technology". It is concerned with ethical aspects of technology, with the nature of technological knowledge, and with fundamental issues regarding the impact of technology on the human condition. Sociologists have likewise taken an increasing interest in technology, studying technology as a social activity and how social issues influence the development and application of technology. However, in both philosophy and sociology there has been a tendency to confuse technology with science, and engineers with scientists; many publications on the philosophy of technology make no mention of engineering at all, and such concepts as 'technoscientist' and 'technologist' tend to confuse the issue even more. One reason for this is probably

E. W. Aslaksen, *The Evolution of Society*,
https://doi.org/10.1007/978-3-031-23054-7_8

that the philosophy of science was already well established and provided the point of departure for work on technology. This is reflected in the implicit or explicit view of most philosophers and sociologists that technology is driven by scientists, rather than engineers, as evidenced by the common reference to 'Science and Technology' and 'technoscience'. A typical example of this is the proceedings of the 21st Nobel Conference, which is entitled *Responsible Science—The Impact of Technology on Society*. It considers questions of judgement and values that permeate issues relating to science-based technology, and states that these demand the attention of the entire society: business leaders, politicians, scientists, and citizens (Byrne, 1985). What happened to engineering and the engineers?

The origin of this confusion may be traced to the period following the First World War; a period of great advance in both science and engineering, and of a growing interest in social and philosophical questions relating to both. However, while the activity related to science was international in character, the activity related to engineering was concentrated in Germany and published in German (an important example is the book *Streit um die Technik*, by Dessauer (1956), first published under the title *Philosophie der Technik* in 1926), and as a result, when the interest in the philosophy of science and of engineering started to pick up under US influence following the Second World War, the philosophy of engineering, such as it was (and has largely been), adopted much from the philosophy of science, while disguising this by talking about "technology". A number of papers discussing and demonstrating this situation can be found in the publications resulting from the forum on Philosophy, Engineering, and Technology, such as fPET 2012 (Mitcham et al., 2018), fPET 2014 (Michelfelder et al., 2017), and fPET 2016 (Fritzsche & Oks, 2018).

One example of how philosophers and sociologists view technology is contained in the book *Autonomous Technology* by Winner (1977), where most of the Introduction is dedicated to a discussion of the history, use, and meaning of 'technology'. He points out that the concept is used so frequently and in such diverse contexts that it has become amorphous in the extreme, to the point where it has come to mean everything and anything; it therefore threatens to mean nothing. He consequently proposes instead to employ the term *apparatus* for the physical objects, such as tools, instruments, machines, etc.; the term *technique* for the activities involved, and the term *organization* for all the related social arrangements. However, while this provides a differentiation of the components of technology, it does little to sharpen the definition of the concept itself, or to replace it, as the title of the book, as well as the frequent use of 'technical', 'technics', and 'technological' throughout the book, demonstrates.

A useful perspective on the everyday use of the concept is given by Marx (1994), where he shows that the character and representation of 'technology' changed in the nineteenth century from discrete, easily identifiable artefacts (e.g., a steam engine) to abstract, scientific, and seemingly neutral systems of production and control. As a result, the newly refurbished concept of 'technology' became invested with a host of metaphysical properties and potencies that invited a belief in it as an autonomous agent of social change, attributing to it powers that bordered on idolatry.

Throughout this book, 'technology' is generally used as is common in daily speech, with the meaning more or less clearly implied by the context in which it is used. However, there will also be many instances where a precise definition is required, and that definition shall be the following:

> Technology is the resource base developed and applied by engineers to meet needs expressed by groups or all of society, and consists of a *construction base* (construction elements, tools, as well as the facilities within industry for fabricating and constructing plant) and a *knowledge base* (textbooks, publications, standards, data bases, heuristics, etc.). Students study technology in order to become engineers.

The resource base consists of the millions of standard construction elements, ranging from reinforcing steel bars to microprocessors, that engineers and the technical workforce can draw on in executing projects, as well as the facilities within industry for fabricating and constructing plant. The knowledge base is comprised of textbooks, standards, published papers, operating manuals for tools and instruments, etc., and spans a continuum from advanced research to Tables for everyday use. It also comprises the tacit knowledge embodied in individual engineers, generated through their personal professional experience. Both of these bases are dynamic: new construction elements are continually being added and older elements are being phased out; new knowledge is being generated through research and experience, and what was advanced knowledge yesterday is tomorrow's accepted practice, often embodied in standards.

With technology as a resource base, a phrase such as "the influence of technology" is, strictly speaking, nonsensical, as technology, just like any form of knowledge, has no influence, and its value is like that of money; it is only realised if and when it is used. It is only applications of technology that can have an influence. An important aspect of this definition is the relationship between technology and engineering, and the central role of engineering is the subject of Sect. 8.2.

8.1.2 Understanding Technology

In Chap. 2 we introduced two complementary views of society—the information view and the action view—and in Sect. 3.4.3 we briefly pointed out how technology has been the most obvious and pervasive manifestation of the evolution of society within the action view. Until now, the focus of the narrative has been on the information view, and will remain so throughout the book, as is appropriate considering our assertion that the primary description of society and its evolution should be in terms of information, and that the driver of evolution is the processing of information. But given the importance of technology in supporting and providing the infrastructure for the information processing, as well as the close coupling between the economic cycle and the transformation cycle, as mentioned in Sect. 3.4.1 and illustrated in Fig. 5.11, we now need to take a closer look at technology in order to understand it and the nature of the relationship between technology and the evolution of society. But what is involved in understanding a complex entity like technology? A description of how it came to be? A description in terms of constituent parts? A description as a process and its products? A description of its relationships to other entities? We shall need to consider all of these aspects, and one of technology's main constituents and a main component of technology as a process is engineering, so Sect. 8.2 is dedicated to developing an understanding of engineering and its role in the relationship between technology and society.

Our most direct experience of technology is in our daily exposure to its applications; we are completely immersed in such applications, and we need to understand the process that produces these applications. Who decides what gets produced? Who is involved in the production process, and what are their roles? These and other questions are considered for technology in general in Sect. 8.2, and for that segment of technology of particular interest to our view of society as an information-processing system—information technology—in Sect. 8.4.

8.1.3 The Influence of Technology

As noted in Sect. 3.4.3, the involvement and importance of technology in all aspects of society is so obvious and ubiquitous that we no longer think of technology and its applications as something separate from us as humans.

We are particularly interested in the relationship between technology and the stability of the evolution of society, and as an extension of the two cycles we considered in Sect. 3.5, we can see that there are two aspects to this relationship: the relationship between technology and society's belief system, and the relationship between technology and our ability to put changes of the belief system that we identified as fluctuations into effect. A main contributor to that ability is weapons technology (or, as it is often euphemistically called—defence technology), and as it is a subject matter of which I have very little knowledge, it will not be considered in this book, except to note that an industry with an annual revenue exceeding 530 billion US dollars (https://sipri.org/databases/armsindustry, accessed 20.06.2022) will be looking to create applications for its products.

The first relationship is how technology influences our view of the world and our position in it. In particular, our view of Nature as something to be exploited by us and of our ability to overcome any limitations Nature might present on our creativity. In this sense, there is a connection between technology and the perception of a divine destiny for humanity; a curious connection, as one would have expected the advance of technology to eliminate the belief in a divine presence. This and related aspects of the relationship are the subject of Sect. 8.3. The ability of information technology to influence the operation of the collective intelligence and to pervert the collective identity is considered in Sect. 8.4.

8.2 The Role of Engineering

8.2.1 A Short History of Engineering

8.2.1.1 Introduction

From the definition of technology in Sect. 8.1.1, we see that engineering is a core activity in developing and applying technology. It is an activity that is poorly understood in the general public, and the same might be true of some of the readers of this book. Engineering can be characterised as the activity of, on the one hand, combining scientific knowledge with practical experience to develop new technology and, on the other hand, applying this technology to meeting needs expressed by society. Therefore, given the tremendous increase in both scientific knowledge and accumulated experience in the period over which we have written records, say, the last 5000 years, it is not surprising that the nature and extent of engineering has changed dramatically in this period. And due to the accelerating nature of this development, it is also not surprising that engineering as we know it today, as a profession with an extensive scientifically based body of knowledge and a defined academic profile, is of relatively recent origin, certainly less than 200 years. However, throughout the whole historic period, the work of the people we can loosely identify as engineers (even if part-time) has exhibited the dual characteristics of using acquired knowledge to develop solutions to construction problems, and then organising and directing the effort required to carry out the construction.

Our purpose in delving briefly into the history of engineering is to obtain a better perspective on the profession as it is today, therefore we shall confine ourselves to developments in Europe, the Middle East, and North America, as it is from these origins that modern engineering arose. Certainly, there were engineering activities in other parts of the world, such as the building of the Great Wall and the numerous canals in China, the mining and smelting of copper in the Andes, and the building of the temples in the Yucatan, but these activities had little or no effect on developments in Europe and often occurred later than corresponding activities in Europe. An exception to this were certain inventions, such as the manufacture of gunpowder and paper, but even in these cases the industrial exploitation of these inventions through engineering took place in Europe.

8.2.1.2 The Beginning

At the outset of the 5000-year period we are considering, the construction activities were centred around irrigation, simple structures such as compacted earth walls, and the production of simple tools and weapons. But from then on, the rate of development increased, both of technology and of the application of it. More sophisticated tools, such as metal chisels and mallets, appeared, and new devices, such as levers and pulleys were constructed and combined in various forms to allow various tasks, such as lifting water or grinding corn, to be carried out more efficiently and to allow animals to substitute for humans in providing muscle power.

As an aside, it is interesting to note that while the invention of the wheel is always cited as a major technological breakthrough, it was really the invention of the bearing that was most significant. Wheels, in the form of logs, had been used to reduce the friction in moving heavy objects, such as massive blocks of stone, from early times, but attaching the wheel to the moving object only became practical by using a bearing with a much smaller diameter than the wheel to reduce the work due to friction, showing a basic, or intuitive, understanding of the concept of work as the product of force and distance.

Simple processes for mining and processing ore appeared, and with them increasing skills of working metals, first gold, then copper, tin, lead, and then, from somewhere around 1000 BC, iron. Metals became the underpinnings of some of the ancient societies, such as Egypt (gold), the Second Hittite Empire (iron), and the city-state of Athens (silver) (Gregory, 1980).

These developments in technology were paralleled by the increase in the size of the construction projects undertaken. The pyramids in Egypt, each of which might have taken 20 years to build with a workforce of 10,000 or so, are the most well-known of these large projects, but there were many others, including the city of Memphis itself and large dams and canals in Mesopotamia (Turner & Golden, 1981). Such construction projects involved a considerable skill in planning, organising, and logistic support, and so we see that what today is loosely called "project management" was always an intrinsic part of the engineer's capabilities. Not only that, but to exercise these capabilities, the engineer had to be a person of considerable standing and authority, often an official such as a governor or administrator, so that the engineering was often only one part or aspect of his work.

In the period up to about 300 AD, engineering made great progress, mainly in the region around the Mediterranean and the Middle East (Hill, 1984). Water wheels driving increasingly sophisticated plant for grinding corn and raising water became commonplace, and around 250 BC a piston pump made of bronze, with the piston turned on a lathe, was designed and produced by Ctesibius in Alexandria. Other well-known engineers in this latter part of the period were Philo of Byzantium, Hero of Alexandria, and Vitruvius in Rome (although the latter would be called an architect today).

More complex building structures appeared, supported by technological advances in the production of such construction elements as mortar, bricks, tiles, quarried stone, etc. Outstanding examples of this are the Roman aqueducts and the Coliseum.

A central development in this period was no doubt the increasing skill in mining and processing of iron ore, and in the processing of iron into tools and devices. The process of hardening by carburisation and heat treatment was fully understood and developed in this period, and such tools as saws with set teeth became available.

Throughout this period, engineering was basically a craft, in that it was based on experience handed down from generation to generation; there was little or no scientific basis for the designs and processes. An exception to this was the understanding of the workings of such basic components as the lever, the wheel (or pulley), the wedge, and the screw, based on geometry, and masons used a few geometric figures as the basis for their designs. For example, Archimedes could calculate leverage on this basis.

In the period from 300 to 1100 AD Europe, or the Christian part of the western world, saw a marked decline in intellectual activity, and a falling-off of standards in such public services as sanitation and water supply. Paris and London in the year 1100 were worse off than Rome in 100 AD with regard to street maintenance and sewers, and would remain so for several hundred years. But in the part of the world conquered by the Arabs and generally identified as the Islamic world, technology and science had a golden age, and the resurgence that started in Europe around 1100 and led up to the Renaissance owes a great debt to Arab and Persian culture and peoples for preserving and continuing the scientific and engineering tradition from classical times. From northern Mesopotamia to Spain, there was a thriving scientific and technological culture, and much good engineering was carried out to meet the demands of such large cities as Baghdad and Cordoba.

8.2.1.3 The Renaissance

The Renaissance marked a turning point and the beginning of a great increase in the speed of development of engineering and the science on which it depended, and the reasons for this were principally.

a. the end of the stifling and autocratic control of education and intellectual activity exercised by the Church throughout the Dark and Middle Ages;
b. the rediscovery of classical science through the contact with Arab culture;
c. the discovery and exploration of new parts of the world (partly as the result of advances in shipbuilding);
d. the development and rapid growth of printing; and
e. the gradual acceptance of the vernacular as a means of written communication, replacing Latin.

The period from 1450 to 1750 saw a proliferation of machinery and equipment to support the demands of a growing and increasingly wealthy population (even if this wealth was very unevenly distributed). Water wheels powered sawmills producing the timber for buildings, ships, and various structures, bellows for smelting furnaces, trip-hammers for forges, and pumps for water supply. Windmills, in use since the start of the millennium, proliferated and were used both for grinding corn and pumping water. The casting and machining of iron produced a variety of equipment, from pumps to cannons, and such "standard" construction elements as shafts, bearings, gear wheels, and pulleys.

Perhaps the most significant feature of this period, in the context of engineering, was the formalisation of the activity of design and its separation from the work involved in implementing it. While the engineer had to have a first-hand understanding of how things were constructed, so that his knowledge was still largely craft-based, the details of the object to be constructed were defined prior to any construction taking place in the form of drawings and written instructions (often as notes on the drawings). The constructed objects became an expression of the intent of the engineer rather than of that of the craftsman; the creativeness involved in an object was transferred from the craftsman to the engineer, and the measure of the craftsman was changed from his ability to create new objects to his skill in producing what the engineer had designed and his ability to do so efficiently. It was the beginning of the process of turning crafts into industries and of the Industrial Revolution, and with it the beginning of engineering as we understand it today.

However, before we go on to considering the Industrial Revolution and what we might call the modern era of engineering, some central features of engineering, both as it was then emerging as a distinct activity and as it is as a profession today, can be illustrated by looking at the work of one man, Leonardo da Vinci (Wikipedia). The historical facts are well known: Born in 1452 as the illegitimate son of a notary in the town of Vinci, near Florence, he served an apprenticeship in the workshop of the famous artist Verrocchio in Florence, became an accomplished draftsman, painter, and sculptor, skilled in all the processes and techniques underlying these arts, and became acquainted with science and engineering through interaction with some of the leading practitioners of the day, such as the physicist Toscanelli. At the age of 31, with confidence in his own knowledge and skills in a wide range of fields, including both arts and science, he went to Milan to work for the ruler, the Duke Lodovico Sforza, and stayed there for 16 years. Following the French invasion of Milan in 1499, Leonardo fled to Venice, but in the next 16 years he alternated between Florence, Milan (after the French were driven out again), and Rome, until settling down in France in the service of Francois I for the last three years of his life until his death in 1519.

Leonardo da Vinci is best known as a painter, although only a relatively small part of his efforts was spent on painting and only 15 paintings are definitely attributed to him. By far the greater part of his efforts were directed at science and engineering; he had an insatiable curiosity and desire to understand how things, both mechanical and animate, worked, and his broad and very solid training, coupled with his inventiveness, allowed him to be at the leading edge in a number of areas, including the foundations of painting (as in light and perspective), astronomy, anatomy, mechanics and machinery, structural design (e.g. bridges and domes), and civil construction (e.g. fortifications and canals). He was a universal genius, often referred to as the archetype of the Renaissance Man, but from our perspective the most interesting and significant characteristics of his work are the following:

a. His studies and his understanding were based on observation; form was everything. If he could see how something was made up of its parts, and could describe, in the form of drawings and notes, what those parts looked like and how they fitted together, he was satisfied he understood it. He did very little experimentation, and his grasp of mathematics was not sufficient to allow him to create abstract models or theories. His few ventures into mathematics were confined to geometry, as the theory of form.

b. His approach to engineering was that of an artist. His main objective was to express his ideas and inventions, which he did in the form of brilliant drawings and sketches. Once the problems were solved in his own mind, turning them into working reality, which involved such tedious and time-consuming activities as obtaining funds and organising and supervising the work force, was of much lesser interest, and so most of his projects and inventions were never realised (although this was sometimes also due to circumstances beyond his control). Even documenting his work adequately was seen as a diversion from the constant activity of creation, and many of his valuable studies remained as unfinished collections of notes.
c. To the extent that he was obsessed with form, he neglected functionality. Take as an example his flying machines; none of them exhibited the functionality of actually being able to fly (which could have been demonstrated by simple calculations), but they were elegant and ingenious depictions of how various components would have to be formed and interrelate in order to represent observed characteristics of flight (e.g., the flapping of wings). In short, they were works of art.

Leonardo da Vinci exhibited, to an extraordinary extent, one of the dual characteristics of an engineer; that of having the creative ability to see how technology could be applied and further developed to provide solutions to given problems. But he lacked the complementary one of planning, organising, and supervising over the long run the activities required to convert a design concept into a successful reality. Not for him the long slog of placing block upon block over 20 years to build a pyramid. Due to his fame, his work as an engineer is perhaps the most striking example of the dilemma that is inherent in engineering (and engineering education); how to resolve the conflict between creativity and inventiveness on the one hand and the purposeful application of existing knowledge and technology to ensure a cost-effective realisation on the other. Somewhat crudely, what is needed in a successful engineer is a balance between the genius and the plodder, and Leonardo was all genius. As a result, his influence on engineering was almost zero.

The period from about 1550 to 1750 was, as far as engineering is concerned, dominated by three developments; the increasing reliance on reason instead of on belief and superstition, the increasing reliance on experimentation and measurement as the arbiters of theoretical propositions, and, of course, the advances in papermaking and printing, which made books more affordable and greatly accelerated the spread and influence of this new knowledge. Based on the advances in science, exemplified by such names as Simon Stevin, Galileo Galilei, and Isaac Newton, engineering started to develop a substantial body of knowledge and a curriculum for an engineering education. The first engineering schools appeared in France at the end of this period, starting with the artillery school at Fère-en-Tardenois near Aisne in the early part of the eighteenth century, then the Ecole National des Ponts et Chaussées in 1747, and finally the Ecole Polytechnique in 1794, and a certain degree of specialisation emerged (Channell, 1989). In particular, mechanical engineering and the design of machinery became distinct from civil engineering.

These developments in the foundations of engineering were accompanied by equally significant developments in their application. New and improved ways of forming and machining materials, as well as advances in mining and metallurgy, resulted in both more sophisticated designs and more efficient production, so that cost-effective machinery and implements found widespread application in factories of all kinds as well as in agriculture and construction. All that was needed for the Industrial Revolution to take off was an efficient source of power (not subjected to the restrictions of wind and water power), and this was provided by James Watt and his inventions, which saw the steam engine transformed from the very inefficient Newcomen engine into its modern form within a couple of decades.

8.2.1.4 The Industrial Revolution

The work of James Watt (1736—1819) is the best, and certainly most famous, example of the change that was taking place in engineering at that time (Hills, 2002). Until then, the engineer had been the creator of individual pieces of equipment or individual constructions, much as an artist created individual works. He either had his own workshop, or he worked closely with workshops that produced his designs. And indeed, this was at first the case with James Watt, who had his own, small workshop attached to the University of Glasgow until he made the first significant improvement, the separate condenser. But then, in partnership with Matthew Boulton, he turned the manufacturing of steam engines into a process which showed all the characteristics of a modern industrial manufacturing process. He broke the plant down into separate components, standardised these components to the extent possible, designed separate processes, tools and jigs for their manufacture, and considered the lifecycle of his products by providing some level of spare parts.

The Industrial Revolution, which is generally considered to have occurred in the period between 1750 and 1900, give or take a few years at either end, contained, besides its enormous social and economic evolution, two developments that were particularly significant for engineering. The first of these was the increased understanding of the properties of material beyond the purely mechanical, in particular chemical and electrical properties, and the exploitation of these properties, through inventions, for practical purposes. Electrical conduction was discovered in 1729, and the first primary cell (an electrochemical process) was developed by Volta in 1799. The telegraph had a rapid rise in the first half of the nineteenth century, with the first transatlantic cable in operation by 1858, and the commercial production of induction motors commenced in Germany around 1890 (Atherton, 1984). Chemical processes, exemplified by the production of alkali for making soap, emerged on an industrial scale (Furter, 1980), with the accompanying necessity for understanding and handling such material properties as corrosion resistance.

The second development was the rise of an industry dedicated to manufacturing engineered products, essentially various forms of machinery and components, which were then used in a second level of manufacturing to produce end user goods, such as textiles. Engineering became a main pillar of the national economy, resulting in a new level of professionalism including design for reliability, maintainability, and constructability, and an increasing level of standardisation. The importance of the latter cannot be overstated; it resulted in a hierarchical structuring of the industry, from the level of materials, such as sheets, rods, and profiles for various metals, through the level of components, from the simplest, such as all types of fasteners, up to more complex ones, such as bearings and valves, and then up to the level of equipment, such as a steam engine or a weaving machine. This development was, in effect, the beginning of systems engineering, and the processes and procedures that made up this structured approach to engineering were expanded, refined, and given a certain intellectual foundation within the academic engineering curriculum, and became an intrinsic part of what was considered good, professional engineering.

A person that straddled this period and the following one, and who in many ways exemplified the transition from craftsman to professional engineer, was Thomas A. Edison (1847—1931). With no formal education, his keen, enquiring intellect and his capacity for hard work made him one of the great inventors, and his good business sense ensured that these inventions (there were about 1000 of them) became the cornerstone of an industrial empire. His preferred method was experimentation and trial-and-error; he was good at building apparatus, both for experiments and for production, and his West Orange laboratory was full of every conceivable material collected from all over the world (Baldwin, 1991). As a craftsman he was not concerned with the theoretical foundations of his inventions, but he would not give up until they had been developed to the stage where they were useful and cost-effective products, and in this important aspect of engineering he was without peer. As engineers, he and Leonardo da Vinci were complete opposites.

A contemporary of Edison's who epitomised the modern, professional engineer was Gustave Eiffel (1832–1923). A product of the excellent French education system, this structural engineer is best known as the creator of the Eiffel Tower in Paris and for his structural design of the Statue of Liberty in New York, but he and his construction company designed and built numerous bridges and buildings all over the world. His designs were the result of state-of-the-art structural calculation methods, and his manufacturing of structural components was so precise that very little adjustment had to be performed when they were assembled on site. His professionalism was also reflected in the attention he gave to workplace management and safety; the Eiffel Tower was constructed without a single fatality. He was a pioneer in advancing the understanding of wind loading on structures, and developed and maintained his own research facilities, including wind tunnels, until his death at 91, but his intellectual curiosity also drove him into exploring other technologies, above all wireless telegraphy, for which he used his tower as a base station (Harvie, 2004).

8.2.1.5 The Last Century

The following period, from 1900 to 1950, saw the emergence of technology-based inventions and innovations, such as the vacuum tube, synthetic materials and plastics, and, finally, the semiconductor, and their conversion into mass-produced products at an increasing rate, among these the automobile, the aeroplane, and the radio. Engineering now had two increasingly distinct aspects: the development of technology and the application of this technology. Engineered products were found not only in the industries that produced end products, such as the textile industry, but they were now end products in themselves. In 1850 nobody had a steam engine in their house; by 1950 many people had an internal combustion engine in their garage. And all this despite two World Wars and the Great Depression. (Or perhaps not despite of, but because of; demonstrating that peace and social stability do not necessarily promote creativity, as Orson Welles remarked, somewhat ungraciously, with regard to the Swiss and the cuckoo clock.)

With a view to our interest in systems engineering, two developments are particularly interesting. In chemical engineering, the concepts of unit processes and unit operations emerged and quickly became accepted as the basis for engineering chemical plants. Complex processes, converting raw materials into finished products, were shown to consist of combinations of the members of a relatively small set of processes; only the values of parameters of each process, such as throughput and temperature, would vary from plant to plant. A plant was viewed as a system of elements arranged in a particular manner, i.e., forming a particular structure. However, it was recognised that despite the simplification this afforded through specialisation and standardisation, maximising the overall plant cost-effectiveness required a holistic view, and this was achieved through a model of the overall process, expressed in terms of an accepted symbolic modelling language. We could therefore say that chemical engineers were the first to embrace systems engineering; they just did not call it that.

The second development came through the rapid spread and general acceptance of the telephone. While the telephone itself is a fine piece of equipment, it is useless on its own; its value lies in being able to be connected to a vast number of other telephones, i.e., in being part of a system. As a result, the theory of networks received a great deal of attention, with such associated subjects as graph theory and queuing theory. Many of these ideas about the flow of information within a network would then be applied to the flow of information within the large defence projects that followed the transition from hot to cold war. The partitioning of the work into a set of work packages, the structuring of this set so as to optimise the probability of a successful outcome, and the use of network concepts and a corresponding graphical language (CPM, PERT, or Gantt chart) are generally seen as the beginning of systems engineering as a separate activity. Because this took place within the defence industry and was heavily promoted and supported by the US Department of Defense, this early formulation of systems engineering focused on project management within the Department's acquisition framework, and that set the direction of systems engineering development for the next several decades.

This brings us to the final period in the history of engineering, from 1950 until today; a period characterised by the space race and arms race, by the PC and by mobile communications, and by the increasing interaction and overlap between engineering, biology, and genetics, and all of it underpinned by an almost incredible acceleration in the development of new technology. In ever smaller and faster semiconductors and integrated circuits, in electronics at optical wavelengths, in new materials and substances and their manufacturing methods; a list almost without end. But perhaps the most significant feature of this period was the emergence of software and software engineering as a new technology, and today it is claimed that more than half of all engineering hours are spent on software engineering. For various reasons, among them the close connection between software and mathematics, especially formal logic, and the fact that the non-material nature of software seemed to free its practitioners from the previous constraints and discipline of engineering, software development took off in a euphoria of uncritical trust in the capabilities of the individual brain, with "gurus" leading the way into the promised land. Only after numerous failures and the prospect of a crisis with serious economic consequences was it generally realised that software development required a controlled and structured approach just as much as engineering did, and so all the features of professional engineering, such as modularisation, standardisation and reuse, structuring, the use of formal, graphical languages to support the design process, etc. were reinvented by the software community as part of the new discipline of Information Technology (IT), albeit in a format tailored to the non-material nature of software.

For systems engineering, the rise of IT was a mixed blessing. On the one hand, the use of computers with the appropriate software allowed many of the systems engineering processes to become cost-effective for smaller projects and so gain wider acceptance. A significant proportion of the vast effort that went into advancing and structuring IT also benefited the further development of systems engineering, particularly in formalising processes and approaches that, until then, had rested on a heuristic foundation. On the other hand, software engineering came into engineering out of left field, so to speak, with very little connection with the tradition and discipline of engineering, and so, by associating systems engineering with software engineering, systems engineering tended to drift away from mainstream engineering. An example of this is the 2008 version of ISO/IEC 15288 (ISO, 2008), which, in my opinion, was a step backward from the 2002 version.

Finally, in coming to the end of this brief historical overview, we must identify a development that has had an increasing influence on all technology-based industries and their associated engineering disciplines, and that will continue to do so in the future, and which may turn out to be the most important driver of systems engineering. For most of the last century, the development of new technology through research was seen as an imperative for developed nations, and there was such an appetite for new technology that almost any new development found an application and a market somewhere. Only in the latter quarter did we start to notice some serious concerns about where all this technology was leading to, and whether its application was always in the best interest of society as a whole. The question started to shift from "can it be done?" to "should it be done?", and the increase in knowledge, both through travel and television, of what was happening in the world outside our own local community made us aware of the fact that we are all sharing the same limited resources and influencing a common environment. It is becoming clear that it is not just a matter of having better technology, it is also a matter of knowing how to apply this technology in the most appropriate manner. This requires an understanding of the interrelation of the application with its environment. While this was always within the scope of engineering, the immediate and direct benefits of introducing a new technology were usually so major that other effects appeared relatively insignificant. Sometimes they were actually insignificant because the scale of the application was initially so small that the side-effects, which are generally dependent on the scale in a very non-linear manner, were also small; at other times they were simply assumed to be small because no methodology existed to handle the increase in complexity involved in a proper assessment. The former reason no longer holds in many cases; for technologies such as the internal combustion engine, irrigation, and power generation the applications have grown to such a scale that what was earlier side-effects have become major effects. The latter reason is no longer acceptable to society, and the legislative framework in which engineering takes place is continually being tightened to ensure that a holistic approach is being taken to determining all the effects of every project over its life cycle. The result is that a whole new dimension of complexity has been added to engineering, creating a strong demand for adopting systems engineering as an integral activity within the engineering process.

8.2.2 The Process of Engineering

To place engineering within the domain of human activities, we recall briefly some work presented in Aslaksen (2017b). Ontologically, engineering belongs in the upper level category of *processes*, and within this category we can distinguish a sub-category of *purposeful processes*; these are processes that have a purpose formulated by humans, as opposed to other processes, such as the change of seasons, erosion, and the processes taking place within stars, as well as the numerous processes carried out by all other organisms. The category of purposeful processes contains two sub-categories. One is what we shall call *realization processes*; these are processes that convert the results of intellectual work into *services* useful to society or to groups of society. The other sub-category is what we shall call *professional processes*. These are distinguished by the extent of the knowledge bases and the intellectual effort required by the practitioners to acquire and apply this knowledge; in effect, by the investment in education and experience. This sub-category includes the class of engineering processes, but also a wide range of processes outside of engineering, such as medicine, dentistry, architecture, and economics. Professional processes are further distinguished by the fact that each one is related to and, to some extent, embedded in, a corresponding realization process. Law is related to law enforcement processes (without which law would be without value), medicine is related to health care processes, and so forth. Engineering is related to the *industrial processes*, which are processes that involve a conversion of natural resources to fulfil their purposes, but the distinguishing feature of engineering is that *every instance of the engineering process is embedded in an instance of the industrial process* (this is also true of architecture, which is embedded in a subset of the industrial processes: the building industry). This means that we cannot consider any effect of engineering on society

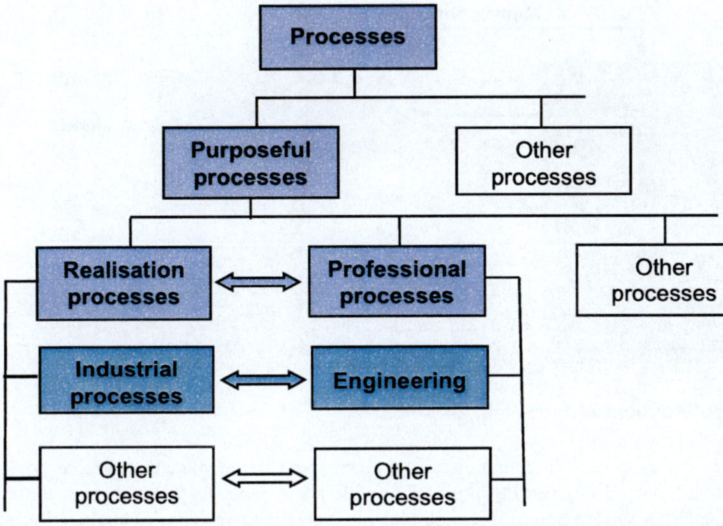

Fig. 8.1 The framework of processes within which engineering is defined (Aslaksen, 2017b)

without considering the associated industrial process. Engineering without industry is just like dreaming.

The foregoing framework is illustrated in Fig. 8.1.

The *process of engineering* (or simply *engineering*) is defined within the class of professional processes by its knowledge and construction bases, and the combination of these two bases is what we defined as *technology*.

A result of the continuous transformation of technology, as well as the current exponential increase in volume, is that, in the sense of understanding, maintaining, and being competent in using, various actors relate to different parts of technology, as was discussed in Aslaksen (2015) and illustrated in Fig. 8.2. The technical workforce includes technologists, technicians, drafters, and trades persons; all persons who, in addition to engineers, require access to the combined knowledge and resource bases, that is, technology. This structuring is defined formally, and to a large extent also in practice, by education and training, but experience and individual interest and aptitude can result in a significant blurring of the boundaries. Initial results of a study of how this structure is perceived by faculty and students is contained in Murphy et al. (2012). For the moment, we shall define an engineer as someone with a degree from an accredited four-year university course and meeting certain requirements for Continuing Professional Education (CPE). Engineers are the practitioners of the professional process of engineering, and the engineering disciplines, such as civil, chemical, electrical, and mechanical engineering, are distinguished by the subdivision of the resource and knowledge bases reflected in their education.

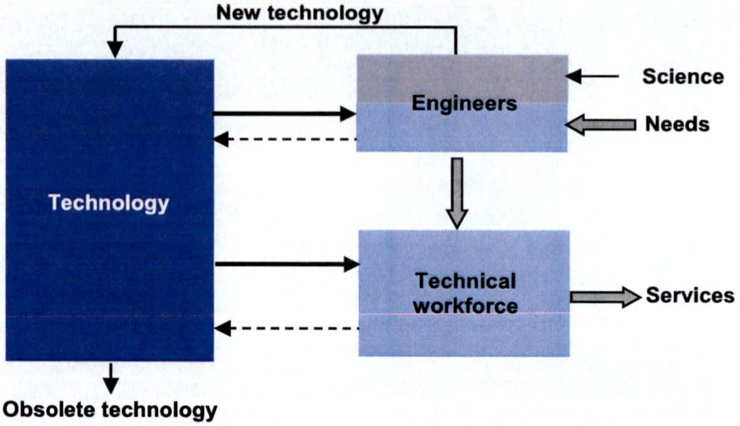

Fig. 8.2 The interaction with technology by engineers and the technical workforce. The dotted arrows indicate that all engineering projects provide some input to technology in the form of experience, and the subdivision of the engineers illustrates the two types of engineering projects (Aslaksen, 2015)

Engineering consists basically of individual *projects*, each on an instance of the engineering process. Each project has a defined purpose, and this allows us to distinguish two groups of engineering projects according to the nature of their purposes, as originally suggested at fPET 2012 (Aslaksen, 2018a):

- projects that utilise the existing resource and knowledge bases to meet a *need* expressed by all or a part of society; and
- projects that increase the resource and knowledge bases.

Or, in other words, projects in the first group *apply* technology in order to meet requirements imposed by entities or people who are generally not engineers, and it is these stakeholders that are the judges of project success; whereas projects in the second group *develop* technology, often using that part of the knowledge base that is provided by science, but sometimes also based on heuristics or arising from trial-and-error, and their success is judged generally by other engineers. Let us agree to call these two groups of engineering projects *application projects* and *development projects*, respectively; this grouping was implied already in Fig. 8.2. There is not a sharp boundary between these two groups, and there will be many projects that contain sub-projects of both types. In any case, every application project also leads to an increase in technology, if by nothing else than simply by acting as an example for later projects, as was also indicated in Fig. 8.2. The usefulness of this grouping and the distinctiveness of the two groups were discussed in Aslaksen (2013a); in particular, as the group of application projects is very much greater in number and direct importance to society than the development group, it allows us, by focusing in the following exclusively on the former group, to make general statements.

8.2.3 Applications of Technology and Our Model

Already in the creation of early settlements there was a level of technology and the ability to apply this technology; extracting and processing the materials required for tools and structures, the knowledge required to use the tools and to design and construct walls, roofs, water supplies, irrigation and drainage canals, streets, and clearing land for agriculture. Similarly, the enabler of the increase in both the flow and the storage of information has always been technology and its applications. For the interaction between settlements, the means of transport, in the form of horses, carriages, roads, and boats, would have been determining factors; again, reflecting a level of technology and its applications. With the advent of writing, somewhere around 5000 years ago, technology was applied in the production of the substrate on which the writing was recorded and in the production of inks and pens, etc., marking the beginning of an accelerating increase in information transmission and an increase in the ratio of technology-mediated transmission versus word of mouth. Thus, referring to our model in Chap. 4, it appears that the level of technology and its application is a main determinant of μ, and the same is true of μ_a, in that the development of more capable and precise instruments drove the increase of our knowledge about Nature. However, the case of μ_0, which is exemplified by the scientist researching some aspect of Nature, makes us realise that it is not enough to have good instruments and facilities; there has to be the background knowledge and training which enable the scientist to work efficiently, as well as the time to do the work. And the same is true, to some extent, of the transmission of information; effective communications requires both knowledge and practice, and if you are scratching around from morning till night just to exist, there is not much time for acquiring either. So, we have arrived at the conclusion that, to a first approximation (or high level of abstraction), the dynamics of Θ is determined by the extent of technology application, and that there are two distinct effects involved. One is the effect of technology on the processing (e.g., printing) and transmission of information, i.e., on the rate of information presented to the individual. The other is the effect of technology on the conditions that allow the individual to process the information, effectively the standard of living in general. The former is the subject of Sect. 8.4; the latter can be seen in the measures discussed in Chap. 6.

8.2.4 The Industry/Business Framework

The industrial processes indicated in Fig. 8.1 provide the means of converting the work of engineers into products and new technology. Seen from the engineering side, these processes provide the industrial framework within which engineers work, a framework that also has a social component. Besides design of the items (products and services) produced by industry, engineers are instrumental in the design, planning, and management of production processes, they are involved in contract

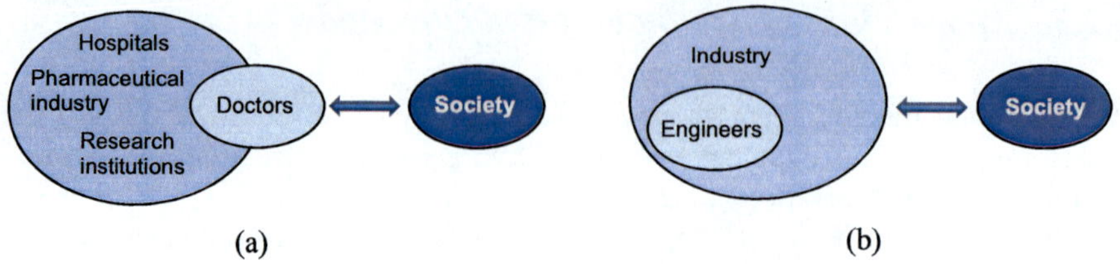

Fig. 8.3 Illustrating the indirect interaction between engineers and society, as opposed to the direct interaction between doctors and society (their patients) (Aslaksen, 2017a)

negotiations and management on both the input (procurement) and output (sales) side, and in numerous smaller specialised roles. Engineers are dependent on this industrial framework, an engineer on his or her own can accomplish very little, and so what society sees is the work of the engineer through an industrial interface in which numerous people play a part, such as workers, tradespeople, marketing and salespeople, business managers, financiers, etc. This is a major difference to other professions, such as medicine, where there is a direct interface between the profession and society, as illustrated in Fig. 8.3.

In the context of this book, perhaps the most important issue with regard to understanding the role of engineers and their position in society is the difference in the environments in which scientists and engineers work. While a scientist may be embedded in an organisation, such as a university or an industrial research department, the research activity itself, and its outcome, are determined by the scientist, and the outcome is published under his or her name. Engineers are embedded in industry, where they work in multidisciplinary teams, and where the result is usually attributed to the organisation; the engineer remains anonymous and invisible to society. The work of the scientist is judged by what new knowledge it produces; the work of the engineer is judged by how beneficial and useful it is. And if the scientific result later turns out to be incorrect; well, that is just further knowledge. Whereas if the engineer's work is incorrect, there is immediately a question of blame and liability. These, and many other differences between science and engineering make it clear that there must be fundamental differences between the philosophy of science and that of engineering, to the extent that the latter is even sensible.

However, there is a further interface between engineers and society, and that is business. A common definition of *business* is "the activity of buying and selling commodities, products, or services". Now, if every household in a society were completely self-sufficient, i.e., produced its own food, clothing, utensils, dwelling, etc. there would be nothing to buy or sell, and no business. Therefore, implicit in the definition is a division of labour and a corresponding *view* (in the sense of systems engineering) of society as a functional structure, a view originally developed by Durkheim (1893). Under this perspective, business appears as the interaction that converts that structure into a system—the economy. Each structural component has a dual nature, as producer and as consumer, and we can then create a high-level view

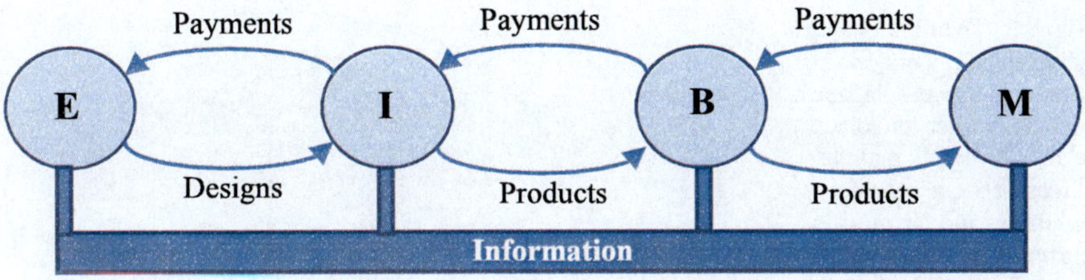

Fig. 8.4 Simplified relationships between the four entities engineering (E), industry (I), business (B), and market (M). In this figure "designs" encompasses all the artefacts and services provided by engineering to industry, and "products" includes commodities and services (Aslaksen, 2017b)

of the economy consisting of two interacting parts: *industry*, as the collection of all the producers, and the *market*, as the collection of all the consumers.

Industry forms the link between engineering and business, and Fig. 8.4 shows a highly simplified view of the relationships between the four entities engineering, industry, business, and market. The main point of this illustration is to emphasize that, in addition to the obvious flows of products and funds, there is a significant interchange of information between all four entities. The relationship between engineering and industry is, as Fig. 8.4 indicates, a very direct one; to the extent that one might say that engineering is *embedded* in industry. The relationship between engineering and business is quite different; indirect as far as the flows of products and payments are concerned, but possibly quite direct as far as information goes.

At this point we need to eliminate a semantic issue: It has been common to refer to the whole process of providing a service that meets an expressed need as "the process of engineering", both because engineers are, to varying degrees, involved throughout the process, and because the general public does not see inside the process and so uses convenient labels like "industry" and "engineering" without reflecting on the precise meaning. This common usage needs to be made more specific in the current context for two reasons: Firstly, because we identify business as a separate process, whereas when industry is considered as one of the segments of the economy, business (i.e., the buying and selling) is included as its commercial (or non-technical) aspect, just as it is in other segments, such as education and healthcare. Secondly, because we shall consider the relationship between engineering and business, as opposed to the role of engineers in the wider context of the interface with the market or, more generally, society. That is why we have identified engineering as a separate process, albeit embedded in an industrial process, and to avoid misunderstandings, we should now refer to the technical content of the process of providing a service as the *industrial process*, in accordance with the taxonomy shown in Fig. 8.1.

This is illustrated in Fig. 8.5, where engineers perform the process of engineering, which is part of the industrial process performed by the technical workforce and which interacts with the market through business. All three entities have an interface to the market, but it is dominated by that of business. What society experiences as "technology" is largely determined by business.

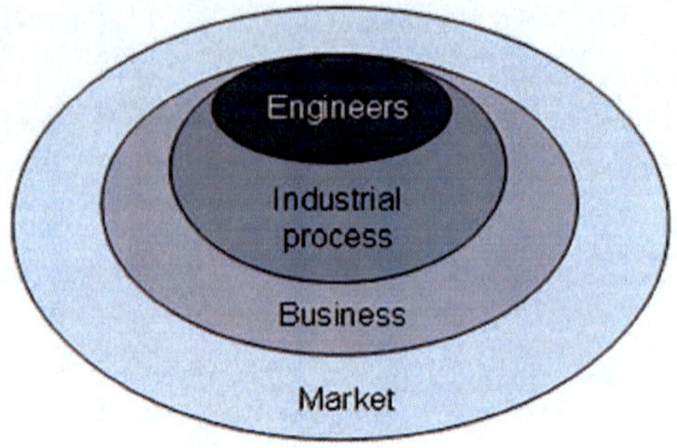

Fig. 8.5 The relationships between engineers, the industrial process, business, and the market (or society). It also illustrates that the direct interface between engineers and business is currently quite limited. From Aslaksen (2017b)

Engineering and business are both situated within the framework of society. They are products of its history and subject to its laws and customs, and, as processes, interact with all the other entities in that framework. But we can also view the two of them under a narrowly focused perspective, which highlights that they operate in quite different domains—engineering in the domain of creation bounded by rationality, continuity, and the laws of Nature; business in the domain of human desires, perceptions, and irrational behaviour—and that the existence and success of each of them are closely linked. In that link technology plays a major role and, somewhat paradoxically, it is the context-dependency of the concept that enables it to play that role in that in the context of business, technology means applications of technology.

In the business domain, the use of "technology" is not sharply defined; on the contrary, it takes on a certain magic quality, representing a promise of overcoming the currently experienced limitations placed on us by our physical environment. It is used to express a general desirable quality arising from the previous successes associated with this concept, such as air transport, high speed rail, space flight, telecommunications, the Internet, computers, mobile phones, advances in medicine, new synthetic materials, and so on. In this domain, "an application of high technology", "based on a recent advance in technology", and similar phrases express the fact that for business, "technology" is one of the parameters characterizing the saleability of a product or service. A symbol of man's triumph over Nature as, indeed, is business itself.

In the engineering domain, technology is a resource; it is something that is used by engineering. The purpose of engineering is to meet needs expressed by society (or parts of society) by creating applications of technology, and so technology, while indispensable, is secondary to the core of engineering, which is the creative activity.

However, as different as these two meanings of "technology" are, they are united in that each is essential to success of the activities in its domain. For business, "technology" is a main driver of profitability and growth; for engineering it is the physical and intellectual foundation on which its existence is based. This Janus-nature of "technology" is the core of the engineering-business nexus, and we shall return to that in the third section of this chapter. But first it is useful, and makes the chapter more self-contained, to consider some fundamental aspects of engineering, based largely on some earlier work, that will support the arguments in the third section.

From a narrow, economic point of view, engineering and business are tied together in a feed-back loop. In looking for opportunities to invest capital and generate a return, business finds needs that are either explicitly expressed by society or perceived by business as latent needs that it believes could be met by applying technology. Engineering creates the required applications, and in so doing, generates new technology through the continuous improvement process that is part of every engineering project. These applications of technology generate further capital in the form of the return on the investment; business looks for new opportunities using the new technology; and so on. In this loop, the interface between engineering and business is technology, but the driver is the need to generate a return on the exponentially increasing amount of capital. The implications of this structure for the evolution of society are discussed in some detail in Aslaksen (2015); for our purpose in this chapter, it is only important to recognize the strength and centrality of this economic relationship in which engineering and business are embraced.

However, aside from the economic aspects, the relationship has undergone some significant changes. In the early days of engineering as a recognized profession, say, up until 1900, engineers and businesspeople had close personal relationships, and leading engineers were often also leading businessmen. Then, with the rise of mass production and the importance of promoting consumption, there was an increasing tendency to isolate the engineering from the business process, and engineers became technical problem solvers and "backroom boys" in an overall stove-piped process with "over the wall" interaction between functions. Since that time, there has been a growing realization of the drawbacks of this type of organization. Not only as inefficiency through misunderstanding and lost information, but, increasingly, because non-economic issues, such as ethical, sustainability, and environmental issues became important and needed consultation between business and engineering.

8.3 Technology and Our View of the World

In addition to the direct influence of technology on our daily lives through its applications, there is another, more subtle aspect of how technology influence us—how we think of ourselves, of the world around us, and of our place in it, and how this again influences our behaviour. This was one of Martin Heidegger's concerns with technology, that the ease with which applications allow us to exploit our environment changes our relationship to it (Heidegger, 1977). We consider it solely as something to be exploited, and value it only as a commodity and for how we can use it, rather than recognising it as an element of the larger system—Planet Earth—of which humanity is just one of the other elements. It is the evolution of that system, with its emergent properties, that should be our concern. Another well-known investigation of how technology can influence our behaviour is Michel Foucault's study of the Panopticon; a prison design by Jeremy Bentham that allowed any of its inmates to be observed at any time (Foucault, 1977). The knowledge that they might be observed would enough to modify the prisoners' behaviour; a psychological effect enabled by technology. Though Bentham's design was never realised, the idea behind it has been influential in prison design ever since, and is, of course, alive and well in our society today. And the simplest example of technology modifying our behaviour is the speed bump.

Furthermore, in a recent book, Cohen (2009) discusses how the appearance of several technologies in the period around the Renaissance had a significant effect on how people changed both their perception of themselves and of their relationship to Nature. The development of mechanical clocks introduced a discipline into our behaviour, which Cohen exemplifies by the rituals of monastic life, as well as an increased understanding and appreciation of the value of time, as exemplified by Darwin's dictum "a man who dares waste one hour has not discovered the value of life" (Darwin, 1887). The development of optics, in the form of telescopes and microscopes, allowed people to see features of Nature, both very small and very large, that not only furthered their knowledge and understanding of Nature, but made them see their own position in the world under a new perspective, realising that our concepts of small and large are not absolute, but relative to our own size. Finally, the development of printing with movable type changed our perception of our personal "self" as a component of society and made us aware of the importance of reflecting this "self" in print, giving rise to literary ambition.

This indirect influence of technology should be contrasted with the direct influence arising out of interacting with, or using, applications of technology, about which there is an extensive body of philosophical as well as sociological/empirical knowledge, and which interaction leads, in its most intimate extent, to the concept of a techno-human hybrid. The indirect influence is a significant factor in what a human is—how we perceive the world around us, the criteria we use in processing the information we receive, how we perceive ourselves, and the like—but it is also a factor that is difficult to define and quantify. And it is certainly something that is way outside my competence and background, hence it will not feature in the further developments in this book, except in Sect. 8.4, where I look at the influence of information technology, and perhaps in the form of an occasional peripheral comment. However, I should say that I am very aware of this influence and experience it every day in the interaction with my wife, who grew up with minimal exposure to technology. There are so many areas where the presence or absence of an exposure to technology makes a very definite difference to how the two of us understand them, with art being perhaps the most obvious, but more generally with a spiritual and emotional view of the world as opposed to a more mechanistic one.

And it is exactly this ubiquitousness that is the core of the problem when it comes to considering both technology and the environment. But while there is a great deal of discussion about the extent to which we are responsible for changes to the environment, there is no doubt about that we are solely responsible for every development and application of technology. Technology does not harbour any inherent force that determines its development and that makes its increasing influence in our lives inevitable. We control that development, and it is becoming increasingly clear that the question should no longer be "can it be done?", but "should it be done?", as we noted earlier. The problem is that our view of the effects of technology have been superficial, in the sense of concentrating on the immediately visible effects, such as the economic effects, or the effects on power projection, or the effects on physical health, and so on. What has had much less scrutiny is the effect on the structure and fabric of society itself. As an example, we have now seen several cases where military superiority is used as a means of suppressing a problem, starting perhaps with Palestine/Israel, followed by Vietnam, Iraq, Afghanistan, and now the whole Middle East and Ukraine. (And the use of technology as a means of suppression is now creeping into our own societies in the form of surveillance, enforcement, and an associated legal framework.) Technology is increasingly allowing one side to inflict grievous losses on the other side with minor own losses, and so this is seen as a relatively easy and, unfortunately, also popular approach to suppressing a problem without having to address the much more difficult task of reducing or eliminating the cause of the problem.

The lack of understanding of, and concern with, the influence of the application of technology on the evolution of society displayed by the general population does not mean that there has not been a significant amount of thought and effort expended on this issue. As we mentioned briefly in the previous chapter, the concern with what might be broadly subsumed under the concept of technology goes back to the Greek philosophers, and from about 1800 onwards, there was a slowly increasing awareness of the applications of technology as significant factors in society. A good review of this developing awareness and of the major issues under discussion is given in the already mentioned book *Streit um die Technik*, by Friedrich Dessauer, most of which was first published in 1926 under the title *Philosophie der Technik*, and in its final form in 1956 (Dessauer, 1956). It is, unfortunately, only available in German, as is much of the literature on this subject in the years before World War Two. The main point, as far as our investigation is concerned, is that most of this work was concerned with the effect of mechanisation on the role of workers, transforming artisans and craftsmen into operators of machinery, without addressing the impact of technology on the development of the fundamental nature of society. There were, certainly, exceptions, such as Marx (1976) and Veblen (1921), but it was only after the appearance and explosive growth of information technology (IT) in the years following World War Two that the influence of technology on the evolution of society became an object of study in its own right, with contributions from philosophy and various branches of social science. A substantial body of work has evolved; here we shall only give a brief overview.

The interaction between technology and society is a complex subject, with numerous components and aspects, and one on which the view has changed significantly over time. Much of the early work on the influence of technology regarded it as taking part between two separate spheres of existence; a genuine (or intrinsically, or unsullied) human sphere and a sphere in which technology is prevalent. Technology was seen as developing under its own imperative, and so the interaction was a one-way process, with conflicts arising at the interface between the two, and with humans sometimes seen as the "victims" of technology. More recent work sees the interaction as a process that is both two-way and so dynamic that it is not possible to make a clear-cut distinction between humans and technology. Humans are always hybrids of supposedly human and technical aspects, and what is of interest are the different kinds of human-technology interactions. This is treated in an article by Dorrestijn (2012), and in the present context it is interesting to note how this two-way process is reflected in the system introduced in the beginning of this section. All the actors in this system (or actor-network, as it is also called) become hybrids, and so there is a feed-back between technology and society that makes the relationship between them take on a dynamic character.

This can be seen, for example, in the importance of a collective readiness to accept and try out new ideas when they become available. This sensitivity to invention is a compound of many social, political, and cultural factors, sustained by tradition and passed on by education and training. A well-known example is the stagnation of technological development in Chinese civilization under the control of the mandarins (see e.g., Buchanan, 1992, Ch. 11). Up until, say, 1400, Chinese technology and its applications were on a par with, if not superior to, technology in Europe, but in the following centuries European technology developed rapidly, whereas in China a veneration of tradition and ritual by a centralised government stifled development. Some measure of political liberty, a degree of freedom from the constraints of class and conformity, a tolerance towards unfamiliar and even apparently bizarre points of view are all parts of the "social package" required for technology to develop. This was again demonstrated in the relative rates of industrial development in France and England during the period 1600–1800, when an entrepreneurial middle class in England had considerable freedom to develop new industries, whereas in France such development was mainly the prerogative of the nobility, under the control of a powerful monarch.

A distinct body of research is what is identified as the social shaping of technology (SST), and a seminal work here is the book *The social shaping of technology*, edited by MacKenzie and Wajcman (1999). The point of departure of SST is to acknowledge the much greater complexity of the socio-technical interface than is recognised by either technological determinism, which saw technology developing according to an inner logic, or social determinism, which saw technology development as reflecting a single influence, such as an economic imperative. Central to SST is the concept that there are choices (although not necessarily conscious choices) inherent in both the design of individual artefacts and systems, and in the direction or trajectory of innovation programmes. Different routes are available, potentially leading to different technological outcomes, and they could have differing implications for society and for particular social groups. Rather than merely assessing the social impacts of a given technology, SST examines what shapes the technology which is having these impacts—its artefacts and practices—and draws together views from different areas of sociology and economics to form a deeper understanding of the innovation process and the social factors influencing it.

An important critical strand within SST has highlighted the politics of technology. Technologies can be viewed as "politics pursued by other means" or as the outcome of social conflict; in either case, technologies are not neutral, but are fostered by groups to preserve or alter social relations. Of particular interest to our investigation is the promotion of the Weberian class conflict perspective by proposing that the ability of a society to favour technical change is enhanced by conflicts taking place in a large number of arenas. The nature of the conflicts may be economic competition or social conflicts, and the arena might be industry, a profession, or a neighbourhood, but in any case, in a totally homogeneous society new technology will not be easily introduced, and technical change is more likely to occur in a society or an arena in which power and influence are unequally distributed among a relatively large number of agents.

Closely related to SST is what is known as Social Construction of Technology (SCOT). It is based on an approach called the Sociology of Scientific Knowledge (SSK), which considered scientific knowledge as arising from the socially influenced interpretation of scientific discoveries. SCOT studies technological artefacts and explains how social factors entered into the particular choices among a number of possible ones. One aspect of this social constructivist program, and which is relevant to our story, is that a problem is closed when the relevant social groups see it as being solved. But this is certainly not true in general; in many cases closure simply means that the differences between social groups have been reduced to the point where power relations of a political or economic nature make any further development futile. Thus, closure does not mean the elimination of conflict, and it is the suppression of latent conflicts through power relations that reduce the stability of a society, as we mentioned above.

Science provides the basis for developing new technology, but the main creators of new technology, as well as the creators of new applications of technology, are engineers. And engineering is very different to science, both in its approach and in its relationship to society. A lack of appreciation of this difference, and of the role of engineers in general, has often made publications on the relationship between technology and society seem somewhat artificial. Basically, whereas science is about discovering the truth of our understanding of Nature, engineering is about using that understanding for beneficial purposes. And whereas the paradigm within a domain of science can change relatively rapidly, caused by a single revolutionary new theory, such as the heliocentric view of the solar system, Newton's laws, Darwin's theory of evolution, relativity, and quantum mechanics, changes within engineering are more gradual. In particular, it is not that existing engineering knowledge and works are found to be incorrect and need to be discarded; it is that new knowledge and works are added and then, over time, replace the old for reasons of greater cost-effectiveness.

8.4 Information Technology

8.4.1 Background

The history of technology is well known and was outlined as part of Sect. 8.2, and we might think of a timeline in terms of some of its major milestones, such as fire, tools, boat-building, agriculture, the domestication of animals, the wheel, concrete, metal extraction and processing, water power, chemical processing, steam power, electricity, telegraph, telephone, the internal combustion engine, pharmaceutics, electronics, and the Internet, and for the case of information technology, a simple timeline was presented in Table 2.2. But if we now pursue Fletcher's insight regarding the role of the stress introduced by interaction on the evolution of society and want to identify this as one of the mechanisms by which technology influences evolution, we realise that there are two completely different epochs: before and after the development of telecommunications. In the earlier epoch, the interaction was based on direct, in situ interaction (the determining factor in Fletcher's density-based interaction limit) or on physical movement; either the movement of the persons involved, or by the movement of the physical carriers of information, such as letters, newspapers, and books. Certainly, technology played a significant role, from horse-riding and boatbuilding to mechanised transport as far as the movement was concerned, and with printing as far as the physical medium was concerned, but both physical distance and the weight of the medium were determining factors in the interaction and in its effect on the evolution of society. But once the telecommunications industry started its exponential growth this changed, and today the situation in the developed part of the world is characterised by the following features:

a. Individuals have instant and unlimited access to a global communications network carrying voice, video, and data.
b. The cost of transmission is largely independent of volume and distance (e.g., fixed price contract for unlimited volume).
c. The capital invested in the telecommunications infrastructure is a significant part of the total investment in fixed capital.

As a result, the strictly spatial dependence of the behavioural stress, as in Fletcher's I-limit, is diluted by a dependence on the information environment, characterised by the parameter μ in our model. In an extreme case, a person could avoid all direct contact with other people and interact only via the telecommunications infrastructure, in which case the dependence on population density would disappear.

Within this second epoch a new one has opened, with analogue encoding of the information having been replaced by digital encoding. With this change, the information contained in an interaction can be detached from the interaction and be turned into data that can be stored, processed, and reused, and the media change their nature from a passive channel for the transmission of information to an active participant in the information exchange. This is what is called the *mediatisation* of communications, and it means that we cannot treat the media as a fixed part of the common infrastructure. Not only do they influence and change the way in which we communicate, but they are themselves participants in the public discourse, both configuring and contributing to this discourse. Mediatisation studies have become a separate subject within sociology, and an introduction to it and to some of its applications can be found in the book *Communicative Configurations—transforming communications in times of deep mediatisation*, (Hepp et al., 2018). It is a subject that lies outside the scope of this book, but it will be referred to again in Chap. 11.

There is another consequence of the rise of media based on modern information technology—in particular, the Internet—and that is the fracturing of the public information sphere. In our model, individuals interact via a public sphere that is represented by the circulating information, as shown in Fig. 5.7—an unstructured body of information to which every individual has equal access. In the past, the mediated part of the public sphere was provided by the mass media—newspapers, radio, and TV—and the information exchange was largely one way; "an elite discourse carried out by representative actors on behalf of the wider citizenry in order to facilitate the formation of public opinion" (Bruns, 2019). While some of the public sphere still exists, it is being supplemented and increasingly replaced by a variety of both public and private information exchange spheres, including domain-specific spheres, public spherules focusing on more narrowly defined themes, culturally defined communities, and dynamic and highly specific domains called "issue publics" in Habermas (2006, p. 422). This fragmentation is clearly an issue of great significance for the information exchange within society, and the work of Alex Bruns provides a good introduction to, and discussion of, the issue in several articles, three of which are Bruns (2008, 2015, 2019). At the high level of abstraction of the view of society presented in the current work, the crucial question addressed in these articles is: How does this development of the public information exchange effect the strength of the interaction? Is it increasing or is it decreasing? This is clearly a subject for further research.

8.4.2 The Interaction Channel

Besides having the financial resources for sustaining the processing function, the collective intelligence depends on the interaction taking place between the individuals, which enables the averaging process—society's immune system—and provides the feedstock for the generation of new information through reflection and critical thinking. While face-to-face interaction is still important, in particular in conveying sentiments, the flow of information impinging on the individual is increasingly mediated by technology, as we mentioned in Sect. 8.1.3, and we noted that the associated IT infrastructure represents a very significant investment. An increasing proportion of this investment is non-government, and accordingly it is looking to achieve a reasonable return. The search for that return is the major determinant of how economics influences the flow of information in society, and to develop an understanding of this, it is useful to consider three different types of communications networks:

a. The original person-to-person communications network, exemplified by the postal services, telegraph, and telephone (fixed, now also mobile). The basic business model was that users paid for the traffic generated at a rate *b*, measured by weight and distance (postal service), words and distance (telegraph), and time and distance (telephone). In addition, if one wanted to have a private terminal, this was available at a rental or subscription cost, *a*, so that the total revenue generated by a user, *u*, was of the form $u = a + bq$, where q is a measure of the traffic generated.

   For both the postal service and telegraph (which is practically extinct), $a = 0$. For the fixed telephone network, one has a choice, in principle, between being a subscriber, in which case $a > 0$, or using a public phone, if one can find one, in which case $a = 0$. Today, in many countries, the main telephone network is the mobile network, which requires the user to have a terminal, so $a > 0$. A user must also purchase a plan (recurring or pre-paid), which effectively adds to both *a* and *b*, but with a pre-paid plan, this cost can be kept very low, so that the main economic barrier to basic telephone service (voice call) is the cost of the mobile telephone.

b. Broadcasting, or one-to-many networks, exemplified by newspapers, radio, television, and podcasting, and the users of these networks are readers, listeners, and viewers. Here the business model is quite different because the main part of the business is not the communications network, but the provision of the *content* to be *distributed* on the network. User-generated traffic is not a relevant concept—the traffic is constant irrespective of the users. The relevant measure is the number of users multiplied by the time each user spends on consuming the content; this is the amount of the user's *attention* the network is able to capture. And the revenue does not come from the users, but from advertising, and depends on that amount of user attention. Hence, the business model essentially becomes a matter of optimising the difference between the advertising revenue and the cost of producing attention-grabbing content.

   In the case of newspapers, there is no fixed cost; $a = 0$, and *b* represents the cost of single newspaper. The cost of accessing wireless (or free-to-air) services is the cost of the terminal device; i.e., *a* is represented by the cost of a radio or a TV set, and $b = 0$. However, except for some government-funded services, the user must endure an amount of advertising.

c. Computer networks, networks that can carry any information as long as it is digitised; speech, text, music, graphics, videos, smell, touch—all that is required are the corresponding transducers connected to the computers at the end of a communications path. For the ordinary user, the only network of importance is the Internet, and from the user's point of view there are two groups of entities involved: the Internet Service Providers (ISPs) and the content providers (where content includes service). The former provide access to the network and certain communications services, such as email; the latter provide almost any content or service imaginable, including data bases (e.g. Wikipedia), voice and video communication (e.g. Skype, on-demand video), social media (e.g. Facebook and Twitter), games, booking services, dating services, blogs, chat rooms, and so on, with search engines (e.g. Google) providing the means for navigating through this mass of sites. The choice of ISP does not limit the size of the network; subscribers are subscribers to the Internet, only via a particular ISP, to whom they pay a service cost, *a*, which often includes the transducer.

The economic aspect of the Internet, as experienced by a subscriber, is the result of a combination of two business models: one for the ISP and one for the content provider. The business model for the ISP is very similar to that of the type a. network, except that with the increasing roll-out of free Wi-Fi, it is not necessary for an individual user to subscribe to an ISP at all (but, of course, the entity providing the free Wi-Fi needs to be a subscriber). And an additional source of revenue can be the sale of information about the subscribers to companies engaged in the processing of "big data". A great deal of useful information and services are available for "free", but often with the acceptance of being subjected to advertising and, in many cases, having to register, which allows the advertiser to send emails to the user or pop-ups to the user's terminal device. However, many of these services and content sites also offer the same, or premium, content without advertising, but at a price, so that, if we consider that advertising takes up some of the available attention time, and that time is money, then advertising can be seen as an impost on the poor.

From this brief description of the main communications networks that mediate the interaction between the members of a society, we see that, aside from the fact that advances in information technology allow increasingly convenient and intense communication, our market-based economic system leads to inequality in what should now be a basic social good, along with education and healthcare. But in the context of this book, with its emphasis on the collective intelligence and the role of interpersonal interaction, there is another aspect of the digital age that is of some importance, and it is the extent to which applications of information technology are reducing the time spent on interpersonal interaction. This issue first came to prominence with the appearance of TV in the majority of homes, and raised concerns about people, but in particular children, becoming passive viewers and neglecting such activities as hobbies, sports, and the social intercourse with family and friends. While the dire predictions raised in this regard were not realised, there are now two new contenders for the attention of the younger generation, and the first one is the various applications that go under the heading of social media. Most of these applications can, in principle be used to carry out an ongoing, two-way interaction, but in practice they are used overwhelmingly as a one-way channel of *posting* information, often in the form of opinions, but also just as a message, as in a photo. Feedback in the form of "likes" and "dislikes" does not count as constituting a conversation.

The second one is *gaming*. The published data on usage varies from country to country and between different sources, but the gaming industry is reported to be worth something like US$140 billion in 2018, and rapidly increasing, with such applications as esports and virtual reality games. Games are, of course, by their nature interactive, but what the player is interacting with is not another person, but a program that is responding according to fixed rules. It may be good mental stimulus to try and figure out what the rules are and then outsmart them, but it does not contribute to developing the identity or to the effectiveness of the collective intelligence. And there is, as with so many of the information technology applications, the opportunity to utilise games to present a particular ideology, sometimes openly, but more often covertly or even inadvertently.

The above presents one factor in the influence of economics on the interaction between the members of society; it is what we might consider a direct factor, in that it shows how the cost of IT services constrains a person's access to those services, and thereby to the information required for the operation of the collective intelligence. There is another, more indirect, economic factor that interferes with the information flow, and it is caused by the concentration of wealth in a very small proportion of the population. On a global scale, 47% of household wealth is owned by 1% of the adult population, and 150,000 people each own more than $US 50 million. (https://www.credit-suisse.com/about-us/en/reports-research/global-wealth-report.html, accessed 20.06.2022), With this wealth comes control of a large part of the IT services, and with control comes the ability to modify the information in various ways, ranging from simply deleting unwanted information to actually producing "fake news". An added factor here is the degree of privatisation of the IT services. In many Western nations privatisation is part of the economic ideology, and the flow of information is slanted towards the interests of the owners, whereas in nations with stronger government ownership and control, such as in China, the information is modified to reflect government policy. In either case, the control of the IT services, together with the ease of modification provided by IT, pose perhaps the most serious threat to the operation of the collective intelligence.

We should also recognize that, given the connection between identity and adaptive actions, there is a fine line between persuasion and enforcement. Being a member of society necessarily involves some restrictions on the individual's actions, and every society has some means of enforcing these restrictions. With the rapid development of information technology new possibilities of enforcement are opening up. Basically, instead of forcing people to behave in a certain way by physical means, we are able to influence their behaviour by manipulating the information they receive. For example, instead of controlling street protests by means of massive police presence, barricades, etc., and incurring the inevitable damage in the form of people injured, looted stores, cars set alight, etc., a much more cost-effective approach would be to suppress the information that led to the protests in the first place. But such manipulation of information as suppressing selected parts to avoid certain behaviour is really at the primitive end of what is becoming possible with information technology.

8.4.3 Manipulating Information

While face-to-face interaction is still important, the flow of information impinging on the individual is increasingly mediated by technology. And it is useful to divide the information and the mediation associated with it into two groups. One is the one-to-one interaction, mediated through writing (letters and email), telegraphy, telephony, and video; the other is the one-to-many interaction, mediated through printing, radio, TV, and the Internet, and with the mobile phone, or tablet, as the iconic user device in both groups. In the first group, the role of technology is to make the transmission of information more convenient (mobile, small device), capable of a great variety of content (speech, pictures graphics, video), and inexpensive (certainly on a bit-kilometre basis). In the second group, in addition to this same role in transmitting information, technology has a role that is becoming by far the most important one, and that is in providing the ability to manipulate the information in real time. This manipulation takes on several well-known forms, including

- Storage and retrieval. Until recently, the main means of this was to store the information as printed matter in libraries and other repositories. Now the collection and recording of information by electronic means, and the proliferation of these relatively low-cost means, such as surveillance cameras, mobile phones, and Internet-enabled devices and equipment, combined with rapidly decreasing cost of storage, is resulting in data bases of various kinds; both publicly available, such as the data stored on web sites and accessible via search engines, as well as specialised ones, available by subscription or for a fee, for practically every conceivable subject. However, many access channels, such as Internet browsers, mix the retrieved information with advertising, and also attach an evaluation to the information by influencing the order in which items are presented.
- Editing. By reorganising the sequence of items making up a message, the meaning of the message can be changed. And in this manner, it is also relatively easy to imply an attribution that is not correct. But it is important to realise that it is not necessary to modify or falsify information to change its meaning; equally effective is simply deleting or ignoring part of the information or certain subjects that contradict the desired message. This is very noticeable in the Australian media when it comes to subjects that reflect adversely on Western interests or behaviour. However, this approach is not without its risk of backfiring, as was noted by Robert Louis Stevenson: "The truth that is suppressed by friends is the readiest weapon of the enemy" (Stevenson, 1973).
- Commenting. By interspersing the factual information with the provider's comments, the impact, and even the meaning, of the information can be changed. It is often difficult to discern what is fact and want is opinion.
- Packaging. The processing and combination of information into consumable "stories" is more convenient than ever, as evidenced by a plethora of newsletters and blogs. The use of the network is moving from interpersonal communication to the provisioning and accessing of *content*, at the same time displacing printed matter, and the distinction between μ and μ_a is becoming less a matter of the type of information than of the individual's *intent*, as "being immersed in society" is becoming more like "being immersed in the web".

The realisation that such manipulation of information is going on is nothing new; it was the subject of numerous publications in the 1980's, with perhaps the best known being the book by Edward S. Herman and Noam Chomsky, *Manufacturing Consent: The Political Economy of the Mass Media* (Herman & Chomsky, 1988) and the two books by Herbert I. Schiller, *Who Knows: Information in the Age of Fortune 500* (Schiller 1981) and *Culture, Inc.: The Corporate Takeover of Public Expression* (Schiller 1989). It was also noted by Jürgen Habermas in his article on deliberative democracy, where he wrote:

> The political public sphere is at the same time dominated by the kind of mediated communication that lacks the defining features of deliberation. Evident shortcomings in this regard are (a) the lack of face-to-face interaction between present participants in a shared practice of collective decision making and (b) the lack of reciprocity between the roles of speakers and addressees in an egalitarian exchange of claims and opinions. Moreover, the *dynamics* of mass communication are driven by the power of the media to select, and shape the presentation of, messages and by the strategic use of political and social power to influence the agendas as well as the triggering and framing of public issues. (Habermas, 2006)

8.4.4 IT in Numbers

Some fairly fragmented data on the growth and current extent of IT, defined as all applications of technology that support the generation, acquisition, transmission, processing, and display of information, is provided for reference. To get a feel for the magnitude of information flow and of its rate of increase (average yearly rate shown in bold font), here are some values presented in Hilbert and López (2011):

World's technological capacity to store information (in exabytes = 10^{18} bytes):

1986	2.6	**25%**
1993	15.8	
2000	54.5	
2007	295.0	

World's technological capacity to receive information through one-way broadcast (in exabytes):

1986	432	**7%**
1993	715	
2000	1200	
2007	1900	

World's effective capacity to exchange information through two-way communication networks (in exabytes):

1986	0.281	**30%**
1993	0.471	
2000	2.2	
2007	65	

The above data values assume optimally compressed data, but that does not need to concern us here. What is of interest is that the yearly increase, shown in bold for each group of data, in the broadcasting of information, as in radio and TV, is much less than for interactive communications, a reflection of the decreasing cost of transmission. And monthly internet traffic, which grew from 1 exabyte in 2004 to 21 exabytes in 2010 (an estimate provided by Cisco), or a yearly increase of about 66%, is an indication of the importance of being able to access content. But irrespective of the accuracy of these estimates and of their detailed interpretation, they demonstrate the fact that the flow and availability of information, represented by μ and μ_a in our model, are increasing exponentially.

A different view of the application of IT emerges if we consider it from a business point of view, as the IT industry, and characterise it in economic terms. A problem with this is that the term "IT industry" is not very sharply defined, but we might start with a research report provided by CompTIA (https://comptia.org, drawing on various sources), which views the IT industry as consisting of the following components:

IT Hardware (29%)

- Servers
- Personal computers
- Storage
- Smartphones
- Tablets
- Network equipment
- Printers and other peripherals.

IT Services (18%)

- Planning and implementation
- Support services
- Operations management
- Training.

Software (12%)

- Applications
- System infrastructure software.

Telecom Services (41%)

- Fixed voice
- Fixed data
- Mobile voice
- Mobile data.

The percentages given for each major component are out of a total global revenue of US $3.8 trillion in 2016, which could be compared with a global GDP of US $78.6

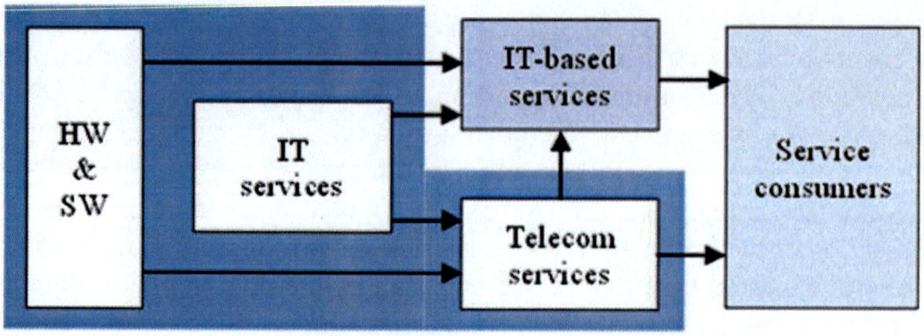

Fig. 8.6 The entities involved in IT-related activities

trillion. However, it is important to be clear about what the 3.8 trillion refers to, as the following figure, Fig. 8.6, demonstrates:

In Fig. 8.6, the block "IT-based services" is by far the most significant one, and it is composed of a number of different types of organisations. One is the online retailer, of which Amazon is the most prominent representative, with a revenue of 100 billion and a capitalisation of 482 billion US$. The company sells directly from its own warehouses, but does also provide an online marketplace for some other retailers. The online marketplace is the second type of organisation, of which Alibaba (capitalisation of 455 billion US$) and eBay are prominent representatives; these companies act as middlemen between buyers and sellers and get their revenue mainly from advertising on their various web sites. A third type are search engines, such as Google (revenue 70 billion and a capitalisation of 600 billion US$), that provide a free service to its users, paid for by advertising. The fourth type are the various social media, such as Facebook (revenue 35 billion and a capitalisation of US$ 500 billion), which get their revenue from advertising. A fifth type consists of a wide range of specialised information providers, often in the form of newsletters or reports, but also as online versions of journals and newspapers; partly free, partly by subscription or purchase, and mostly with additional revenue from advertising.

There are, of course, a whole host of companies providing specialised online services, such as email marketing (4.5 billion US$ yearly revenue, according to Transparency Market Research), accounting, tax returns, and the like. And, finally, there are a number of companies providing online payment services, obtaining their revenue from a percentage of the payment amounts.

The three blocks in the dark area of Fig. 8.6 are the ones accounting for the 3.8 trillion US$, and out of this sum the 41%, or US $1.56 trillion, of the hardware and software block presumably represents the gross fixed capital formation of the IT infrastructure of the telecom and IT-based services industry.

The three main points to take away from this are that, through their business activities, these companies provide a vast amount of information, a considerable part of this information is in the form of advertising (that is, its purpose is to persuade), and their combined capitalisation makes them a significant part of the economy.

A final indication of the present role of IT-mediated information exchange in our society is the amount of time we spend on IT-based activities, including fixed and mobile voice and data, TV, and radio. There are various estimates, e.g., from eMarketer (www.emarketer.com/ accessed 20.06.2022), and while they are fairly poorly defined, it would appear that in the US people spend several hours per day watching TV and as much as 4 h per day on mobile devices, of which 1.5 h on social media and 2.5 h on apps. This is certainly considerably more than they spend on face-to-face interaction.

In summary:

a. For the average person in the developed world, information via some form of technology-supported medium has become the dominant form of information input—the dominant component of μ in our model—and the time spent in interacting with these media is taking up a significant and increasing part of the time available to us for processing information.
b. Advertising is a large and increasing part of the information input.
c. The ability of advertising, and other forms of information aimed at persuading the individual, to catch the attention of the individual is increasing due to personalised targeting, with the data required for this targeting mined from the individual's activities with electronic media.
d. A further effect of the personalised targeting is the increased persuasiveness of the messages.

Chapter 9
The Role of Education

9.1 A Brief Review

9.1.1 A Simple Taxonomy of Education

The importance of education in supporting both the maintenance and further development of society has been recognised for a long time—at least since the ancient Athenian society. But the understanding of what the word "education" means, including such aspects as its purpose, how it realises this purpose, the degree of its importance, who is involved in it, and what activities it comprises, has varied greatly over time and in different societies, as exemplified by Athens and Sparta. For our purpose it will prove useful to characterise education according to its purpose and the framework in which it is delivered, resulting in four main types, as shown in Fig. 9.1.

In the classification of purpose, "Social" signifies that the purpose of the education is for the benefit of society; for the recipient to become a productive citizen, in the sense of providing an economic return, as well as in the sense of providing an input to the stability and orderly operation of society. "Personal" signifies that the purpose is to develop the individual's sense of uniqueness and that the nature of the individual's relationship to society is more than just as a cog in the machinery; it should be a conversation with society. Thus, the purpose could be expressed as developing the individual into a valued partner in this conversation, and it should enable self-fulfilment and the enrichment of the person, allowing the individual to appreciate and contribute to the culture in which it is embedded, without necessarily having any direct economic effect. In the classification of framework, "Institutionalised" indicates that the education is provided within a framework institutionalised by society and with attainment resulting in a formal degree or qualification recognised by society, and "Open" encompasses all other forms of information input with the potential to influence a person's behaviour.

The two "coordinates"—framework and purpose—were chosen here because the framework can be related to how technology mediates the education process, and the purpose is a determining factor in how education influences the evolution of

E. W. Aslaksen, *The Evolution of Society*,
https://doi.org/10.1007/978-3-031-23054-7_9

	Purpose	
Framework	Social	Personal
Institutionalised	A	B
Open	C	D

Fig. 9.1 The four types of education, characterised by the purpose of the education and by the framework within which it is delivered

society. This choice is by no means unique; in particular, it completely ignores the whole issue of *how* education is provided, i.e., pedagogy. Nor are the two coordinates completely orthogonal, but they will provide a useful space in which to analyse the relationship between education and the evolution of society. A different classification is one according to the content of what is being transmitted in the education process. At the basic level it is *facts*—anything from the names of things to the values of their parameters. At the middle level, it is knowledge about how to *apply* the facts, e.g., how to design or create something—from a cake to a space station. At the top level, it is knowledge about how to *evaluate* the facts; *critical thinking*, to which we shall return shortly.

To relate this high-level taxonomy of education to the existing body of knowledge on education and its place in society, a good place to start is with the overview provided by four essays by Emile Durkheim, published in the early part of the twentieth century, and published in English by The Free Press under the title *Education and Sociology* (Durkheim, 1970). As a sociologist, Durkheim approached education as a characteristic of society, the means by which it secures its own existence, and as a result, each society has its own form of education. However, while he recognised that education could be extended throughout a person's life by social interaction, his focus was narrower, and he defined education thus:

> Education is the influence exercised by adult generations on those that are not ready for social life. Its object is to arouse and to develop in the child a certain number of physical, intellectual and moral states which are demanded of him by both the political society as a whole and the special milieu for which he is specifically destined. (*ibid* 71)

Durkheim elaborates on this by saying that we can differentiate between two sources of our behaviour: one is made up of the mental states that apply only to ourselves and to the events of our personal lives, the other is made up of the ideas and practices that reflect the society in which we exist, such as religious and moral beliefs, and generally collective opinions of every kind. It is these latter items that form the social being, the development of which is the goal of education; "It is society that draws for us the portrait of the kind of man we should be, and in this portrait all the peculiarities of its organization come to be reflected." (*ibid* 123).

This emphasis on education reflecting the needs of the particular society, which he illustrates with the development from primitive, tribal societies, through Greek, Roman, Middle Age, and Renaissance societies, up to modern society, is compelling. But he does not enter into why society developed, what drove this development, and when he describes "the egotistical and asocial being that has just been born", he seems to miss the fact that it is this being that has created our present society. In the view expressed in our model, besides its physiological features and capabilities, this being is born with an information processing capability that, generally speaking, is capable of evaluating externally supplied information based on stored knowledge and a single criterion: survival. What constitutes survival, and the strategies for achieving it, is developed through interaction with the individual's environment and, by our unique ability to communicate, this knowledge has been transmitted from generation to generation, continually enriching that environment until arriving at today's complex society, in which survival is not only survival of the individual, but increasingly the survival of society itself.

However, Durkheim then goes on to effectively reduce the importance of this initial nature by stating "Now, if one leaves aside the vague and indefinite tendencies which can be attributed to heredity, the child, on entering life, brings to it only his nature as an individual. Society finds itself, with each new generation, faced with a *tabula rasa*, very nearly, on which it must build anew." (*ibid* 72). And he comes back to this point in several places; e.g., "But fortunately one of the characteristics of man is that the innate predispositions in him are very general and very vague." (*ibid* 82), and "But only very general, very vague dispositions, expressing the characteristics common to all individual experiences, can survive and pass from one generation to another." (*ibid* 84). With these qualifications, his view essentially conforms to our simplification in Sect. 1.3.3 and affirms the pre-eminence of education over heredity.

With regard to the types of education defined in Fig. 9.1, Durkheim's concept of education is definitely as an institutional activity, provided by professional teachers. However, it is when it comes to the purpose of education that the relevance and usefulness of his analysis to our model requires an amount of interpretation. He recognises two components of education—one with general content applicable to all, and one with specialised content, subdivided according to professions, or to "the special milieu for which he is specifically destined"—but the purpose of both is social, in the sense of providing a dividend to society, and together they create the social being. The specialised component is clearly in the social column, i.e., of type A, but the purpose of the general component is open to some interpretation; in particular, given the context in which it was provided (France around 1900). Durkheim states that the purpose of the general content is to make a person conform to the sentiments and institutions of the particular society, to make the person a good citizen. "Education is, then, only the means by which society prepares, within the children, the essential conditions of its very existence." (*ibid* 71). This places the general component also in type A, but if we reformulate the purpose as "make the person a valuable contributor to the development of society", the component would be classified as type B. With this, we could interpret Durkheim's general component of education as concerned with the strategy for survival and, as the information processed by the two components of education corresponds roughly to our classes 2 and 3, we can say that the purpose of the general component is to create the individual's identity. Throughout this collection of essays, Durkheim focuses on the importance of the general component, and it is this that makes his observations so relevant to our model.

9.1.2 The Interaction Between Education and Society

The first thing to note is the mutual interaction between society and the individual, with education as the most important mediator:

> Whereas we showed society fashioning individuals according to its needs, it could seem, from this fact, that the individuals were submitting to an insupportable tyranny. But in reality, they are themselves interested in this submission; for the new being that collective influence, through education, thus builds up in each of us, represents what is best in us. Man is man, in fact, only because he lives in society. (*ibid* 76)

> Thus, the antagonism that has too often been admitted between society and individual corresponds to nothing in the facts. Indeed, far from these two being in opposition and being able to develop only at the expense of the other, they imply each other. The individual, in willing society, wills himself. (*ibid* 78)

Except for the phrase "what is best in us", which seems to be in conflict with the *tabula rasa* view of the newborn child, this expresses the idea of the ongoing interaction between individual and society. The role of education is not to make the individual conform to a fixed set of rules, but to ensure that the accumulated wisdom provides the foundation for each individual to contribute to society as it evolves from generation to generation. This evolution is not progressing toward some externally given ideal state; it is a progress without limit. We could therefore interpret "what is best in us" as the ability to discern what is most likely to support the survival of society; i.e., the collective intelligence.

Durkheim emphasized a historical approach to understanding education, recognising the changes that have taken place in societies over time, and the many distinct societies that have existed, and how each one placed different requirements on education. In particular, on how to ensure that the individual developed into a social being that was able to contribute to the maintenance of a stable society. But there is much less, if any, emphasis on what characteristics the individual should have in order to be able to contribute to the *dynamics* of society. Society does not present the individual with a static environment; the individual is continually exposed to fluctuations, which can be of economic, political, and social nature, and the manner in which it responds to the fluctuations determines the evolution of society. This issue is taken up by Richard S. Peters in the essay *What is an educational process* (Peters, 1967), where he says ".... activities like history, literary appreciation, and philosophy, unlike Bingo and billiards, involve forms of thought and awareness that can and should spill over into things that go on outside, and transform them.", and "All forms of thought and awareness have their own internal standards of appraisal. To be on the inside of them is to both understand this and to care. Indeed the understanding is difficult to distinguish from the caring; for without such care the activities lose their point." (*ibid* 78), and "(For an educated man) his knowledge and understanding must not be inert either in the sense that they make no difference to his general view of the world, his actions within it and reactions to it *or* in the sense that they involve no concern for the standards immanent in forms of thought and awareness, as well as the ability to attain them." (*ibid* 79).

Peters then proceeds to develop this issue in term of *principles.* "Understanding principles does not depend on the accumulation of extra items of knowledge. Rather it requires reflection on what we already know, so that a principle can be found to illuminate facts. This often involves the postulation of what is unobservable to explain what is observed. So it could never be lighted upon by 'experience'." (*ibid* 18). Principles are close to what we have identified as the information items in the identity, and the salient concept here is *reflection.* We introduced this in connection with our model, in Chap. 4, and it is central to the evolution of society. It is the process that allows us to make sense of information about changes to society and to take adaptive action, where required. The process of reflection has two sides to it; one, establishing relationships between information items and forming principles, and two, assessing the applicability and correctness of existing principles and assumptions, both as they stand and in the light of new information items. Regarding this latter activity, *critical thinking*, Peters makes two observations. The first is that while it is desirable to acquire a critical attitude, critical thinking is no substitute for action; only the two together lead to effective improvement. The second is that critical clarification of principles is a very different exercise from applying them in concrete circumstances, which requires judgement, something related to a critical attitude, but developed mainly through experience, rather than through a thought process about principles. These issues, as they relate to education, are also treated in the last chapter in the same book, *On teaching to be critical* (Passmore, 1967).

9.1.3 The Concept of Bildung

A more recent book, *Educating Humanity: Bildung in Postmodernity*, (Lövlie et al., 2003) treats reflection and critical thinking under the perspective of the concept of *Bildung*, a product of the neo-humanist tradition that flourished in Germany between 1770 and 1830, and expressed the ideals of the Enlightenment. It is related to the concept of an educated person, as contrasted with a trained person, but *Bildung* has two sides to it. One is the content, in the sense of knowledge—what constitutes "a good education"—as in a list of what one should have read and studied; the other is a capacity for discerning the reality behind what presents itself as necessary, natural, general, and universal; an understanding of how the world operates. As such, *Bildung* as a process was very much one of self-*Bildung*, one of a personal quest to achieve rational autonomy, or, in terms of our model, one of developing one's identity. It could be seen as developing a capacity for critical thinking, except for the fact that the concept of *Bildung* was based on the assumption that there were concepts and principles that were general or universal, and also enduring, and the assessment of the world was tied to this measure. As this assumption came under increasing scrutiny in the nineteenth and twentieth centuries and the interest shifted to more down-to-earth behaviouristic concepts of capabilities, the concept of *Bildung* lost much of its relevance, but the authors of this book examine to what extent and in what form it can be resurrected, given that it contains much of value.

Relaxing the requirement for the core concepts and principles of Bildung to be enduring, we can still ask if there are, at each point in time, concepts and principles that are characteristic of society, in the sense of forming a basis for the evaluation of everyday experience. This is the question addressed by Gert Biesta in *How General can Bildung Be? Reflections on the Future of a Modern Educational Ideal* (Biesta, 2003), and he approaches it by considering two different answers. The first is the *epistemological* answer, in which the general is understood as universal; the second is from the *sociology of knowledge*, in which the general is understood as a social construction. In the first, knowledge is considered to be general if it is valid and can be applied everywhere and at any time; it is an intrinsic quality of knowledge and is taken as proof that it represents reality 'as it is'. The existence of such knowledge is demonstrated by the omnipresent success of technology based on science; what Biesta calls *the technology argument*. In the second, the general is only an expression and product of social relationships, and hence an expression and reinforcement of the way in which power is distributed at a certain moment in time, so that the general is in fact a manifestation of the particular at a specific time.

Biestra provides valuable insight into problems associated with both of these answers, as well as with an extension of the second one—the *critical theory* of *Bildung*. But, in view of these problems, he then goes on to propose a different understanding of the general, based on work by Bruno Latour. As with some of the other work of philosophers of technology, this work of Latour's appears highly artificial to an engineer and reflects a poor understanding of the relationships between science, technology, engineering, and industry. In the context of our model, sociology of knowledge (all knowledge in class 3 is a sociological product) together with critical theory, as described by Biestra, provide an adequate definition of the process side of *Bildung*. The involvement of critical theory is further elaborated in *Bildung and Critical Theory in the Face of Postmodern Education* (Ilan Gur-Ze'er, 2003). "This article attempts to show that the prospects for today's resistance to the current process of dehumanisation, which flows from normalising education, are closely bound up with the project of *Bildung* and its re-articulation by the critical thinkers of the Frankfurt School." (p. 75). A central point in this re-articulation is the autonomy of inwardness of the individual and an emphasis on the importance of reflection as a precondition for, and in a certain sense the first manifestation of, the resistance to the hegemonic order of things. Transcendence here takes on almost the opposite to its religious meaning; it is the possibility of overcoming the normalisation and hegemonic order.

The chapter *On Irritation and Transformation: A-telelogical* Bildung *and its Significance for the Democratic Form of Living* (Reichenbach, 2003) contains further insight into the meaning and relevance of *Bildung* in today's society (p. 93). He starts out by saying "The question of the meaning of the (old) idea of *Bildung*—as the refinement of intellect, sensibility and judgement—in a postmodern society can be approached by starting with a characterisation of postmodernism as a tired, even exhausted modernity; a modernity that has lost faith in its own dreams and promises (although it may nevertheless be able to evaluate these positively)." (p. 93). This may be a somewhat pessimistic view; the fact that the "dreams and promises", in the sense of a blissful end-state, has been abandoned, does not mean that the process part of *Bildung* has lost its relevance and importance. And Reichenbach acknowledges this when he asks "Could it be that there is only one plausible goal left to postulate by a late-modern theory of *Bildung*? I am referring to the individual's ability to initiate processes of self-transformation." (p. 95). That he sees self-transformation as a process without any hope of success, in the sense of not reaching an end-state, is really irrelevant; it is one of many processes whose success is measured by their ongoing productivity rather than by any change. Indeed, most, if not all, of the processes, both physical and mental, in the human body are of this type. The productivity of self-transformation can be seen as the opposite to acting through habit and "everyday thoughtlessness", a term he attributes to Eugen Fink (p. 100).

Reichenbach introduces the concept of *self-deception*, and stresses the importance of accepting this possibility in two directions. In the inward direction, the individual needs to accept that, despite its best and sincere efforts, the view of life arrived at may be subject to valid objections; otherwise *Bildung* degenerates into dogmatism. In the outward direction, every individual, and thereby society as a whole, needs to accept and make allowance for self-deception and see it as an expression of society's vibrancy rather than as something resulting in unnecessary or disproportionally severe social sanctions that only serve to cause the self to harden (p. 100). This dual acceptance becomes a prerequisite for the proper functioning of what we have called the averaging process within the operation of the collective intelligence, and thereby for the democratic form of living.

As a conclusion to this very brief and selective foray into the sociology and philosophy of education, we find that the process part of *Bildung*, in its postmodern form, is the component of education that has the greatest impact on the evolution of society; it is the component that constitutes the major part of type B (and, as we shall see, type D) education, and that enables the process of reflection. So far, we have, more or less by implication, considered education in an institutional framework. However, the fact that the individual's interaction with society throughout its lifetime includes elements of education, and that, as a consequence, the identity keeps evolving, has always been the case. Indeed, before institutionalised education became significant, it was the main form of education. It was always recognised as a factor in *Bildung*, and from the introduction of printing it has gained in importance due to the enabling effect of technology, as we shall consider in Sect. 9.4.

9.2 Education as a Process

9.2.1 Lifecycle of the Individual

In order to be able to focus on the role of education in the evolution of society, which, as a transfer process from society to the individual, involves the dynamics of the information processing within each person, we need to distinguish between a macroscopic (social) view and a microscopic (individual) view. So far, our model has not explicitly distinguished between the two views, and the parameter α was a characteristic of both the individual and the social identity; now we must distinguish between what we have called *society's belief system* and the identity of the individual, with the former being the average of the latter. Even if society is unchanging, each individual progresses through its lifecycle, and to look at the information processing that accompanies this progress at the individual level, we need to identify individuals by their age. Denoting the life expectancy of persons in the society by T, we can, as far as education is concerned, distinguish three periods within this lifetime, as shown in Fig. 9.2.

The duration assigned to each of the periods in Fig. 9.2 should be understood as approximate and representative of the lifecycle in a developed society. The significance of each period is as follows, and the assumptions made in these descriptions constitute further simplifications of our model presented in Chap. 4.

PI is the period in which the individual's identity is initially formed through the acceptance of information received from authoritative sources. First from parents, and then through a formal, institutionalised education, i.e., schooling; both based on the existing state of the collective identity. The size of the identity rises to its full value, *w*; the value of α will depend on both the socio-economic structure of the society and on the schooling system and the content of its curriculum but would normally be high throughout the phase.

During PII and PIII, the individual is exposed to interaction with other individuals, with a resultant value of the level of conflict in the identity, β, but is also engaged in making its own observations and evaluations. This is the wider scope of education;

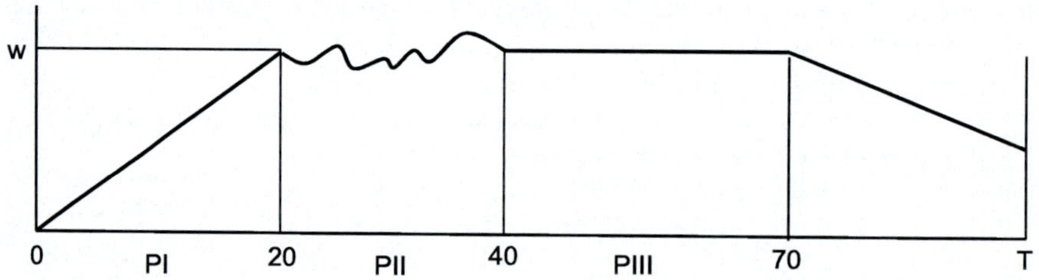

Fig. 9.2 The development of the identity over the individual's lifetime. PI is the period of schooling, PII is the period of adult development, and PIII is the period of intergenerational interaction. The timescale is indicative only

the individual develops a personal character and, in doing so, creates new information items, with the result that the value of α decreases. The individual provides the input to the schooling and development of the next generation, transferring both old and new information items developed during these phases, and thereby contributing to the evolution of society's belief system. However, in these phases the individual performs another, very important task: evaluating input received from other individuals and commenting negatively on those items that conflict with the individual's identity (which, of course, contains, in its α-part, the items common to all individuals). The effect of this, together with the averaging performed to create society's belief system, is to dampen any sudden, large changes to this belief system. The change to the individual's identity is less in PIII than in PII; this is due to (at least) three factors: The attention factor, x, introduced earlier, becomes smaller with age; a matter of *déjà vu.* The persuasiveness of new information items, our parameter p, is also reduced with age; we become more sceptical. And, finally, our personal questioning and investigating activity, our parameter μ_0, decreases; we become more resigned. As a consequence, and with reference to Chap. 4, we shall assume that α remains constant in PIII, and that the value of β is small.

In the process we have developed, the evolution of society is continuous, but in daily language we sometimes measure society's evolution in *generations*, implying a certain discontinuity or "bunching". In its daily usage the concept is not well defined; it is sometimes used to define a time period, and as such it is generally taken to mean 30 years, but it is now also used, particularly in the US, to describe persons borne in a specific time period, such as "generation X" (1965–1980). The concept may also be related to major events, such as in 'the inter-war generation", or to disruptive technologies, such as "the computer generation". In the essay *The Problem of Generations*, Karl Mannheim discusses various aspects of the concept; in particular, that besides this unilinear quantitative meaning, it can also have a qualitative meaning, in the sense of intellectual evolution and personal experience (Mannheim, 1927).

However, it is obvious that the process of education introduces a time constant into the evolution of society, and we can easily discern two contributors. One is the duration of PI; as the extent of our knowledge increases, the time needed for reaching a level of competency that will allow the individual to perform satisfactorily as a member of society has been increasing, despite greater specialisation.

The other contributor is the duration of PII; a period in which the information acquired during PI is put to the test and modified by experience and new information arising from the interaction with other individuals. This process decays with time from the end of PI as the identity "hardens"; at some point becoming negligible and thus marking the end of PII. But the decay is counteracted by the increase in leisure time, combined with the improved access to information through information technology, with the result that there has been an increase in the duration of PII. It is interesting to consider that as the duration of the phases increased, due to both increasing life expectancy and complexity of society, the relative duration of the three phases remained largely unchanged, which would be interpreted to mean that society needs a balance between the functions performed by people in the different phases in order to stay functional.

9.2.2 *Inter- and Intra-Generational Information Transfer*

In the essays we referenced in Sect. 9.1, Emile Durkheim emphasized the nature of institutionalised education as a transfer of knowledge from one generation to the following one, and while this is being somewhat blurred by the decrease in teacher/student interaction in schooling, the availability of information for ongoing non-institutionalised education, and the educational role of social media, the delay inherent in this transfer is still an important factor in the dynamics of society. To get a feel for the effects of this delay, we can construct a highly simplified model of the information transfer, based on the individual lifecycle defined in Fig. 9.2 and, as we are only considering information in class 3, on the concepts of the individual identity and the social cohesion, α, as introduced in Sect. 3.4. In Fig. 9.2 we gave some indicative durations of the three phases; for simplicity, we shall now assume that the three phases have the same duration, L. This allows us to divide time into segments of duration L. During one segment, the institutionalised education transfers the content of the α-part of the identity of the individuals in the phases PII and PIII (society's belief system) into the identity of the individuals in phase PI. Let U(T) be the information received by individuals in PI, V(T) the information transferred from individuals in PII, and W(T) the information transferred from individuals in PIII. If there is no new information generated *within* a phase, then $V(T+1) = U(T)$, and $W(T+1) = V(T) = U(T-1)$; that is, education only preserves the *status quo*, and society does not evolve. But in Sect. 4.1.1 we introduced the activities that result in the information flow we labelled as *discovery* in Fig. 4.1. Let us denote the information generated from these activities during the time segment *n* in phase *i* by $D_i(n)$, with $i = 1, 2, 3$, but assume that the *amount* of this information is the same in each phase and time segment, and denote it by D. Furthermore, as the size of identity, *w*, is constant (as we explained in Sect. 3.3), and, if the evolution is in a stable state, α is also constant, it follows that the amount D *replaces* a fraction of the information from the previous phase, reducing that amount of information by a factor *r*. Now, let the information in each phase in the initial time segment be identical and its amount be X, so that $r \cdot X + D = X$, and let the phase-internal activities be switched on at the beginning of time segment 1; we then get the temporal development shown in Fig. 9.3.

The temporal development displayed in Fig. 9.3 can be characterised by how the original content of the identity, i.e., of society's beliefs, in the material that is propagated by education decays over time; this is the proportion of X in U(T). If we denote that proportion by R(T), Fig. 9.3 allows us to deduce that

$$R(T) = \frac{r^T}{2 - r^{(T-1)}}. \tag{9.1}$$

T	U(T)	V(T)	W(T)
0	X	X	X
1	$rX+D_1(1)$	$rX+D_2(1)$	$rX+D_3(1)$
2	$r(rX+D_2(1)+D_3(1))/(2-r)$ $+D_1(2)$	$r(rX+D_1(1))+D_2(2)$	$r(rX+D_2(1))+D_3(2)$
3	$r(r^2X+r(D_1(1)+D_2(1))+D_2(2)$ $+D_3(2))/(2-r^2)+D_1(3)$	$r^2(r^2X+r(D_2(1)+D_3(1)))/(2-r)$ $+rD_1(2)+D_2(3)$	$r^2(rX+D_1(1))+rD_2(2)+D_3(3)$

Fig. 9.3 The evolution of the α-part of the identity from an initial content of X under the influence of the phase-internal processes of reflection and discovery

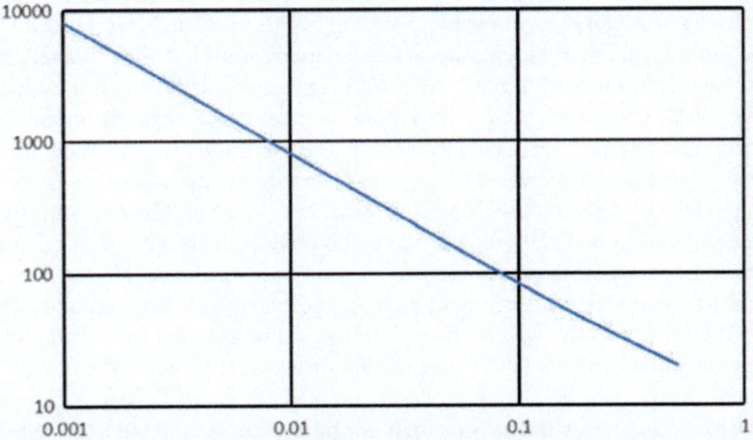

Fig. 9.4 The half-life of the α-part of the identity—society's belief system—in years, as a function of the intensity of the generation of new information during a phase through reflection and discovery, measured as a fraction of the existing information. The duration of a phase, L, has been chosen as 20 years

We can then define the half-life of society's belief system as the value of T that makes R(T) = 0.5, and this is shown in Fig. 9.4 as a function of the intensity of the generation of new information, 1 − r, measured as a fraction of the existing information, for a choice of the unit of T (i.e., the duration of a phase) of 20 years.

9.3 Education and the Model of Society

In order to relate the evolution of society's belief system through the institutional, or schooling, part of education depicted in the previous section to the highly simplified model of society and its evolution we developed in Chaps. 4–6, we need to consider two issues. The first one is that the ageing process, which transfers information from one generation (value of T) to the next, is a continuous process, but in Fig. 9.3 it is shown as a batch process. For example, in the cell V(2), the information rX +

$D_1(1)$ is transferred from U(1) at the beginning of the time period T = 2. Then, during that period, the discovery and reflection activities that are performed by the individuals in phase PII of their lives result in new information, and the exchange and evaluation of this new information between the individuals during the phase results in the creation of new information in the α-part of the identity, $D_2(2)$, which replaces the same amount of information in the information present at the beginning of the phase. Thus, this updated information, which becomes part of the basis for education in U(3), has already undergone the evaluation and averaging process we identified as the operation of the collective intelligence, and it is this process, which we can consider to be society's immune system, which has a stabilising influence on the evolution of society. Any new idea, or item of information, even if meeting with approval in a local environment, needs to win approval of the wider society before it enters the curriculum in PI; the "price" of this process is the delay it introduces, in the order of L. The rate at which the averaging takes place depends on the flow of information, μ_1, and on the persuasiveness, *p*. But our model is concerned only with the steady state, so that the value of *p* is what is needed to maintain a constant value of α in the face of a given flow of new information, μ_0, as we shall see shortly.

The second issue is that, while the education process deals with persons and their lifecycles, the model considers society to be composed of identical individuals, with the value of a parameter of an individual to be the average of that parameter over all adults in the real society, which allowed us to assume that the size of the identity, *w*, stays essentially constant. This means that, if we want to relate the model to the three-phase view of the education process presented above, we have to divide society into two sub-societies. The one, to which our model can be applied as it stands, is a society comprised of individuals based on persons in PII and PIII, and this is where the evolution of society takes place, where new information is absorbed into the identity. The other sub-society, based on persons in PI, is dedicated to education, and while the model concepts still apply here, the model would have to be completely changed to account for the behaviour in this phase. Also, the amount of evolution taking place in PI, although assumed to be identical in volume to that in the other two phases, i.e., D in all three phases, only becomes effective once it transitions into PII.

This process taking place in the PII/PIII society, in which α stays approximately constant, is what we consider to be the normal evolution of society. If α = 0 there is no society, and if α = 1 there is complete stagnation; essentially, a dead society. A healthy society would have a value of α somewhere around 0.5, and any significant fluctuation from this value would indicate a society in distress. If we now go back to Sect. 4.3 and Eqs. 4.1 and 4.3, and, as in Sect. 5.3.2 set ν = 1, we have

$$\alpha\mu_0 = p\mu_1(1-\alpha), \tag{9.2}$$

and then, with

$$\mathrm{D} = L\cdot\alpha\cdot\mu_0 = L\cdot\mu_1\cdot p(1-\alpha) \tag{9.3}$$

and with $X = \alpha \cdot w$, we obtain

$$1 - r = \frac{D}{X} = \frac{L\mu_1}{w} \cdot p\frac{1-\alpha}{\alpha}, \tag{9.4}$$

which is what we would expect. For a constant value of α, the product of the volume of new information produced in the duration of one phase, L, relative to the size of Θ, and the persuasiveness, *p*, determines the rate of change of the society's belief system.

Perhaps the most important aspect of these relationships becomes evident if we consider what we have been calling persuasiveness in more detail, and which we introduced in Sect. 5.2.3 as a defining characteristic of an information item. It is not a sharply defined concept, and while it is always related to the process of changing people's minds, it may be a measure of the effort that goes into it, but it may also be used as a measure of the efficiency of the process, or of its success rate, and so on. In a face-to-face situation, persuasiveness is comprised of many personal factors, such as tone of voice, facial expression, body language, and the use of humour; generally, what is included in the concept of charisma. But as information transmission is increasingly mediated by numerous applications of technology, it is the means technology offers for increasing persuasiveness that becomes the dominant factor. There are numerous physical channels, both fixed and mobile, that allow the same message to be received by a significant proportion of society. And while the most recent developments are in the area of telecommunications, other means, such as printing (newspapers, flyers, billboards, etc.) have been around for some time. The format of messages is expanded from voice only to voice, images and videos, and music, as well as any combination of these, and most recently it is also possible to deliver the message via a two-way, or interactive, process. And the cost of repetition, which is a common means of increasing persuasiveness, is reduced. But perhaps the most significant aspect, as far as persuasiveness is concerned, is the ability to modify or edit the message in order to, on the one hand, make it resonate with characteristics of a particular audience, and, on the other hand, modify original content so as to change its meaning, e.g., by selecting only part of the content, or by rearranging the sequence of components of the content to reverse cause and effect, as was described in Sect. 8.4.3. We can therefore ascribe a main part of the exponentially increasing rate of churn within society to the role of technology in increasing not only the flow of information, but also its persuasiveness.

There is another characteristic of the information flow (μ in the model) that is related to its persuasiveness, and that is its ability to capture the *attention* of the individual (the factor *x* in the model). Only a relatively small proportion of the information impinging on the individual is noticed and processed in any meaningful way, and it is a main purpose of advertising and any other form of thought-manipulation (including teaching) to increase this proportion; only then can persuasion commence. Here again, information technology plays a major role, often by the ability to combine different information items in the same message, e.g., creating a captive audience through apps that combine information with advertising, and so on.

Turning to the other sub-society—the one formed by the persons in PI of their lifecycles—the situation is quite different, in that rather than a small part of the information, D, being new, all the information is new. And instead of the main process being a circulation of information, it is mainly a one-way path from teacher to student, steadily increasing the size of the identity, w. At any point in PI, α will have a particular value, and traditionally the aim has been to achieve a relatively large value, which, we now know, requires greater effort. However, the nature of the persuasiveness in this phase is somewhat different from what it is in PII/PIII. There persuasion meant changing an existing belief; in PI it means getting the individual to accept the importance of having a belief in the first place, and then to understand the importance of a shared belief structure in maintaining a viable society.

9.4 Education and Technology

As we did in the previous section, we can start our consideration of the relationship between education and technology by introducing a structure to the relationship, and the first thing to note is that the relationship is a mutual one. Education influences technology (E → T), and applications of technology influence education. This latter influence has two distinct aspects: One is direct, in that Information Technology (IT) *enables* many forms of providing education (T → Ee). The other is indirect, in that technology *supports* the type of society and standard of living that allow people to engage in education (T → Es). Hence, we can identify twelve elements of the relationship, as shown in Fig. 9.5.

A brief description of each element is as follows:

(a) All applications of technology require some form of knowledge and skill, and the earliest institutionalised form of transmitting this from one generation to the next would have been workshops, often organised into various forms of guilds. Today, the development and application of technology is completely dependent on a suitably educated workforce, from engineers to skilled tradespeople, and the dependence of technology on education is reflected in the current emphasis on STEM subjects.

(b) Depending on the exact definition of the Personal purpose of education, we could say that this element includes the influence of ancient beliefs on technology, in the form of temples, pyramids, and other structures and artefacts expressing personal beliefs as propagated through religious education. But otherwise, the

Fig. 9.5 The twelve elements of the interaction between education and applications of technology

	A	B	C	D
E→T	a	b	c	d
T→Ee	e	f	g	h
T→Es	i	j	k	l

significant influence only arose in the last century or so, with institutional education starting to create a critical awareness and assessment of the role and effect of various technology applications until, today, attitudes to technology applications have become part of people's belief structure and effect a corresponding influence on these applications (e.g., nuclear energy, GM foods, global warming, etc.).

(c) One might think that this element would simply be a reduced version of element *a*, to the extent that Open education is (still?) a very much smaller component of education than the Institutional one, but this would be overlooking the fact that they are located in different periods of the individual's life. While the Institutional component is located predominantly in PI, the Open component is located predominantly in PII; that is, a period where experience is starting to reveal shortcomings in the state of technology. The effect of education in this period is therefore likely to be biased towards the development of new technology rather than towards the aims of element *a*.

(d) Much the same situation as in element *c* above, except that the subject matter is now not technology itself, but a critical examination of the influence of technology applications on society, and also that the experience and maturity of the individual is likely to amplify the effectiveness of this element.

(e) This is the element most immediately associated with the concept of technology in education, and it has undergone a rapidly accelerating growth in the last decade. It offers the possibility of tailoring the education to a diverse student population (instead of e.g. having selective schools), of a more interactive delivery mode, and of flexibility of timing and location. An overview of these and other possibilities, as well as of some associated problems, are contained in (Herold, 2016), and some thoughts on the current enthusiasm are presented in (Selwyn, 2016). Furthermore, the increase in the time spent on education noted in Sect. 3.4.2 is to a large extent the result of the demands of technology.

(f) This is probably the element least influenced by technology; if anything, it is a negative influence. It is the element where the personality of the teacher has the greatest effect.

(g) This element is very significantly influenced by technology; the Internet is presenting a plethora of sources of educational material that can be accessed outside of any institutional framework.

(h) Besides a great amount of low-cost printed material for self-education and self-improvement, there are innumerable websites in the form of blogs, discussion groups, wikis, etc. that purport to be concerned with developing the individual's capability for critical thinking, but they are often promoting a particular point of view or degenerate into a free-ranging chatter with little, if any, educational value. It is difficult to say to what extent technology, beyond the level that printing reached fifty years ago, has had an influence on this element.

(i) Institutionalised education depends on the ability of society to generate a surplus above what is needed for the existence of the individual, and this surplus is a consequence of the development and application of technology.

(j) There is no in-principle difference between this and the preceding element of the relationship; in practice, the economic aspect of these two elements may tend to place more importance on the former.
(k) In addition to the economic aspect (e.g. allowing people to pay for information and for IT services), there is the important role of technology in creating more free time (leisure time) in which educational activities can be pursued. The increase in leisure time is noted in Sect. 3.4.2.
(l) As for element j.

In the description of the interaction between education and technology above, it appears as a mutual interaction between two equal partners. But a moment's reflection shows that this cannot be quite true; education is a process of transmitting information, and that information must be created before it can be transmitted. That creation takes place within what we characterise as *industry*, in the widest interpretation of that concept; through scientific research flowing into the development and application of technology by engineers. Education then responds to a *demand* for the information created to be transmitted, both within the generation that created it as well as to following generations. The influence of education on technology is reflected in how well education responds to the demand, and that, in turn, depends to a certain extent on how well education utilises the means put at its disposal by technology. So, while the interaction between education and technology is complex, it is technology that is the driving force; this is just the current extension of Durkheim's analysis.

While the above is true in general, there is a difference in the relationship between the two frameworks in which education is delivered. Within the institutional framework, the institutions have a choice in what, if any, technology they apply in providing the education, and they can, at least in principle, resist attempts by the technology providers to pervert the education process for their own commercial or ideological purposes. (The use of technology by an autocratic government to pervert the education process is a different matter.) Within the open framework it is much more difficult to differentiate between what is the provision of information for educational purposes and what is, essentially, advertising or propaganda, and the information technology offers more opportunities for presenting a message rather than factual information. Nevertheless, information technology is making the open framework increasingly important for providing education in phase PII, and within that offering are unprecedented means of focusing on the Personal purpose of education. A search for enlightenment and personal development that would, even as recently as a century ago, have been possible for only a small proportion of the population, and mostly in academia, is today available to anyone with Internet access. How many avail themselves of this opportunity is a different matter, but it is at least potentially a great force for change.

9.5 Education and Economics

Realising that there is a strong relationship between education and technology, and recognising how production is totally dependent on applications of technology, we would expect there to be a strong correlation between the level of education and the change in GDP *per capita*. This is indeed the case, and more for completeness than anything else, we review some of the available data.

We start out with a paper by E. A. Hanushek and L. Wössmann, *Education and Economic Growth* (Hanushek & Wössmann, 2010), mainly because they open their analysis with the observation that it is not only the attained level of education or the amount of resources spent on education that determine the impact on economic growth, but also the quality of the education, as in its effectiveness in producing cognitive skills. A straight-forward correlation between years of schooling and GDP per capita growth suggests that each year of schooling is associated with long-run growth that is 0.58% points higher. But if the quality of the education is taken into account, which is now possible due to the increasing participation in standardised tests (e.g. Trends in International Mathematics and Science Study (TIMMS) and Programme for International Student Assessment (PISA)), it turns out that the variation in annual growth rates of real GDP per capita is explained mainly by the variation in education quality, with the effect of years of schooling greatly reduced.

Another measure is the education index, which measures a country's relative achievement in both adult literacy and combined primary, secondary, and tertiary gross enrolment, and it is clearly correlated with the per capita GDP, as demonstrated e.g., in (Habermeier, 2007, Fig. 5), which shows an exponential dependence of GDP on the education index. Further analysis in this paper shows that, for developed countries, the technology aspect of the education is dominant, and Habermeier concludes "The education in science therefore is a mandatory prerequisite for sustainable economic performance of a country." (*ibid* 66).

An interesting approach to studying the relationship between education and economic development, and one that is based on the explicit involvement of technology in this relationship, is described in a paper by Yong Jin Kim and Akika Terada-Hagiwara, *A Survey of the Relationship between Education and Growth with Implications for Developing Asia* (Kim & Terada-Hagiwara, 2010). They focus on total factor productivity (TFP)—the part of output not explained by the amount of inputs used in production—as the most appropriate measure for analysing the influence of education on the economy. This is based on Solow's observation that long-run growth in income per capita in an economy with an aggregate neoclassical production function must be driven by growth in the TFP (Solow, 1956). Denoting the TFP by A, the growth rate of the TFP is given by the expression

$$dA/dt = g_y - \alpha \cdot g_k - (1 - \alpha)g_h; \tag{9.5}$$

where

g_y = growth rate of aggregate output;

g_k = growth rate of aggregate capital;
g_h = growth rate of aggregate labour; and.
α = capital share.

As the values of these quantities are all captured in the national accounts, the growth rate of the TFP can be determined for each nation. The results of the analysis are presented using two further concepts: the *technology gap*, which is the difference in the TFP of a nation and that of the most advanced nation (the TFP of the latter is called the *technology frontier*), and what we might call the *level of education*, as the ratio of tertiary education expenditure to primary and secondary education expenditure (the development in the paper goes via the relationship of education, or human capital, to technology, and is quite detailed). The main result is that, while TFP is positively correlated with expenditure on education, the influence depends on the relationship of the technology gap to the level of education. For a large gap, it is expenditure on primary and secondary education that is most effective, whereas for a small gap it is expenditure on tertiary education. But there is a limiting effect of the level of education, in that too high a level (implying too great a degree of specialisation) reduces the flexibility with regard to adopting new technology.

9.6 The Challenge for Education

Emile Durkheim's definition of the purpose of education, which we met at the beginning of Sect. 9.1, is formulated in terms of society's requirements on the individual, something he considered to be both relatively fixed and well-defined in the France of 1900. Today, and in a society like the Australian one, that is no longer the case, and so the question of the purpose of education has been shifted into a question about the purpose of the individual. Durkheim's subdivision of the purpose into two parts, one general and one specialised, is still valid, but they both relate to the maintenance of (a static) society and are contained in our types A and C in Fig. 6.1—the social component of the purpose. We have introduced a second component, the personal component, consisting of the types B and D in Fig. 9.1, additional to what Durkheim saw as the purpose of education. What we now have here is one component relating to society as a process, and one component relating to the dynamics of this process. Crudely put, the former component is driven by our desire to survive, the latter by our view of what should survive.

In Australia, there is currently a great deal of concern, both publicly and in the education community, about the declining proficiency of Australian students in the subjects grouped under the acronym STEM (Science, Technology, Engineering, and Mathematics), and there are basically two reasons for this concern. One is that the world in which we live is increasingly dominated by applications of technology, so that to understand that world and participate, as citizens, in deliberations about its management and development, a good understanding of these subjects is required. Without this understanding, the public discourse degenerates into a fruitless emotional, ideological, and acrimonial exchange. The other, more directly practical and measurable reason is that the performance and competitiveness of our economy is correlated with our performance in these subjects, as we saw in the previous section. Important as this concern is, it inevitably raises the more general issue of the purpose of education, and of the allocation of limited resources to the various subjects and structural elements that make up education. In particular, it raises the issue of the role of the individual in society, and of how education should prepare the new generation for this role. This aspect of education seems to become somewhat lost in the current debate, and in the context of our view of the evolution of society this is a major concern.

This perception is by no means new or original, and it typically appears under the heading of sociology of education. A paper by Julie Matthews (Matthews, 2013), although focused mainly on some specific areas, such as gender, sexuality, ethnicity, and multiculturalism, gives a good overview of sociology of education in Australia up to 2010, and raises the various aspects of this perception. The historical trajectory goes from an initial focus on the practical problems of teaching, via increased interest in the relationship between education and social change, to a bureaucratic preoccupation with narrow utilitarian objectives and with testing and test results as the determinant of success.

Staying for the moment with Durkheim's subdivision of the social component, we realise that there has been a very significant change in the ratio of the information associated with the two parts of this component since his time. The information within the specialised part has been increasing at a rapidly accelerating rate, whereas the information belonging to the general part has changed hardly at all in volume. One popular measure (although of somewhat uncertain accuracy) of the accelerating rate is the time it takes for the amount of information to double; before Durkheim's time it was about 100 years; today it is estimated to be around 12 h. It is clear that this must have a profound effect, not only on education per se, but on our relationship to information in general, and the current situation is characterised by three main approaches. The first one is an increasing degree of specialisation; by narrowing "the special milieu for which he is specifically destined", the growth of the knowledge required to perform a profession is contained. But this specialisation has a downside, in that the communication between specialisations becomes more difficult and less effective; the interfaces between specialisations increase in number and present an increasing overhead. The second one is to increase the time spent on education, prior to entering the workforce, as documented in Sect. 3.4.2. However, as the individual's lifespan is limited, and the energy level declines at some point after the half-way mark, this increase can only have a very limited effect. And there is a further problem, in that the amount of information that can be usefully handled by the brain is limited; with too long an education period, what was learnt at the beginning has been forgotten at the end.

The third, and by far most promising way forward is to utilise the capabilities of the technology that is causing the information explosion in the first case, and to form a hybrid which combines human and artificial information processing capabilities in an optimal way. But there is another change looming for education, and it is, in one sense, the core of what this essay is leading up to. It arises as an integral feature of the evolution of society and is, perhaps, best approached from the perspective of the individual's understanding of its role in society, which will be taken up in Chap. 10. There, we propose that a measure of this evolution could be the ratio of the individual's perception of its purpose as being the advancement of its own interests to it being the advancement of the interests of society, and while the former is inherent in our nature as individuals, the latter is acquired through education (in our wide interpretation of the concept). The problem we are faced with—the effects of which we can observe all around us—is that the rate of change of society, and the increase in its complexity, has led to a situation where the *actual* integration of the individual in society has outstripped our understanding (or acceptance?) of it. That is, the value of the above-mentioned ratio is higher than it should be for a harmonious operation of society, and this is where education comes in. In particular, the types of education we identified as Personal, i.e., B and D in Fig. 9.1. This may at first seem paradoxical: is it not the social types we need to promote? But this apparent paradox disappears once we recall the definition of the types. The social part of education, types A and C, has the purpose of preparing the individual to be a valuable contributor to society, to its maintenance and its growth. It is a view that sees the individual and society as two separate entities; society is something given, with certain well-defined characteristics, such as structures, institutions, and beliefs, that allow education to be fashioned so as to fit the educated individual into it. This is, essentially, Durkheim's view, and one that has maintained much of its influence until today. It is a situation very similar to thermodynamics, which, despite the "dynamics", relies on the change being so slow that it is a good approximation to assume that the system under consideration is in equilibrium at each point in time. For society today, as a complex, dynamic system, this is no longer a good approximation. The rate of change is such that the change within an individual's lifetime is significant, so that a decoupling of the dynamics of the individual from that of society is no longer appropriate.

At this point, let us just recall that what we are considering is the information of class 3—society's belief system. We are not dealing with the change to the rest of our knowledge base and the associated need for ongoing education throughout an individual's lifetime. That is very different; it is the response to a change that has taken place, whereas what we are considering is the individual's involvement in change itself, and education's new role is to prepare the individual for that involvement. It is not about acquiring information relevant to an expected future development of society, as it is impossible to know how society will develop; it is about acquiring and exercising an ability for critical thought, as discussed in Sect. 9.1.

This new role for education can also be seen as a shift in emphasis, from knowing to understanding. Until recently, the measure of intellectual prowess was the amount of knowledge a person could trot out at any time, and a profound thinker was often confused with someone who could produce appropriate quotations; now the knowledge and the quotations are readily available on demand. What we are experiencing is a change in our relationship to information, or knowledge. First came the accumulation of facts, knowledge being simply the knowledge of facts. Then came the application of facts, the knowledge of how to benefit from the accumulated facts, which developed through crafts to industry and the support of modern society. And now we are engaged in developing our understanding of the meaning of information, as something that emerges from the relationships between information items. This change is one of increasing complexity; our society has become so complex that only a holistic view of our knowledge and of the relationships contained within it will allow us to make correct judgements about its further evolution. And the means for operating on this system level of knowledge can no longer be the individual in isolation; it can only be the greatly increased processing power of individuals interacting through the free exchange of information—the *collective intelligence*. To give the individual the ability to participate effectively in this gigantic information processing venture by responding with a critical evaluation to the received flow of information is the new, additional task of education.

Chapter 10
Measures of Evolution

10.1 Some Basic Concepts

10.1.1 The Essence of Society

The problem of finding significant measures of the evolution of society, in the sense of new species of the genus *society*, was raised in Sect. 2.4 as a question: "What do we understand by evolution in the context of society?". That is, we are looking for a characteristic of society such that a (significantly) different value of this characteristic would distinguish different societies, and we would want this characteristic to arise as a feature of our model of society. Section 2.4 introduced social complexity as a measure, and there are other others, based e.g., on legal or technological status, so that it may well be that the concept of a measure of the evolution of society is a multi-faceted concept, like a cube with six sides, and which side we see depends on our perspective.

Section 2.3 discussed the evolution of society and pointed out that, if we define the state of society in terms of a set of variables, **S**, and denote the values of these variables at a particular time t by $\mathbf{S}(t)$, then we can trace the history of the state of society as $(t < t_0)$, where $t = t_0$ is the present. But we cannot find an expression for $d\mathbf{S}/dt$ that would allow us to integrate it and determine the future state of society, $\mathbf{S}(t > t_0)$, because **S** contains stochastic variables representing individual decisions. It is not even clear that a current set of variables would be relevant in the future, not least because of the unpredictability of the development of technology. What we have identified is the process that drives the evolution of society—the *transformation process*—and a criterion for characterising the effect of this process—*survival*.

This was also discussed in Sect. 3.3.1, and when we take it up again now, the purpose is to find a useful subset of **S**, where, by useful, we understand variables for which the values are readily available and that are significant in characterising the evolution of society. A starting point is to recall our concept of generalised energy with its two appearances as free energy and bound energy (see Sect. 2.1), and the increase in the ratio of bound energy to total energy as the general process of

E. W. Aslaksen, *The Evolution of Society*,
https://doi.org/10.1007/978-3-031-23054-7_10

evolution progressed. Reflecting this general idea onto society, where the energies are represented by processes, we recognised, in Sect. 2.4, that the primary driver of the evolution of human society is the interaction between its members, and that therefore the primary measures of social evolution must relate to this interaction. The ratio of bound to free energy, the ratio of commonality to individuality, and, in terms of our model, the value of α as the extent of the collective identity, are all measures of the individual's integration into society, and all related to the strength of this interaction. But, as the discussion about the value of α made clear, in Sect. 4.3, these measures must each have an optimal value; values that are, for the time being, determined by the nature of the species *homo sapiens*. Those optimal values are expressions of the optimal balance of the roles of the individual as an independent being and as an element of society, and finding a set of variables in **S** that allows us to define this optimal point is the Holy Grail in our quest to understand the evolution of society, and would form what we consider to be the *essence* of society.

To illustrate the concept of the essence of society, let the societies of interest be the 200 or so nation-states in which the world's population is contained, and let S be so detailed that the societies are all different, i.e., they differ in at least one of the variables in S. If we now perform a stepwise reduction in the level of detail of the description by ignoring the least significant variable, the set of societies will be subdivided into increasingly larger subsets of societies that are equivalent at this level of abstraction. Finally, all the variables have been stripped away except the ones defining the interaction between the society's members; this is the essence of society that cannot be eliminated without destroying the society and turning it into a collection of non-interacting individuals. At this high level of abstraction, we shall consider two societies to be different if they differ in this essence.

The essence we are searching for is not directly a measure of the state of development of a society, in the sense of going from primitive to a developed modern society. In a primitive society, say, an isolated tribe in the Amazon, the individuals may be highly integrated into society, both through their behaviour and through their beliefs, and the communal life may dominate any private life. It may even be difficult to discern any private life as completely distinct from the communal life. But in the type of society of interest to us—one of the 200 nation-states, or even the world society—we have a strong sense of the difference between our private lives and our involvement in society, and the understanding of what 'private' means and the strength of its influence on the fabric of society is a measure of the state of society. Compared with the primitive society the difference between private and communal is not only one of quantity; it is one of quality, they have a very different character. We see society in fairly abstract terms; we know it is there, we interact with it every day, and we pay to keep it operating, but we do not often feel any direct involvement with it. It is like air; we are surrounded by it, we breathe it all the time, but we do not have any conscious involvement with it, it is simply there. On a day-to-day basis our involvement, our thoughts, and our concerns are with our private lives—family, career, leisure activities; this is where we can have a direct involvement, exercise our will, and where we can achieve a sense of accomplishment. The fact that all of these activities depend, to varying degrees, on other people and on the environment provided by society is largely suppressed. There is a dichotomy here, already mentioned in Sect. 1.3.2: on the one hand we see ourselves as independent individuals, on the other hand we have created a complex society in which we are increasingly enmeshed. The understanding of this dichotomy has varied through history and between different cultures, and this variation can be observed in the world today, with the West displaying the least understanding, clinging to an ideology that favours the independent individual.

A contrasting understanding of the two sides to our existence is the Ubuntu philosophy, which was mentioned briefly in Sect. 1.3.2. It is a normative ethical framework for the relationship between persons, expressed in such statements as "I am because we are and I am human because I belong" and "a person is a person through other persons". Because of its focus on direct interaction between persons, it would only express a desirable relationship between the individual and society in a simple society, with a low degree of structuring and without the interaction being mediated by technology, such as pre-colonial African societies. But it is an important example of the recognition of the two sides to human behaviour in *any* society; it is an inherent feature of the concept of a society. For further information about Ubuntu, see (Ewuoso & Hall, 2019) and (Hailey, 2008).

Society's belief system and the value of α reflect the essence of society, but they are both difficult to measure because they are located in the information domain; it is much easier to describe and measure actions and their results than to determine what information led to these actions. The activities and their results arising out of society's belief system constitute the *common infrastructure*, first introduced in Sect. 1.2. It will be defined in the next subsection; here we recognise it as one measure of the essence of society, as an expression of the extent of the individual's relationship to society.

There is a further expression of the extent of the interaction, arising from the general observation that as the strength of the interaction increases, there comes a point where the structureless state is no longer stable and the individuals form subsets and thereby create a structured society, with the interaction within the subsets greater than the interaction between subsets. The nature of the subsets depends on the properties of the interaction, and as the interaction strength continues to increase the subdivision continues, creating subsets within the subsets, increasing the *structural complexity* of society. This concept will be defined in Sect. 10.1.3 below, but first we note that we encountered a closely related concept—*social complexity*—in Sect. 2.4. The difference between the two is mainly one of focus; social complexity focuses on how the interaction between individuals is structured, structural complexity focuses on the resulting physical structure, in the form of institutions and systems. And this difference leads us to a final comment regarding the essence of society: In Chap. 6 we developed various measures of society, but the purpose of those measures was to define the conditions under which the collective intelligence, as the driving part of the transformation process, operates within society. That is, the conditions determining the change of society, the $d\mathbf{S}/dt$—its direction and rate, as discussed in Chap. 2. The essence of society is only concerned with a particular characteristic of society—the level of integration of the individual.

10.1.2 Common Infrastructure

The concept of a common infrastructure arises quite naturally if we consider our own existence as members of society. Our daily concerns and activities fall roughly into two groups: those that relate to our personal lives, our families, and our friends, and those that relate to what we consider to be *services* provided to us by society, such as education, law and order and, in particular, the ability to participate in the public discourse. The extent of these services varies as a function of a society's development status and depend also on the ideology of the political economy, but intuitively we recognise that this extent is somehow a measure of the importance of society in our lives.

In such a society as a developed nation, we can think of the various activities being performed by elements of society: that is, in the action view we consider society to be a system of *functional elements.* These elements fall into two groups: those that produce the above services, and all the rest, and the elements in the former group constitute the *common infrastructure.* It is a further aspect of the essence of society, and as the evolution of society is the change of this essence, the common infrastructure is both a measure of and an enabler of the evolution of society.

An element in the common infrastructure provides a service, and in all societies, past and present, ensuring that such services are provided is an essential part of what defines *government*, introduced in Chap. 7. The extent to which the services are actually provided by government or outsourced to industry is a matter of the prevailing political economy, but for an element to be considered as part of the common infrastructure, its service must satisfy the following criteria:

1. It is government owned and controlled.
2. The service is provided irrespective of any personal characteristic, such as race or belief, although some services may be based on need (e.g., disability support is provided only to persons with disabilities).
3. The service is provided free of charge. (Any charges are counted as taxes.)
4. It provides a publicly agreed and verifiable level of service.

The common infrastructure typically includes such activities as education, health and aged care, sanitation, transport and communication, law and order, defence, customs and border control, general administration, political infrastructure, and, now increasingly, the access to information. In addition to potentially increasing the cost-effectiveness of the services provided (no profit, no advertising, etc.), the common infrastructure provides a powerful means of promoting social integration and solidarity, and it is essentially the representation of the bound energy in society. But irrespective of the type of service, its purpose must include support of the operation of the collective intelligence; a service such as Facebook does not qualify, as its purpose is to provide its owners with a return on investment by means of advertising. The fact that it is used as an enabler of the public discourse is a different matter.

10.1.3 *Types of Measures, and Social Integration*

The state of society can be measured in terms of many different parameters, depending on what one intends to achieve by the measurement. One measure might be the size of society, measured by national or total world population; another measure might be the Gross National or World Product; a further one could be the number of books published each year, or the level of education world-wide, and so on. The extent of technology could in principle be an obvious measure, even though the definition of this measure is difficult due to the fuzzy meaning of "technology" in daily language. It could also be a combination of measures, but in any case, in addition to defining the state of society, the measures need to be able to express the direction of evolution, its dynamics, as that is the focus of this book.

The operative word here is "direction". Consider, as a simple, but concrete example, the movement of a car. As we have a measure of distance—the metre, and a measure of time—the second, we can measure how far it moved in a given amount of time, and, by dividing distance by time, we can determine the speed of the movement. We can also measure the direction in which it has moved, by comparing with a preferred direction, such as the direction of the road or a compass direction, and the measure is in angular degrees. For none of these measurements is it necessary to know where the journey will eventually end, or if it even has an end. It is a bit like Forrest Gump; the essence is in the journey, not in the goal.

A first perspective on our search for system variables arises from the concept of a generalised energy introduced earlier, and by associating this energy with the activities performed by the individuals. We can then define the bound energy as the energy associated with those activities that require cooperation between individuals and are determined by society's requirements, rules, structure, and other constraints on such cooperation; and free energy as the energy associated with those activities that do not depend on the cooperation of other individuals and are at the individual's discretion. For each individual we can form the ratio of bound to total energy, and average this over all individuals; the resulting quantity is one aspect of a parameter we may think of as the degree of *social integration* (the choice of this name will become clearer further on). If we then consider three states of society, as measured by this parameter, and form an analogy with three states of a physical system—specifically, our earlier analogy with a collection of water molecules—we can obtain the following picture:

Society		Water molecules	
Social integration	State	State	Temperature
Low	Individual hunter/gatherers	Gas (steam)	High > 100 °C
Medium	Nomadic societies	Liquid (water)	Medium 0–100 °C
High	Settled societies	Solid (ice)	Low < 0 °C

The immediate idea you might take away from this picture is that the inverse of the degree of social integration is what we can think of as a *system temperature*. This temperature of societies has been declining from a very high value towards a value of 1, and we might reasonably ask if there is an ideal value of the temperature of a society, and what would determine such a value. For a given society, too high a temperature would tend to cause disintegration; too low a temperature would lead to stagnation; in this sense a society is like a living organism. A society is a dynamic entity, constantly evolving, but there is no goal to this evolution, no target society; it is impossible to know what society will look like in a thousand years from now. What is important is not its form at any one time, but the process of change, and the temperature is an indication of the stability of this process.

Based on our view of society as a collection of interacting individuals, the possible measures fall into two groups: those that relate to the individual, which we might call *micro measures*, and those relating to the collection as a whole, i.e., to the environment which supports the existence of the individuals and provides the infrastructure that supports the interaction between them, which we might call *macro measures*, of which the temperature is one. To illustrate the difference between these two groups of measures, consider again the analogy with a collection of water molecules. The properties of the collection will depend on the properties of the individual molecules, such as dipole moment, but they will also depend on the state of the collection as a whole, with its three distinct phases of gas (steam), liquid (water), and solid (ice). The most common measure of this collective state is temperature; another measure is what we identified as free energy, with the conversion of free energy into bound energy (and vice versa) taking place at the phase transitions. Another measure would be the *freedom* of the individual molecule within the "society"; it ranges from completely free in the gas to completely bound in the solid, but it is obvious that this measure can be viewed as either a macro measure or a micro measure, and so we are led to the realisation that micro measures are further divided into two subgroups: *intrinsic* measures that characterise the individual as such, independently of any relations to other individuals, and *contingent* measures, that depend on the fact that the individual is embedded in a society. Or, in other words, there is a conceptual difference between an individual considered in isolation and an individual embedded in society, where it is interacting with other individuals; this is a reflection of the dual nature of the concept of an individual, as mentioned earlier.

An important micro measure of the second kind, and one that we will consider from several perspectives, is related to *the individual's perception of its purpose*: the advancement of its own interests, or the advancement of the interests of society. The extreme is a person that perceives its purpose to be the advancement of its own interests only, without any other consideration. This is a person that sees society simply as a collection of individuals; as an environment with which it interacts for its own benefit only, much as we used to see Nature (and some people appear to still do). Such a society would be simply the sum of its members, without any emergent properties. Only in the most primitive of societies would this be a reasonable view; in a modern society it must be considered delusional, and the behaviour resulting from it would be completely asocial. In thinking about how we could assign a measure to this dimension, say, as the ratio of social interests to individual interests, we first need to realise that as society evolves in the direction of stronger interactions and greater complexity, not only do society's requirements on the individual's actions increase, but so do the options open to the individual, so that this measure is not the same as what we think of as progress in the form of increasing GDP *per capita*. It is possible (likely?) that in our society today, the options offered to the individual are increasing faster than society is able to introduce requirements, and that the evolution of society, along this dimension, is actually regressing.

The difficulty with, but also the importance of, this measure is that it is defined in terms of the individual's perception, something that is not immediately measurable or directly observable. Any determination of this measure, which is a new perspective on the degree of social integration introduced above, would have to be determined by taking a test, similar to an IQ test, and the design of such a test would be a project within social science. It is useful to compare this concept of social integration with the well-established concept of *socialisation*. The Internet offers a number of definitions (some slightly edited):

Socialization is the process of internalizing the norms and ideologies of society. Socialization encompasses both learning and teaching and is thus "the means by which social and cultural continuity are attained" (https://en.wikipedia.org/wiki/Socialization, accessed 20.06.2022).

Socialization is a continuing process whereby an individual acquires a personal identity and learns the norms, values, behavior, and social skills appropriate to his or her social position (http://www.dictionary.com/browse/socialization, accessed 20.06.2022).

Process by which individuals acquire the knowledge, language, social skills, and value to conform to the norms and roles required for integration into a group or community. It is a combination of both self-imposed (because the individual wants to conform) and externally-imposed rules, and the expectations of the others (http://www.businessdictionary.com/definition/socialization.html, accessed 20.06.2022).

Socialization, the process whereby an individual learns to adjust to a group (or society) and behave in a manner approved by the group (or society), and is a central influence on the behaviour, beliefs, and actions of adults as well as of children (https://www.britannica.com/topic/socialization, accessed 20.06.2022).

Socialization is the process that prepares humans to function in social life. It should be re-iterated here that socialization is culturally relative—people in different cultures and people that occupy different racial, classed, gendered, sexual, and religious social locations are socialized differently (https://en.wikibooks.org/wiki/Introduction_to_Sociology/Socialization, Accessed 20.06.2022).

What all these definitions have in common is that the degree to which a person is socialised is judged by society; it is a matter of the person adapting his or her *behaviour* to the expectations of society. Whereas the degree of social integration is a characteristic of the person's *belief* or world view, a private view that does not necessarily have to be revealed to society. In most cases the two measures will be close, but it is entirely possible for a person, who society judges to be highly socialised, to have a very low opinion of society, seeing it only as an environment to be exploited for personal benefit, and not as something of which he or she is an integral part.

Social integration is not the same as morality; a person with a low degree of social integration can have high moral standards and behave accordingly, as the following example illustrates: Consider persons with high moral standards and a strong work ethic, who are, and have been, working hard and are earning a great deal of money. They employ aggressive tax minimisation schemes, to the maximum extent allowed within the law, but they are compassionate about people in need and donate a significant portion of their wealth to charity, as well as to other causes they deem worthy, such as art, research, education, and sport. They see wealth redistribution by the government as usurping an important personal moral activity and duty, besides, perhaps, having little faith in the government's ability to operate cost-effectively. Disregarding any ulterior motives, such as the desire for recognition and influence that tends to accrue from this behaviour, many of these persons would be convinced that they are acting in the best interests of society.

The issue here is not one of morality or of selfishness; it is one of understanding. The most immediate understanding, as individuals, of our relationship to society is as a relationship between individuals; between me and other individuals like me, with no reference to society. And so, if each one of us, individually, does what is right, society will be right. This is the personal morality that has been delivered to us from antiquity onwards, and as long as the intensity of the interactions between individuals was relatively low, this was a workable approximation, resulting in a zeroth-order view of society as the aggregate of individual contributions. But as the intensity of the interactions increased, societies displayed behaviours that could not be explained in terms of characteristics of individuals; the interactions introduced emergent features of the behaviour. It is then no longer enough to have a code of personal behaviour; we need a code of behaviour for society, and as the only active elements of society are individuals, it follows that each individual has, in addition to a responsibility for its own behaviour, a responsibility for the behaviour of society. The prerequisite for the individual taking on such a responsibility is that it has an adequate understanding of how the behaviour of society arises out of the behaviour of individuals and their interactions, just as we understand how the macroscopic properties of a substance arise out of the properties of its atoms (or molecules) and their interactions. This is not a simple relationship, nor is it one that is intuitive, given our origin as individuals; it is an understanding we have started to develop through experience and analysis over thousands of years.

What about macro measures of the evolution of society? We mentioned temperature earlier, although that is more a qualitative measure than a quantitative measure, and we introduced the concept of social integration. And in Sect. 10.2 we shall develop the concept of social integration further. But in the context of our view of society as an information-processing system, the significant macro measures will be measures of how the other activities taking place in society interact with and influence the information processing; these measures are already well defined and understood, and data for most of them is readily available; this was the subject of Chap. 6.

10.1.4 Structural Complexity

'Structure' and 'complexity' are both related to each other and to the concept of a system, as was alluded to already in Sect. 1.3.1, and the description of society as a system is central to the view of society developed in this book. Accordingly, this might be the appropriate place to define the system concept in more detail; it will allow us to get a better understanding of the nature of 'structural complexity' and of the various approaches to describing society and its evolution.

The starting point is ontology and its role in linguistics, where the use of the word is somewhat different to its traditional use in philosophy. There it is concerned with the question of existence and what exists; the current interest is about explicit specification of conceptualisation; about the vocabulary we can use to speak about a particular domain of interest. An ontological commitment is the agreement to use a shared vocabulary and rules of grammar in the communication between people. Or, conversely, a given ontology defines the objects that can be represented by its concepts, the universe of discourse defined by that ontology. That there is more than one ontology arises from the fact that natural language is not semantically definite; the meanings of words and sentences depend on the context in which they are used. In particular, they depend on the area of professional specialisation, and with the rapid increase in knowledge and consequent increasing specialisation, there is an accompanying need for correspondingly specialised sub-languages.

However, if the specialised meaning of words is ignored, they can be grouped into a few main groups or categories; in what is called a top-level ontology. A number of such top-level categorisations have been put forward, starting with Aristotle's Metaphysics, but a more recent one is one shown in Fig. 10.1; this ontological sextet is from Munn and Smith (2008), but with the formal-ontological relations amended by adding the relations between properties and processes, making properties symmetrical with regard to substances and processes.

However, as it stands, this categorisation is not entirely suitable for a treatment of the evolution of society; the problem being that the concept of "substance" must go beyond the narrow meaning of something that has mass. To understand what this wider meaning is, we need to take a step back and consider an ontological structure that, at its highest level, has only two categories, *continuants* and *occurrents.* A

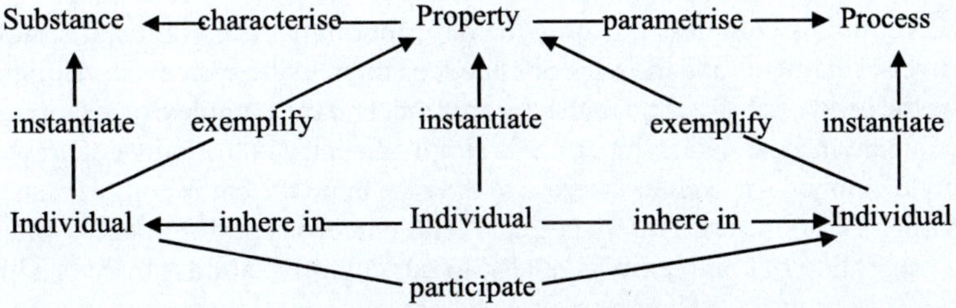

Fig. 10.1 The ontological sextet and the formal-ontological relations

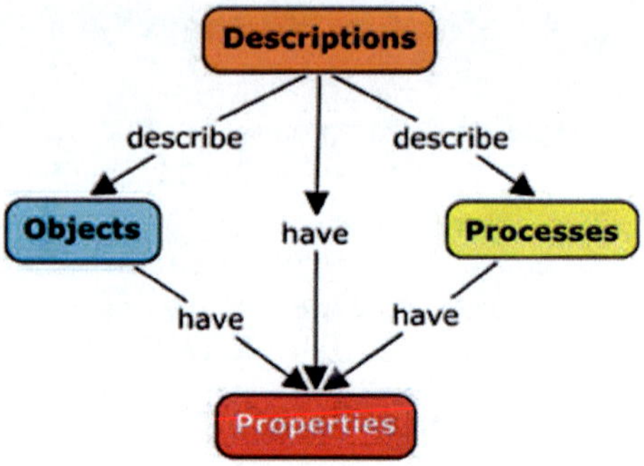

Fig. 10.2 The four top-level categories of the ontology

continuant is anything that exists *as a whole* at any point in time when it exists, which we shall call *object*, as opposed to an occurrent, which only exists as a whole over a period of time. The category of occurrents includes what we shall call *processes*. The category of continuants can be divided into the subcategories of independent continuants and dependent continuants, and the latter category includes properties and what we shall call *descriptions*.

Thus, to the top-level ontology in Fig. 10.1, *mode of description* must be added as a fourth category, as shown in Fig. 10.2, and the system concept is an *instantiation* of this category.

A system is a description of an entity (object or process) in terms of interrelated or interacting elements. Nothing is a system, but anything can be described as a system. An entity can always be described in terms of its externally observable variables only (a black box), and any entity can always be described in terms of a set of interacting elements (down to its atoms, if required, and if that makes any sense).

There is not a single or unique system associated with an entity; there is generally a family of systems that can be associated with an entity. This family is structured in two dimensions: horizontally in subject matter, i.e., what is to be described, and vertically in level of detail of the description (with each level of increasing detail consisting of an increasing number of interacting elements). This was treated already in Chap. 3, where the horizontal dimension was described in terms of *views*.

This brings us to 'complexity'. Intuitively, we feel that something is complex if it needs to be viewed from many perspectives in order to understand it, if it consists of many parts, if it is defined by many variables; in short, if it is described as a system. Nothing is complex, it is only the description that is complex, measured by the number of elements and the relations between them in the system description. Just increase the level of abstraction in the description, and the complexity disappears. To the accountant in a car dealership, a car is simply an entry in the ledger characterised by a single number—its value, whereas to the mechanic the car is complex, a system of many parts with intricate interfaces, and with numerous performance parameters.

The perception of complexity is related to our cognitive ability; there is a limit to how many elements and relations we can keep in our consciousness at one time, i.e., view as describing an entity, as was demonstrated by Miller (1956). The perception of complexity increases sharply when the number of elements and relations surpasses seven, or thereabouts. That fact provides the foundation for applying the system concept as a methodology for reducing complexity; if the description of an entity to a desired level of detail contains too many elements and their relations, one endeavours to divide the description up into more easily comprehended subsystems and interfaces between them. The art of applying this methodology lies in choosing the partitioning of the description so that the interfaces between the subsystems do not introduce so much complexity as to negate the benefit.

It is sometimes stated that, in addition to the number of elements and relations between them, there is a further dimension to complexity, that of uncertainty. Wikipedia defines uncertainty as "epistemic situations involving imperfect or unknown information. It applies to predictions of future events, to physical measurements that are already made, or to the unknown. Uncertainty arises in partially observable and/or stochastic environments, as well as due to ignorance, indolence, or both". Uncertainty arises from a lack of knowledge; complexity arises from increasing knowledge. Uncertainty changes the variables from numbers to distributions; this does not increase the complexity (as defined here), it just increases the effort required and may require the use of computer algorithms instead of a calculator. There is a particular use of "complexity", in "computational complexity", where it is tied directly to the effort (resources and time) involved in completing a calculation, but again, this has nothing to do with uncertainty.

Structure refers to the subdivision of an entity into a set of items and the resulting relationships between the items—the relationships that make them together represent the entity, and from the foregoing we see that structure is a characteristic of the system concept. Depending on the nature of the entity the term 'structural complexity' takes on a particular meaning; one example is the knowledge required to complete a task, see e.g. Gill (2008). Applied to society, structural complexity arises out of the division of labour and a corresponding description of society as a system with the elements being the groups of individuals forming similar functions, and the relationships between them is described within a two-dimensional framework, with a hierarchical/administrative dimension and a discipline dimension. The structure of this description is formed by the relationships between the elements, and the structural complexity is a measure of the number of relationships, or links. For example, one such measure—the connectedness of the elements—is the average number of links at an element; for N elements this measure takes on a value between 2(N − 1)/N and N − 1. However, this is a measure of the description of society, not of the society itself. To make it a measure of the society, we would have to define the elements of the description (i.e., the level of detail of the description).

In the present context the entity of interest is the common infrastructure, and we define *structural complexity* as the greatest number of independent functions performed by that infrastructure, where "independence" means that the performance of a function—the service provided—is not dependent on the performance of any other function. That is, in this case there is no interaction between the functions; the only relationship is 'belongs to' or 'is part of'. Of course, depending on the purpose of a description of the common infrastructure we may choose to aggregate functions, and in the next section we describe the common infrastructure in terms of a single functional unit—government.

This definition of structural complexity is a coarse definition, in that, except for the number of links, they neglect any information about the links themselves, such as duration, intensity and, indeed, the complexity of the information being exchanged. But we can make a comment regarding the dependence of structure on the size of the society; an issue that will be raised in Sect. 10.2.3. The work of Kasarda (1974) shows that it is not so much the structure of society that is dependent on size as the content of the information being exchanged between the structural elements. The majority of today's nations have a similar structure, at least down to some level of detail, but the relationship between the elements, as defined by the interaction between them, depends on both the type of society (autocratic, democratic, oligarchic) and the size (population) of the society. It is also interesting to note that Kasarda's data shows a marked increase with size in the proportion of the economically active population that is engaged in administration of the society, whereas the data provided on the LEI in Sect. 10.2.3 shows that the government expenditure as a proportion of GDP is around 20% for a developed nation, independent of size (e.g., Sweden vs. US). This would indicate that salaries make up an increasing proportion of government expenditure with increasing society size, which can be further interpreted as showing the increase in the information content of government activities (data collection, studies, etc.).

10.1.5 Social Complexity

A good starting point is the approach put forward in (Bergman & Beehner, 2015). Their purpose was to propose a measure of social complexity that could be measured over a wide range of species and so provide the input to testing the hypothesis of a correlation across species between social complexity and individual intelligence, or cognitive ability. To this end they stated that "social complexity should be measured as the number of differentiated relationships that individuals have. By differentiated relationships, we simply mean the number of relationships that can be distinguished by an observer; that is, the number of consistently different interactions that are seen." So, if in a society with N individuals every individual treats every other individual identically, the complexity would be 1, and if they treat all other individuals differently, the complexity would be $N - 1$.

Bergman and Beehner were concerned with correlation across species, and the focus of their development and justification of their definition, as well as their examples, are from the animal world, where the individuals are treated as identical, whereas we are interested in comparisons across (human) societies. And only across recent societies, i.e., a timespan in which the basic processing capacity of the brain has not changed significantly. But their definition of social complexity and the correlation with cognitive ability still offers a useful perspective. In humans, cognitive ability is not an intrinsic ability, it is acquired by growing up in a social environment with a certain complexity, i.e., a society where the individual is exposed to a certain number of differentiated relationships. The thought-processing ability (*ibid*, Fig. 25) is basically an evaluation of thought inputs based on existing information; information that is the result of processing of previously received inputs. Consequently, social complexity is particular to each individual, it is a characteristic of society as perceived by the individual; only its average over all individuals is a characteristic of a particular society.

A slightly different perspective on the issue of social complexity that reflects this close relationship between social complexity and cognitive ability in human is provided by the concept of *cognitive complexity*, a concept attributed to Bieri (1955). It is based on the recognition that our interpersonal relationships are framed by a number of constructs (effectively evaluation criteria) and "a system of constructs which differentiates highly among persons is considered to be cognitively complex". A review of the ensuing constructivist research of cognitive complexity is contained in Kovárová and Filip (2015), and in addition to reviewing both the various approaches and definitions that have appeared, the article also considers the empirical evidence supporting them. A central theme in the article is the relationship between differentiation and integration, and the perceived ambiguities in the conceptualisation of these two components of cognitive complexity. Given the relationship between the system concept and complexity set out in the previous subsection, there is no ambiguity; differentiation and integration are converse aspects of a system and are thus reflected onto the concept of complexity. (For a more detailed discussion of this assertion, although in a slightly different context, see Aslaksen (2013, p. 210).)

Of the relationships an individual has with other members of society, some may be identical, i.e., not differentiated, and these members are perceived by the individual to form a group, so that the differentiated relationships provide a structure of society from the perspective of the individual. For example, parents and siblings may constitute two such groups. This structure has nothing to do with the structure in structural complexity, to be considered in the next section, but the numbers of members of each group provide an additional dimension to the definition of social complexity.

Social complexity is a subject of ongoing research within social science and was the theme of a recent issue of *Behavioural Ecology and Sociobiology*, with an Introduction by Kappeler (2019). As is true of the subject in general, most of the articles in this issue are concerned with animal behaviour and societies; a reflection of the fact that interactions between members of animal species are both more limited and more easily observable, as well as of the desire to use social complexity as a measure of biological evolution. Towards the end of the Introduction, the authors ask: "Do species differences in social complexity actually matter?". If we consider societies to be the "species" in the evolution of human societies, we would expect social complexity to be correlated with the individual's understanding of and integration in society, so the answer is "yes".

Where social complexity is studied as an indication of cultural evolution in human, or hominin, societies, it is mostly concerned with timeframes of 100,000 years or more and with technology as the main component of culture (e.g., stone age, bronze age, iron age). The article (van Schaik et al., 2019) gives a good review of research in this area and provides valuable insights into cultural evolution in that timeframe. In particular, the importance of social transmission and of teaching as its major component in determining the level of technology. Our interest in such knowledge is to see if and how it can be useful in relation to today's society; a subject matter in Chap. 11.

The variation in social complexity, and with it, the variation in cognitive ability, provides a further measure of social inequality, in addition to the inequality measured by economic parameters, as in Sects. 6.4 and 6.5. Establishing the degree of correlation between them would require empirical research, but in the absence of that, what all our measures relate to is the integration of the individual into society—the involvement in society, the understanding and appreciation of this involvement and of the fact that society is the result of interaction between individuals. The inequality measures are measures of various aspects of the inequality of this integration. The evolution of society is the evolution of the relationship between the individual and society.

10.2 Economic Measures

10.2.1 Approach

The evolution of society over the last 10,000 years or so is one of increasing cooperation and mutual dependence of its members, accompanied by an increasingly complex structure and increasing flow of information. This evolution is reflected in various ways—in our shared beliefs (both religious and social), in the restrictions placed on individual behaviour in order to facilitate this cooperation, and in the relationship between private and public ownership. But these and other indicators are all difficult to define and measure. How to measure someone's belief, how to define what people consider "public" in a time of increasing outsourcing and increasing financial *versus* non-financial assets, and so on. Considering all of this, it would appear that a promising approach might be via economics, where the variables are reasonably well defined and there is a wealth of data available; the problem is then 'reduced' to defining significant measures in terms of this data. And it is, above all, a problem because economists tend to view economics as a self-contained description of human activity, largely unrelated to any other considerations, such as social considerations or considerations of stability in anything but economical terms. But the evolution of society is not primarily a matter of economic growth, as in GDP or in the level of technology, or even in changes to such parameters as income and wealth inequality; these are measures of secondary effects. The primary driver of the evolution of human society is the interaction between its members, and therefore the primary measures of social evolution must relate to this interaction, to its strength, timeliness, and quality. It is the purpose of this section to explore to what extent such measures can be extracted from economic data.

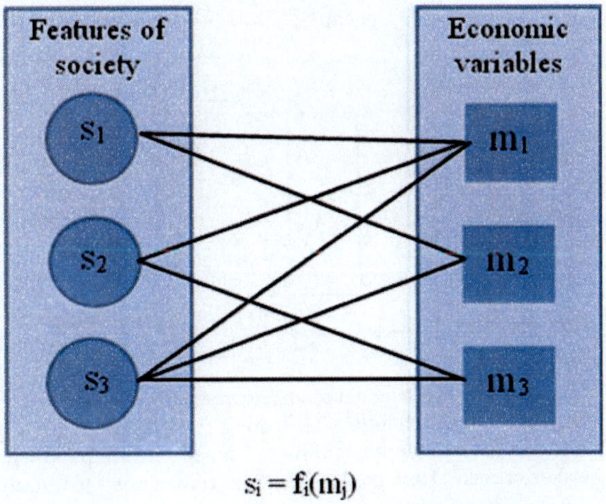

Fig. 10.3 The purpose of this section is to determine functions of the economic variables that provide useful correlations with the observed features of social evolution

Our approach is illustrated in Fig. 10.3, and the first step in this this approach is to define what features of society are significant indicators of its state. Modern society is very complex, and it is easy to think of a vast array of features, related to such functions as education, healthcare, and security, and to measures of life expectancy, wellbeing, application of technology, and of various forms of art, just to mention a few. But which of these are most significant; which ones would we use to say that one society is better than another, or to say that one is further advanced along the path of evolution of society than another? Trying to answer such questions is the purpose of the next section.

10.2.2 Macroeconomic Model

There are various macroeconomic models, from simple aggregate demand—aggregate supply models to complex dynamic stochastic general equilibrium models. Each one defines a set of variables and the relationships between them, and with increasing sophistication the models are able to explain (and predict) the behaviour of the economy with increasing detail and accuracy. For our purpose, which is to use economic data to characterise the evolution of society, a simple model will suffice, depicting society as consisting of only three components—households, industry, and government—as shown in Fig. 10.4.

The model is designed to separate out that part of a society's economic activity that is associated with the operation of the *common infrastructure*, as defined earlier, and there are a number of assumptions/simplifications in this model. The most significant one arises in the definition of the government component; it is defined as the component that encompasses the common infrastructure, but nothing else. That is, if

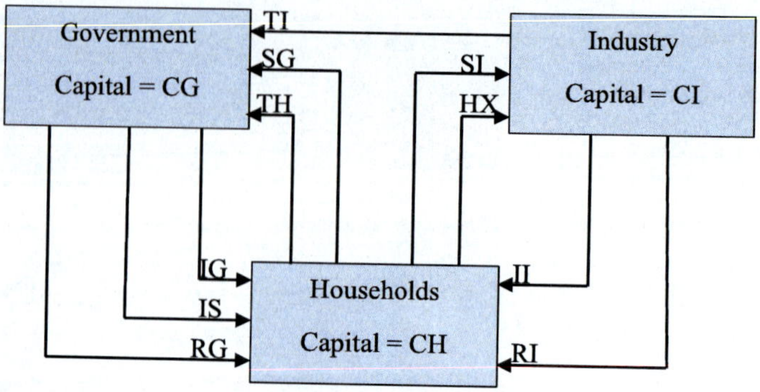

Fig. 10.4 A simple model of society's economy, showing only money flows, not product flows. The flows are labelled as follows: II Income (salaries) from industry, IG Income (salaries) from government, IS Income support (from government), RI Return (on investment) from industry, RG Return (on investment) from gov't, SI Savings (investment) to industry, SG Savings (investment) to gov't, TI Taxes from industry, TH Taxes from households, HX Household expenditure (Consumption)

economic data for government includes the operation of government-owned industries not involved in the provision of common services, as may be the case e.g., in China, then this must be deducted and added to industry. Also, because the services provided by the government, through the common infrastructure, are provided at no cost (paid for through taxes), any user-pays feature of such services, are included in the taxes. The income support includes the various government pensions, disability support, unemployment payments, etc.

Industry comprises all non-government income-earning activities, including housing rental. The savings to industry includes bank deposits, shares, and bonds, the return on investments from industry includes interest, dividends, rents, and capital gains; the investment (savings) in government and the return relate mainly to government bonds.

The data of interest to us is aggregate macro-economic data, and in the first instance we consider two of these measures—the Gross Domestic Product (GDP) and Government Expenditure (GE). They are related to the variables in the model in Fig. 10.4 as follows:

$$\mathrm{GDP} = \mathrm{II} + \mathrm{IG} + \mathrm{IS} + \mathrm{RI} + \mathrm{RG}; \text{ and}$$
$$\mathrm{GE} = \mathrm{IG} + \mathrm{IS} + \mathrm{RG}; \text{ so that } \mathrm{GDP} = \mathrm{GE} + \mathrm{II} + \mathrm{RI}$$

Macro-economic data is generally available on a national basis in the form of National Accounts, which are based on the System of National Accounts developed and maintained by the UN. In this system, there are several approaches to calculating GDP; the one most aligned with our model is the Expenditure Approach (GDP(E)), defined as:

GDP(E) = Final Consumption Expenditure by Households and Government + Gross Capital Formation + Exports of goods and services − Imports of goods and services.

(A related measure is Gross National Income (GNI); it is equal to GDP plus net primary income receivable from non-residents (NPINR); i.e., GNI = GDP + NPINR. For many nations the difference between GDP and GNI is quite small, in which case we can use GNI data where this is more readily available.)

10.2.3 Level of Economic Integration

The previous section introduced the idea of considering the activities performed by the members of society as being of two types—common and individual, and to think of the associated effort as bound and free energy. This led to a first definition of the level of social integration as the ratio of bound to total energy, and if we now translate this into the realm of economics by means of the model in Fig. 10.3, we can give a first indication of the level of social integration in economic terms in the form of a level of economic integration, LEI, as

$$LEI = \frac{GE}{GDP}. \tag{10.1}$$

Time series of LEI only make sense for individual nations, because the difference in the temporal development of the two variables, DGP and GE, between nations would obscure the significance of the LEI once the data were aggregated. Also, the time period for which data is available varies considerably from nation to nation. In the following, thc LEI for a fcw nations for which relevant data is readily available are presented.

Australia. Australia is ranked no. 13 out of 180 nations by GDP (World Bank, 2018), but no. 10 by per capita GDP (US$ 57,305). The data shown in Fig. 10.5 has been extracted from the Australian Bureau of Statistics.

Chile. Chile is ranked no. 40 out of 180 nations by GDP (World Bank, 2018), and no. 58 by per capita GDP (US$ 15,923). The data displayed in Fig. 10.6 is from the World Bank dataset.

China. The PRC is ranked no. 2 out of 180 nations by GDP (World Bank, 2018), but no. 64 by per capita GDP (US$ 9771). The data displayed in Fig. 10.7 is from the World Bank dataset.

India. India is ranked no. 7 out of 180 nations by GDP (World Bank, 2018), but no. 146 by per capita GDP (US$ 2010). The data displayed in Fig. 10.8 is from the World Bank dataset.

Indonesia. Indonesia is ranked no. 16 out of 180 nations by GDP (World Bank, 2018), but no. 118 by per capita GDP (US$ 3894). The data displayed in Fig. 10.9 is from the World Bank dataset.

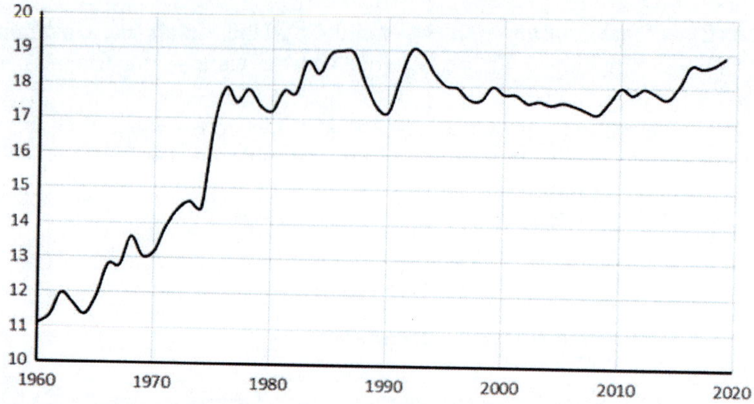

Fig. 10.5 The LEI for Australia for the period from 1960 to 2019 (in %)

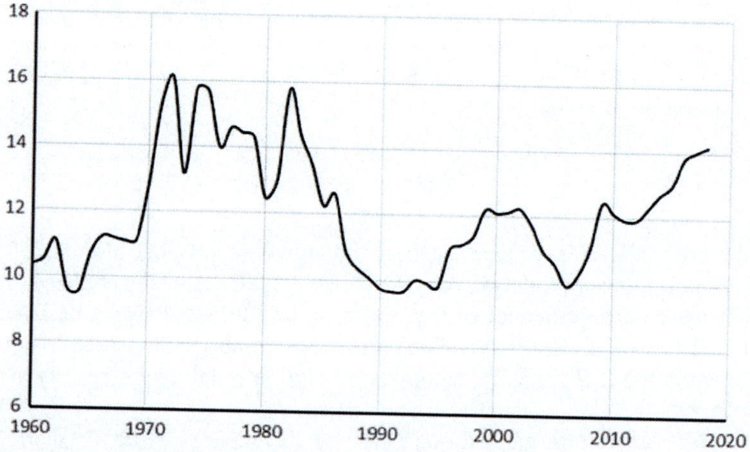

Fig. 10.6 The LEI for Chile for the period from 1960 to 2019 (in %)

Nigeria. Nigeria is ranked no. 30 out of 180 nations by GDP (World Bank, 2018), but no. 145 by per capita GDP (US$ 2028). The curve in Fig. 10.10 is derived from the World Bank dataset.

Sweden. Sweden is ranked no. 22 out of 180 nations by GDP (World Bank, 2018), but no. 11 by per capita GDP (US$ 54,112) The data used was extracted from the database of the Swedish Bureau of Statistics (http://www.statistikdatabasen.scb.se/pxweb/en/ssd/), both from the national accounts (for the period 1980–2020) and from the historical database (Fig. 10.11).

United Kingdom. The UK is ranked no. 5 out of 180 nations by GDP (World Bank, 2018), but no. 21 by per capita GDP (US$ 41,030). The data used here was originally presented in Broadberry et al. (2015), and quoted by the Bank of England in A millennium of macro-economic data for the UK, sheets A5/6/7, available at

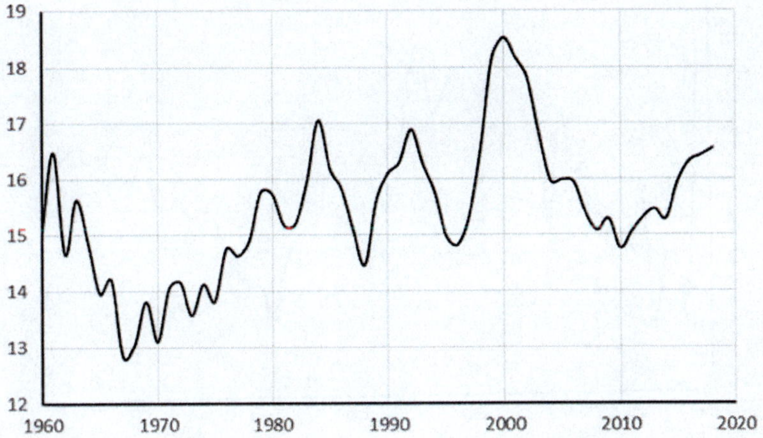

Fig. 10.7 The LEI for the Peoples Republic of China (PRC) for the period from 1960 to 2018 (in %)

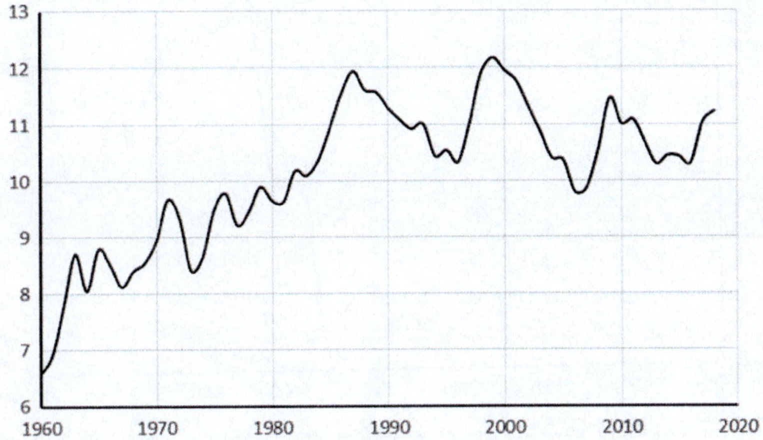

Fig. 10.8 The LEI for India for the period from 1960 to 2018 (in %)

https://www.bankofengland.co.uk/statistics/research-datasets. The resulting values of the LEI for the period 1280–2015 AD are shown in Fig. 10.12.

Besides the progressive increase, the trend line shows the effect of two factors: the increase in times of war (as the GE includes defence expenditure)—the Hundred Years War and the War of the Roses, then the colonial wars, the war against Napoleon, and, finally, the two World Wars—and the decrease in times of capitalist expansion—first mercantile, then industrial. However, for the most recent time period, 1960–2015, the trend indicated is very deceptive, as the more detailed display of the data in Fig. 10.13 shows; the value of the LEI has remained largely constant.

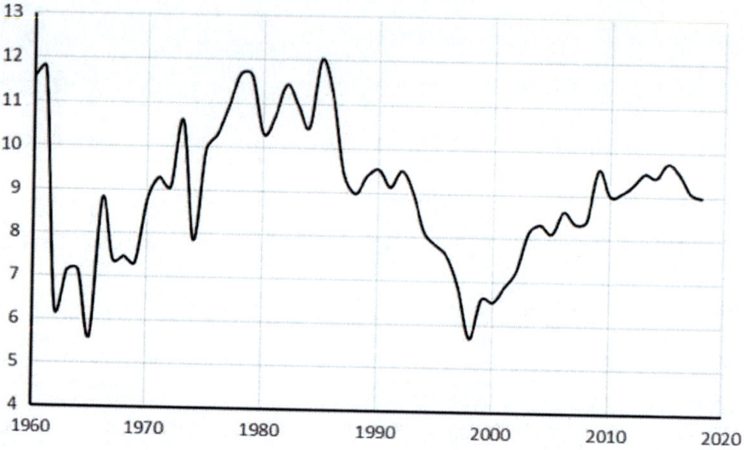

Fig. 10.9 The LEI for Indonesia for the period from 1960 to 2018 (in %)

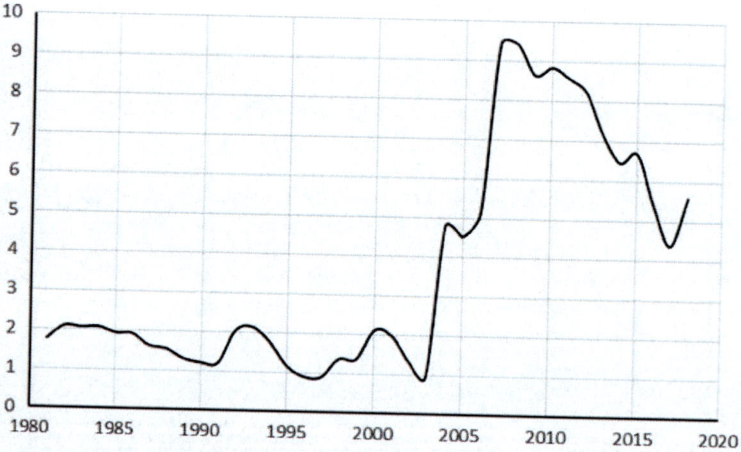

Fig. 10.10 The LEI for Nigeria for the period 1982 to 2018 (in %)

United States. The US is ranked no. 1 out of 180 nations by GDP (World Bank 2018), but no. 8 by per capita GDP (US$ 62,641). The data presented here was downloaded from the website of the Bureau of Economic Analysis (http://bea.gov/data) for the period 1929–2018. For 1820–1913, the GDP data was copied from the OECD publication The World Economy: Historical Statistics, by Angus Maddison, 2003, and converted to current dollars by using the CPI scaling data in http://liberalarts.oregonstate.edu/app/polisci/robert-sahr; the government spending was copied from http://www.usgovernmentspending.com.

The timeline of the US LEI in Fig. 10.14 shows the increase that resulted from the Great Depression and the New Deal, followed by the spike of WW2 and then the cost

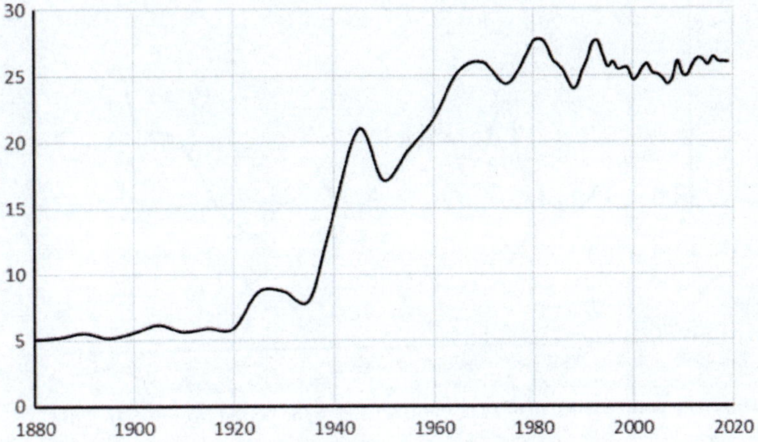

Fig. 10.11 The LEI for Sweden for the period 1880 to 2020 (in %)

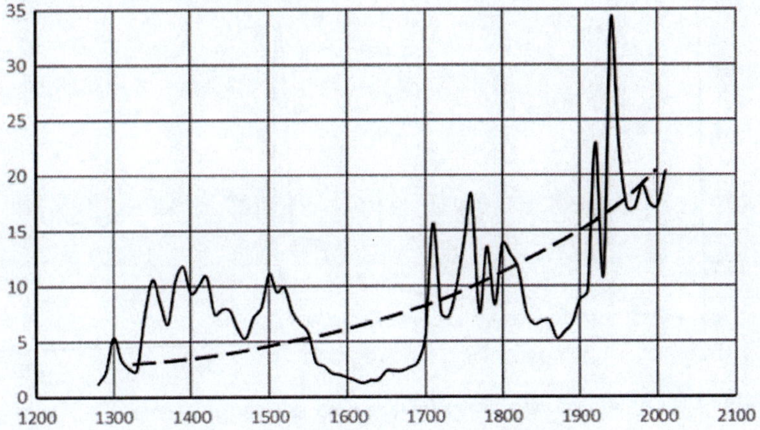

Fig. 10.12 The level of economic integration (LEI) in GB/UK over the time period 1280–2015 AD (in %)

of the Cold War. But since 1970 there has been a steady decline—a manifestation of the increasing power of neoconservatism. The LEI is now well below 20%, and in 2018 the military budget made up 3.2% of the 18% LEI.

World. Regarding the world as the next, and for the foreseeable future ultimate, society, with nations as its individuals, we can identify the very early manifestations of a corresponding common infrastructure and associated LEI in the form of the expenditure on the United Nations, as reported in https://unsceb.org/total-expenses, accessed 20.06.2022. For the year 2018, the expenditure, in $US, was as follows:

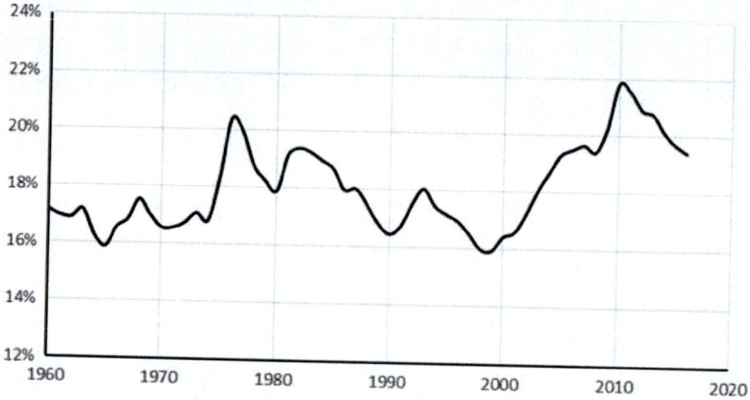

Fig. 10.13 A more detailed display of the UK LEI for the period from 1960 to 2015

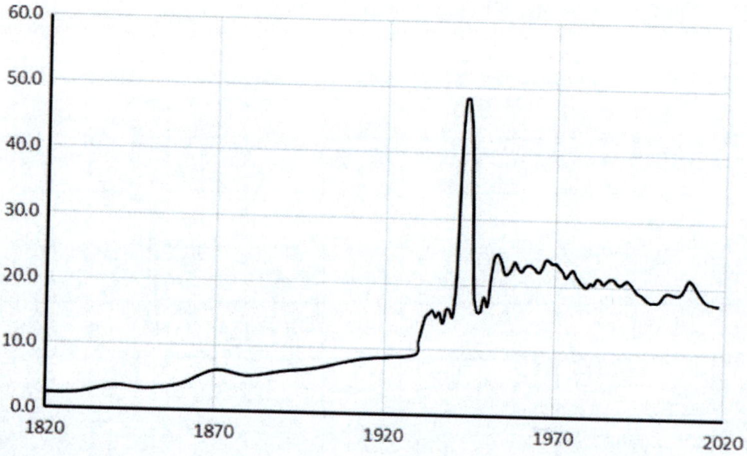

Fig. 10.14 The LEI for the United States for the period 1820–2018

Expense category	Expenditure
Humanitarian assistance	19,228,144,556
Development assistance	16,844,407,512
Peace operations	9,866,214,744
Global agenda and specialised assistance	6,837,061,899
Total	52,775,828,711

Compared with the world's GDP in the same year, $US 85.911 trillion (https://data.worldbank.org/indicator/ny.gdp.mktp.cd?most_recent_value_desc=false), this is only about 0.06%, so by this measure we are still very far from having any significant common infrastructure on a world basis. However, there is strong and increasing interaction between the nations of the world, evidenced by the following data on international trade and travel (Figs. 10.15 and 10.16):

Comments on the LEI. From the time series presented for the selected nations above, it is possible to draw a few conclusions:

(a) For those three nations (Sweden, UK, and US) where earlier historical data is available, the increase in the LEI correlates with the rise in industrialisation and decline in employment in agriculture; that is, an increase in the division of

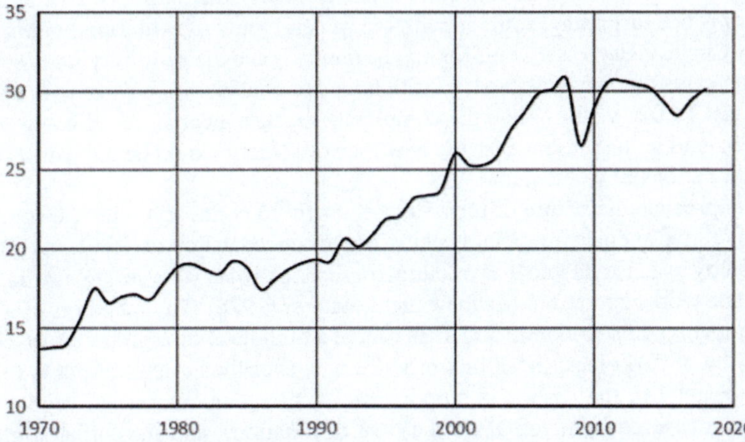

Fig. 10.15 World trade volume as a percentage of world GDP (https://data.worldbank.org/indicator/NE.EXP.GNFS.ZS, accessed 20.06.2022)

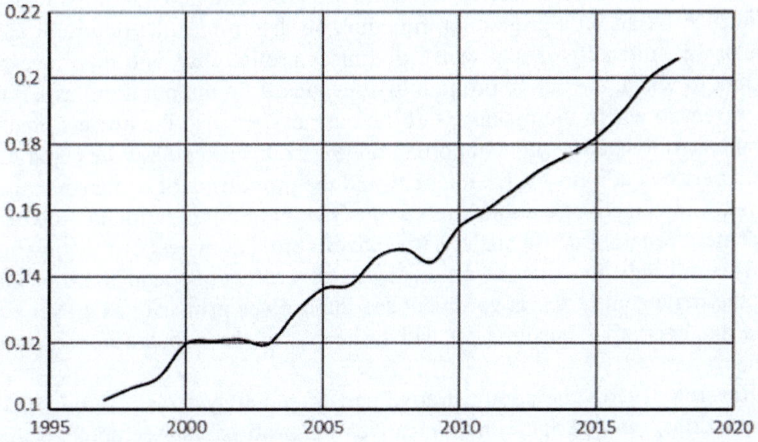

Fig. 10.16 International departures per year and person (https://data.worldbank.org/indicator/ST.INT.DPRT, accessed 20.06.2022)

labour and in the integration of the individual into society. The increase in the LEI can be seen as a response to the increase in the complexity of society.

(b) For a developed nation, the LEI appears to settle around 20%, with the difference around this value indicating the degree of social solidarity and integration, with Sweden at 25% and the US steadily declining below 20%. Australia still has a way to go, and the UK shows the ravages of the Thatcher years.

(c) The developing countries, on a *per capita* GDP basis, are still quite some way off from establishing a common infrastructure and achieve a stable LEI.

The fact that the LEI seems to converge to a value in the range 20–25% as nations develop is not surprising; it fits in with our general view of evolution, within which society can be viewed as a genus (or maybe family?) and the various types of societies as the species that have evolved, of which we are considering only human societies in the last 10,000 years. When the complexity reaches a certain level a new species emerges, and in the present case the new type of society would be the world society, with the nations as its elements.

To appreciate the nature of the LEI, it is useful to contrast it with the concept of administrative intensity in social systems, as introduced in Pondy (1969), and pursued in the study by Kasarda into the structural implications of system size (Kasarda, 1974) and in the work reported in Mayhew and Levinger (1976). The members of a social system involved in economic activities can, at a high level of abstraction, be considered to be of two types: those that produce whatever the system output is (e.g., an actual product in the case of a corporation, or the conglomerate of products and services measured by the GDP in the case of a nation), and those that administer the production process. The latter, expressed as a fraction of the total, is the administrative intensity. Based on a study of 45 manufacturing industries, Pondy finds a definite positive correlation between functional complexity, defined as the number of distinct professions involved in the production process, and administrative intensity. Further, based on a production function, he determines the optimum value of administrative intensity as that which optimises profitability, and then investigates the extent to which the actual profitability deviates from the optimum as a function of the extent to which the managers of the business are also the owners, and states the result as follow: "The most important finding of this research is the negative relationship between administrative intensity and the proportion of owner-managers, or conversely, the positive relationship between size of administration and ownership-management separation." In society, the citizens are the owners of the government, and this result indicates that the efficiency of government depends on the extent to which the managers of the government see themselves primarily as citizens rather than as managers of a business for which the rest of the population constitute the market.

In Kasarda (1974), the administrative part of social systems is subdivided into three functions: managerial, communicative, and professional/technical. The study investigates three types of social systems: institutional, communal, and societies, and for the latter, which comprises a group of 43 non-agriculturally based nations, the results are as follows:

Table 10.1 The administrative component and its sub-components, in percent of the economically active population, as a function of society size (From Kasarda, 1974)

Economically active population in millions	Total administrative component	Managerial component	Communicative component	Professional/technical component
< 0.5 (n = 11)	13.57	1.72	6.49	5.36
0.5 − 5 (n = 20)	16.80	2.18	7.98	6.73
5 − 25 (n = 9)	18.26	2.67	8.27	7.41
> 25 (n = 3)	25.17	4.64	12.56	8.00

With regard to the results in Table 10.1, Kasarda states "The most prominent organizational changes occur in communication. As societies expand, substantially greater proportions of their personnel are devoted to communicative (clerical) functions. It may therefore be inferred that the major role of holding large social systems together rests with those whose primary function is facilitating communication.". This resonates with the primacy of the interaction between individuals in our view of societies and as the driver of their evolution.

Finally, Mayhew and Levinger (1976) develops a model based on graph theory that shows the dependence of the density of interaction in human aggregates (i.e., social systems) on population size, and with it provides a formal rationale for Kasarda's hypothesis relating size and communications in societies, as noted above, as well as for such effects as decreased interaction time and increased anonymity. Leaving aside the authors' attempt to give the explanatory aspects a mathematical rigour, which appears misplaced, what is of interest in the present context is the combination of the relation between size and expected density of interaction with the relationship between size and structural differentiation reported in Mayhew et al. (1972) leading to role strain, to which we shall return in Chap. 11.

Returning now to the LEI, we see several relationships to the above work:

(a) While administrative intensity is an important component of the LEI, through the personnel cost of government, the LEI includes several other components, such as the cost of providing the common services, and the cost of direct payments for various welfare schemes.

(b) Because the LEI is a proportion of the GDP, the dependence on the size of the population is greatly reduced, and the increases shown in Figs. 10.3, 10.4, 10.5, 10.6, 10.7, 10.8, 10.9, 10.10, 10.11 and 10.12 are more due to a change in society's attitude to solidarity than to an increase in the population.

(c) The "saturation" of the LEI at a value in the range 20–25% is very likely linked to such parameters as role strain and decreased interaction time within the government, and while this will be discussed briefly in Chap. 11, it needs to be supported by empirical data.

10.2.4 Education

The second economic measure to be considered is the expenditure on public education. In particular, the expenditure on that part of education that promotes social cohesion and solidarity, which basically means expenditure on public primary and secondary education. Private education, although it may provide excellent formal education in the sense of knowledge transfer, has as a major purpose a segregation of society and the creation or maintenance of a privileged class. And tertiary education, while it is of great importance in developing people's ability to contribute to and succeed in society, does not really contribute much to social cohesion and solidarity, due to the high degree of specialisation.

In the following, *public* education (often called government education) is education provided within an institution controlled by the government and with a curriculum defined by the government; *private* education (often called non-government education) is education provided within an institution controlled by private enterprises, churches, or any other non-government organisations.

Education, as the means by which knowledge is transferred from one generation to the next, is one of the most important factors influencing the evolution of society, but it is a complex influence, with several different aspects. One aspect, and perhaps the main one, is education's purpose of preparing the individual for a good and productive life in the existing society. That means, on the one hand, for the individual to understand and accept the features of that society—its organisation and its laws and customs; on the other hand, to prepare for an occupation, for contributing to the economy of the society. This aspect is the *integration* of the individual into the existing society, but this integration is more than just a "fitting in", like a piece in a puzzle; it is also about the individual's belief that it *belongs* to, and is an important part of, society—a feeling of solidarity and equality. For example, the laws and customs are observed not to avoid punishment, but because they are seen as just and appropriate. This belief is a measure of the cohesion of the society, and under the aspect of fostering this cohesion, there is a difference between public and private education. Public education fosters solidarity and equality on the scale of society, whereas private education fosters solidarity within a small part of the population, but inequality and a lack of solidarity on the scale of society. Hence, there are two parts to measuring this influence: the expenditure on public education, and the ratio of private to public education expenditure.

Given the above comment on the purpose of education, we are only interested in education up to the end of secondary education, as tertiary education does not enhance the belief of belonging to society (possibly the opposite). That is, our measure is government (all levels) expenditure on primary and secondary education in public institutions as a percentage of GDP; a measure we shall call the *Public Education Indicator*, or PEI. There are three things to note regarding this measure: One, funds from private organisations can be contributed to education in public institutions. Two, in many nations, e.g., Australia, significant government expenditure flows to private educational institutions, so that it would be incorrect to simply use government expenditure on education as a measure. And thirdly, in some countries, more that in others, households contribute a significant amount to public education in the form of fees for various activities and cost of educational material (see e.g. *Who pays for what?*, UNESCO Institute for Statistics, 2016).

Another aspect is that education should develop the individual's *critical ability*, for while it is important to understand and accept the existing society, there is no need to believe that it is perfect and impervious to changing circumstances; such a belief is an *ideology*. Society is constantly undergoing small changes, and it is the task of the collective intelligence—the operation of interacting human brains, of which the individual's critical ability is a major component, to decide which ones to keep and which ones to discard, thereby guiding the evolution of society. Capturing the extent to which education is promoting critical thinking by means of any economic parameters is probably impossible, but an indication of the converse, capturing the extent to which ideology finds its way into education in order to suppress critical thinking, could possibly be obtained from economic data (e.g., the budgets of organisations dedicated to this activity). An oblique indication of this is the extent to which education is privatised, as private education is most often related to a particular ideology or religion, and the expression of public vs. private education in economic terms is therefore included as a measure, in the form of the *Public Education Fraction*, PEF, defined as the ratio of expenditure on education in public institutions to total expenditure on education, both relating to primary and secondary education. Clearly, the two measures are far from orthogonal.

As is evident in the following treatment of the individual nations from the previous subsection, data on details of the expenditure on education are, in most cases, very difficult to obtain, and where it has been necessary to make estimates or "educated guesses", the accuracy is uncertain.

Data is sourced from the OECD data base (https://data.oecd.org/searchresults/?hf=20&b=0&r=%2Bf%2Ftype%2Findicators&r=%2Bf%2Ftopics_en%2Feducation&l=en&s=score, accessed 20.06.2022) and from the UNESCO data base (https://tcg.uis.unesco.org/data-resources/, accessed 20.06.2022).

Not all spending on educational goods and services occurs within educational institutions. For example, families may purchase textbooks and materials commercially or seek private tutoring for their children outside educational institutions. All such expenditure outside educational institutions is excluded here.

Australia. The public expenditure on primary and secondary education in Australia in the period 1993–2016, in % of GDP, as provided by the UNESCO data base, is shown in Fig. 10.17.

As an initial example of the components of the expenditure on primary and secondary education, the values for Australia in 2015 were as follows (in $millions) (Rice et al., 2019):

Public education: 35,679, of which 1482 was contributed by private organisations.

Private education: 24,476, of which 14,553 was contributed by government as is illustrated in Fig. 10.18.

Australia's GDP in 2015 was $1,629,282 million in 2015, so that our two measures take on the following values:

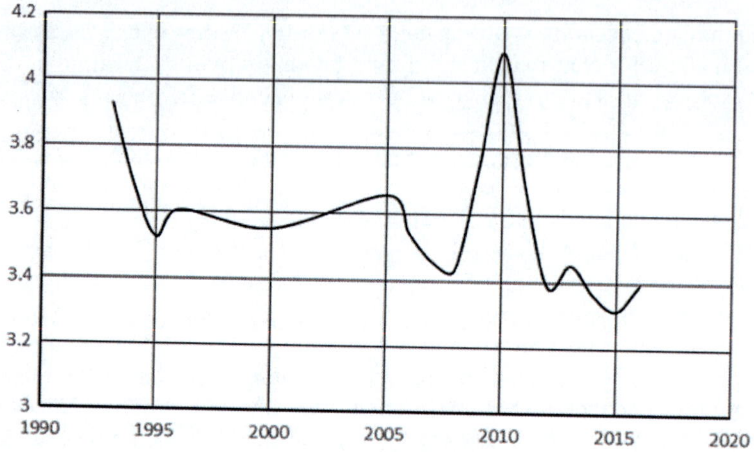

Fig. 10.17 The public expenditure on primary and secondary education in Australia, in % of GDP

Fig. 10.18 The expenditure on primary and secondary education, in $million, in Australia for the year 2015

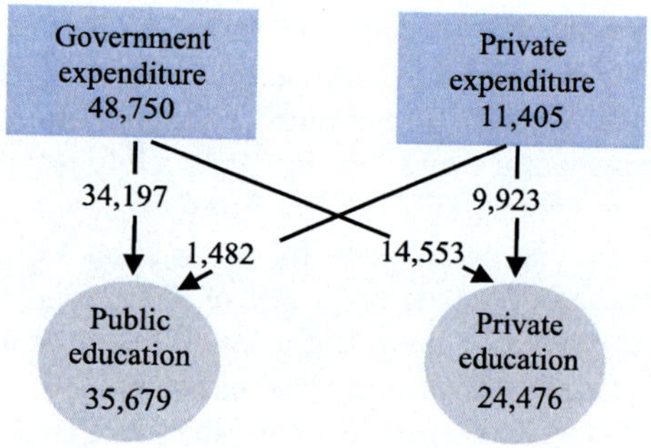

$$\text{PEI} = 2.19\%; \text{ and PEF} = 59.31\%$$

The government expenditure on private primary and secondary education in 2015 is about 30% of the total government expenditure on primary and secondary education; way above the OECD average of 13%. This is also reflected in the OECD figure for public expenditure on primary/secondary education, which is 3.182% of GDP.

Chile. The public expenditure on primary and secondary education in Chile in the period 1974–2017, in % of GDP, as provided by the UNESCO data base, is shown in Fig. 10.19.

For Chile, the OECD data base gives the following values for 2016, in millions of CLP (Table 10.2):

In 2016, Chile's GDP was 167,227,448 million CLP, with which we obtain

$$\text{PEI} = 1.49\%; \text{ and PEF} = 40.92\%$$

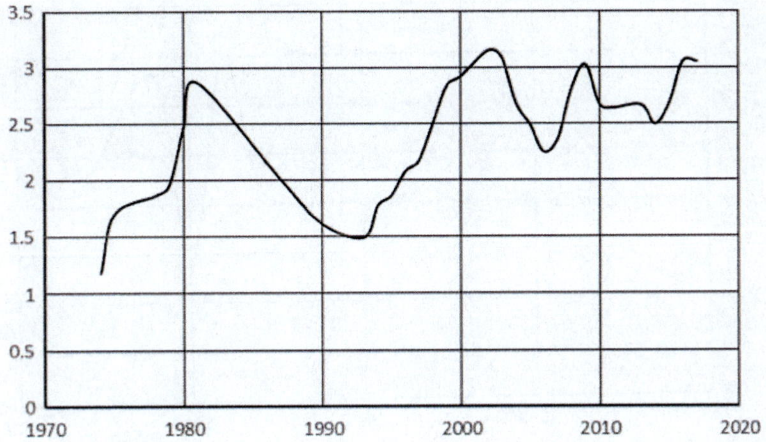

Fig. 10.19 The public expenditure on primary and secondary education in Chile, in % of GDP

Table 10.2 Expenditure on primary and secondary education in Chile, by source and institutions

Source	Institutions	Primary	Secondary
All	Public	1,237,127	1,252,101
	Private	1,807,925	1,785,055
Government	Public	1,237,127	1,252,101
	Private	1,293,443	1,274,202

China. The public expenditure on primary and secondary education in China in the period 1980–1999, in % of GDP, as provided by the UNESCO data base, is shown in Fig. 10.20.

A more recent article (Li, 2017), provides the following information regarding the expenditure on education in the year 2011 (somewhat approximate, as the information is only provided in the form of curves):

Total expenditure on education: 5 % of GDP, of which 3.9 % was public (Fig. 2)
Of the total expenditure, 57.5 % was spent of primary/secondary education (Fig. 6)
Of the public expenditure, 53 % was spent of primary/secondary education (Fig. 8)

From this it follows that the total expenditure on primary/secondary education was 2.88% of GDP, of which public expenditure was 2.07% of GDP and private expenditure was 0.81% of GDP. However, while almost all of the public expenditure was on public institutions, there is no information on how much of the private expenditure was spent on public institutions. From data presented on the website (tradingeconomics.com/china/education) it can be determined that the proportion of primary/secondary students enrolled in private institutions was about 8%, and taking into account that cost per student is higher in private than in public institutions, and also that a significant amount of the private expenditure on students in public institutions is for tutoring (Yuan & Zhang, 2015), it is likely that only a small amount,

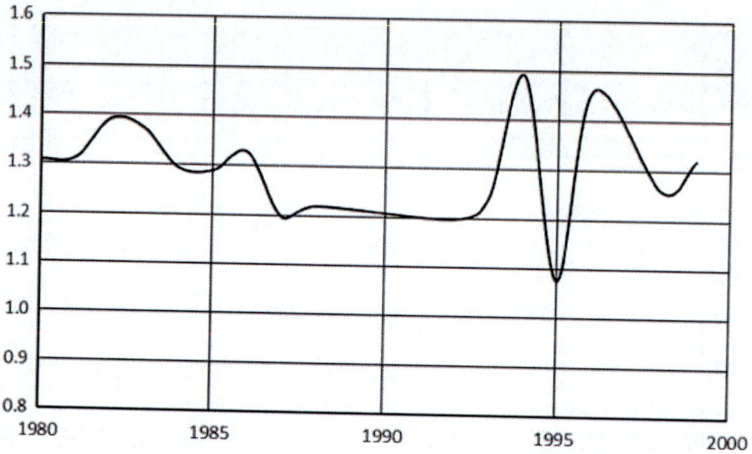

Fig. 10.20 The public expenditure on primary and secondary education in China, in % of GDP

say, ¼, of the private expenditure should be added to the public expenditure. With this, we obtain

$$\text{PEI} = 2.27\%; \text{ and PEF} = 78.8\%$$

India. The public expenditure on primary and secondary education in India in the period 1999–2013, in % of GDP, as provided by the UNESCO data base, is shown in Fig. 10.21.

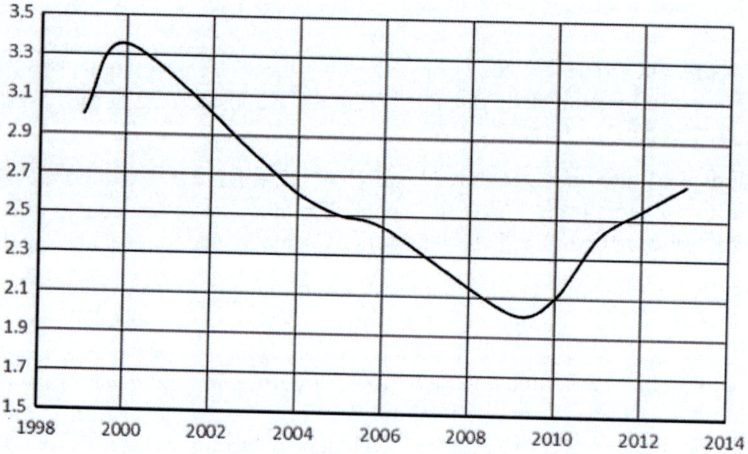

Fig. 10.21 The public expenditure on primary and secondary education in India, in % of GDP

The Indian Ministry of Human Resource Development, in their data base at (https://mhrd.gov.in/statistics-new?shs_term_node_tid_depth=378), under *Educational Statistics—At a Glance—2016*, show expenditure for primary education in 2011–12 at 1.86% of GDP, and for secondary education at 1.03% of GDP (the corresponding values in the UNESCO data base are 1.05% and 1.5%). In order to stay consistent, we stay with the UNESCO values. In the data provided in (tradingeconomics.com/india/education), the enrolment in private institutions is given as 35.21% of total enrolment in primary education, and as 50.51% for secondary education. If we then take expenditure to be proportional to enrolment, the total expenditure on primary/secondary education in 2011–12 equals 4.65% of GDP, and

$$\text{PEI} = 1.05\%; \text{ and PEF} = 54.8\%$$

Indonesia. The public expenditure on primary and secondary education in Indonesia in the period 2007–2015, in % of GDP, as provided by the UNESCO data base, is shown in Fig. 10.22.

For Indonesia, the OECD data base gives the following values for 2015, in millions of rupiahs (Table 10.3):

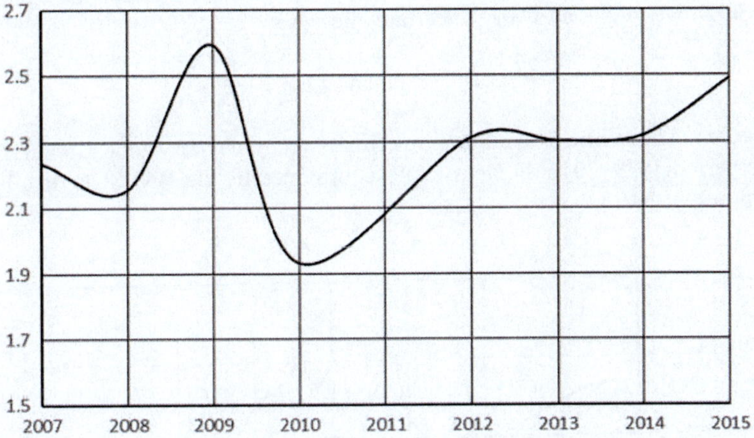

Fig. 10.22 The public expenditure on primary and secondary education in Indonesia, in % of GDP

Table 10.3 Expenditure on primary and secondary education in Indonesia, by source and institutions

Source	Institutions	Primary	Secondary
All	Public	167,100,000	110,100,000
	Private	14,257,762	24,682,038
Government	Public	–	–
	Private	–	–

The Indonesian education system and its financing are quite complex. The report (World Bank, 2013) gives the proportion of private education (by enrolment, in 2009) as 8% in primary, 19% in junior secondary, and 32% in senior secondary, so, given the reduction in enrolment with increasing education level, we might estimate the effective proportion of private expenditure in the overall primary/secondary expenditure of 316 trillion rupiahs to be around 20%.

In 2015, Indonesia's GDP was 11.526 quadrillion rupiahs, with which we obtain

$$\text{PEI} = 2.19\%; \text{ and PEF} = 80\%$$

Nigeria. The availability of data regarding the expenditure on education in Nigeria is very limited; the UNESCO data base gives the amount of government spending in 1975 as 3.06% of GDP. The Nigerian bureau of statistics states that the expenditure on education in Q3 of 2019 was 2.12% of GDP, but the Nigerian budget for 2019 shows an allocation of $1.7 billion (620.5 billion naira) to education, which, with a GDP of $446.543 billion, would only be 0.38% of GDP. The enrolment numbers given in (tradingeconomics.com/nigeria/education) indicate that education in private institutions amounts to about 15% of the total of primary and secondary education.

Taking the amount of 0.38% of GDP as an upper limit on expenditure on primary and secondary education, and the fraction of public primary and secondary education to be 85%, we might, as a very uncertain guess, set

$$\text{PEI} = 0.38\%; \text{ and PEF} = 85\%$$

Sweden. The public expenditure on primary and secondary education in Sweden in the period 1979–2016, in % of GDP, as provided by the UNESCO data base, is shown in Fig. 10.23.

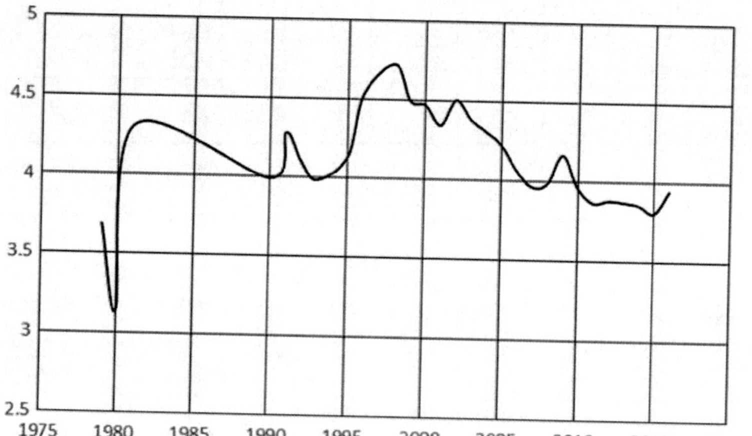

Fig. 10.23 The public expenditure on primary and secondary education in Sweden, in % of GDP

Table 10.4 Expenditure on primary and secondary education in Sweden, by source and institutions

Source	Institutions	Primary	Secondary
All	Public	72,446	69,517
	Private	8637	15,053
Government	Public	72,446	69,517
	Private	8637	15,053

For Sweden, the OECD data base gives the following values for 2016, in millions of kronor (Table 10.4):

In 2016, Sweden's GDP was 4,404,802 million kronor, with which we obtain

$$\text{PEI} = 3.23\%; \text{ and PEF} = 85.70\%$$

United Kingdom. The public expenditure on primary and secondary education in the UK in the period 1971–2016, in % of GDP, as provided by the UNESCO data base, is shown in Fig. 10.24.

For the UK, the OECD data base gives the following values for 2016, in millions of pounds (Table 10.5):

In 2016, the UK's GDP was 1,963,311 million pounds, with which we obtain

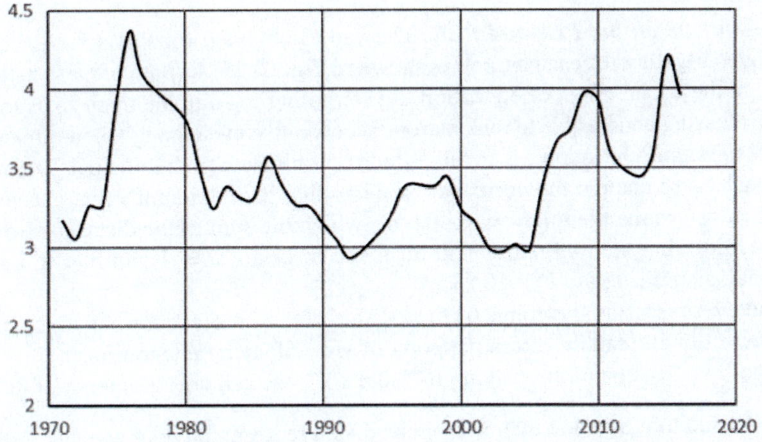

Fig. 10.24 The public expenditure on primary and secondary education in the UK, in % of GDP

Table 10.5 Expenditure on primary and secondary education in the UK, by source and institutions

Source	Institutions	Primary	Secondary
All	Public	27,284	13,161
	Private	10,066	34,084
Government	Public	25,729	12,544
	Private	7363	25,982

Table 10.6 Expenditure on primary and secondary education in the US, by source and institutions

Source	Institutions	Primary	Secondary
All	Public	288,186	313,187
	Private	19,679	28,882
Government	Public	282,375	307,936
	Private	1856	2724

$$\text{PEI} = 2.06\%; \text{ and PEF} = 47.81\%$$

United States. For the US, the UNESCO data base only gives four values for the public expenditure on primary and secondary education, as a percentage of GDP:

2013	2014	2015	2016
3.268	3.251	3.247	3.254

However, the OECD data base gives the following values for 2016, in millions of dollars (Table 10.6):

In 2016, the US's GDP was 18,624 trillion dollars, with which we obtain

$$\text{PEI} = 3.23\%; \text{ and PEF} = 92.53\%.$$

Comments on the PEI and PEF. The values obtained for PEI and PEF can be displayed together for each nation, as shown in Fig. 10.25. As both of these measures relate to the social effect of education, it is tempting to combine them so as to get a single measure and be able to rank nations accordingly, even though their dimensions are not the same. One approach to this, which is reminiscent of a Principal Component approach, is to change the horizontal scale so that its maximum value is also 100, and then determine the line $y = ax + b$ for which the sum of the distances from the points to the line is minimal, which turns out to be for $a = 1$ and $b = 0$, i.e., the diagonal, as indicated in Fig. 10.25.

Then, representing the points by the x-value of their projections onto this line, we find the following ranking from the point of view of social integration:

From Fig. 10.25 and the ranking in Table 10.7, we can note a number of points:

(i) We see that Sweden, which is viewed as a representative of socialist policies and the welfare state, and the United States, which is viewed as a promoter of capitalist and neoconservative policies, are both together at the top of the scale. This illustrates the point that while a common education promotes social integration and a common belief system, it does not determine what that belief is; that depends on the content of the education.

(ii) The size of the society is not a determining factor in the ranking.

(iii) The ranking of Australia demonstrates that the level of social integration and the attainment level, as measured e.g., by PISA testing, are different, and need

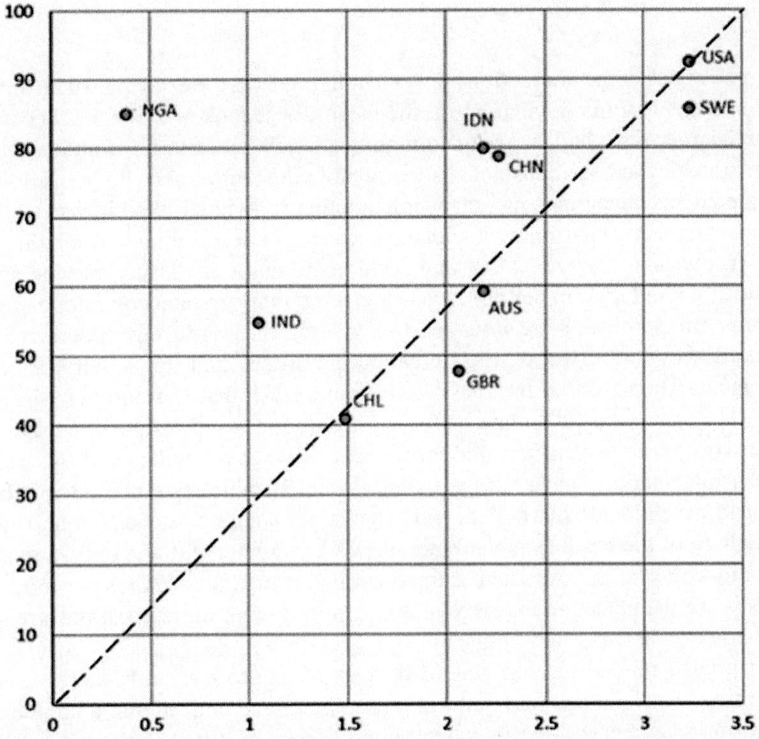

Fig. 10.25 Horizontal axis: PEI, in % of GDP; vertical axis: PEF; in %

Table 10.7 Ranking of the nine nations considered, according to the impact of the education system on social integration

United States	0.92
Sweden	0.89
China	0.72
Indonesia	0.71
Australia	0.61
United Kingdom	0.53
Nigeria	0.48
India	0.42
Chile	0.42

to be considered as such. (For example, in the 2018 tests, the average scores were Australia 499, Indonesia 382, and UK 504.)

(iv) Developed countries lie on or below the line, whereas developing countries lie above, indicating the rising affluence tends to promote increasing privatisation of education.

10.2.5 Art and Culture

When we speak of the interaction between individuals as the exchange of information, we immediately think of language as the means, be it spoken or written, because it is by far the most prevalent means of communication. We receive information at a much greater rate through the constant observation of our environment, but most of it is not interpreted as conveying a message from another individual. Walking on a crowded street I observe my environment with the purpose of not colliding with anyone; going for a bushwalk I observe Nature and admire and enjoy its beauty and the sense of tranquillity it induces; in neither case is there any interpersonal communication. But some manmade objects are intended to convey a message from the creator to the observer; they are artworks, their creators are artists, and the activity of creating them is art. The nature of the message is complex, in that it usually requires many words to convey the full range of feelings, emotions, and insights conveyed by a single work of art. Art is a major component of what we understand by "culture", and the information conveyed by art is in what is defined as class 3 in Chap. 3; information relating to our beliefs and capable of changing our behaviour. If we use the analogy of the brain as a computer, artworks interact with the operating system via the machine language, whereas interaction via language requires a compiler, and in this sense the extent and variety of art is an important measure of the strength of the interaction between individuals.

Art is also related to, and should be part of, education, and "there's a strong relationship between arts and cultural engagement and educational attainment. We see an improvement in literacy when young people take part in drama and library activities, and better performance in maths and languages when they take part in structured music activities." (Arts Council chief Sir Peter Bazalgette, The Guardian, 27.04.2014).

A report by the Australian Academy of the Humanities (A New Approach/Insight research series/Report One/2019), states "It is well-established that engaging in cultural and creative activities helps us develop a sense of belonging, forges social cohesion, stimulates curiosity and the ability to engage with different perspectives, and can have a range of beneficial effects on health, wellbeing and education outcomes."

Obtaining a measure of the strength of interaction in terms of economic data is difficult for at least two reasons. One is that the value of an artwork, as the price it would fetch on the market, has little to do with its ability to convey a message; art has become a commodity, and price is determined by scarcity. Secondly, if we want to reduce the effect of the market and instead obtain a measure of the importance society places on art by looking at government expenditure on the arts, it is difficult to obtain comparable values (or, indeed, to determine exactly what a particular expenditure value covers). Some include expenditure on libraries, others do not; some include expenditure on art in schooling, others do not, and for many nations no relevant data is available at all. Developing this measure must therefore remain a work in progress, but as an initial input, the following Table gives the per capita expenditure, in US$,

Table 10.8 Government *per capita* expenditure on the arts, in US$, for one year in the range 1993–96

Australia	25	Ireland	9
Canada	46	Netherlands	46
Finland	91	Sweden	57
France	57	UK	26
Germany	85	US	6

Table 10.9 The government expenditure on the arts as a proportion of GDP—the Arts Appreciation Factor AAF—based on the data in Table 10.8. The *per capita* GDP is as per IMF estimates for the year 1995

Nation	Table 10.8	per capita GDP	AAF (10^{-4})
Australia	25	20,690	12.1
Canada	46	20,642	22.3
Finland	91	26,350	34.5
France	57	27,898	20.3
Germany	85	31,908	26.6
Ireland	9	19,083	4.7
Netherlands	46	28,911	16.9
Sweden	57	29,883	19.1
UK	26	22,759	11.4
US	6	28,763	2.1

for a year in the range 1993–96 (Government Expenditure on Arts and Museums, Table 1—International Data on Government Spending on the Arts, US National Endowment for the Arts, Research Division Note #74, January 2000) (Table 10.8):

That a wealthy nation spends more on arts per capita does not in itself give a good indication of how important the society considers art to be. A more appropriate measure is the government expenditure on the arts as a proportion of GDP—a measure we might call the *Arts Appreciation Factor*, AAP—and Table 10.9 shows the corresponding conversion of the data in Table 10.8 (with 1995 as the year).

In the case of Australia, the aforementioned AAH report states "Australia's creative and cultural activity is a significant component of our national economy, contributing more than $111.7 billion, or a 6.4 per cent share of Australia's Gross Domestic Product (GDP), in 2016–17. More than half a million Australians work in the creative economy, which employed 593,830 people in 2016, representing about 5.5% of the national workforce." Comparing this with the value of government expenditure given for Australia in Table 2.9, even though this is for art alone, it is clear that government expenditure is just one (small) measure of a nation's appreciation and engagement with art and culture. Nevertheless, government expenditure is a measure of the importance society assigns to art.

10.2.6 *Significance of the Economic Measures*

What can we say about the significance of the economic measures introduced above with regard to our view of the evolution of society? A starting point for answering that question is to reiterate what our view is: The evolution of society is the current stage of a general process of evolution; a process that at the highest level of abstraction can be seen as a process of converting free energy into bound energy, with the realisation of the generalised concept of energy taking on a different form in each successive stage. On Earth, this sequence of stages is divided into two parts: before and after the advent of life. Life meant that the individuals of the species were no longer only substances; they embodied processes, so their existence became dependent on time, and with it arose the concept of survival. This could not be just the survival of the individual (although that was obviously essential), but the survival of the species, and as evolution went through a progression of increasingly complex species, so did the concept of survival. For the last tens of thousands of years, the species have been the instantiations of the genus society, and the progression of evolution has been determined not by random mutations and sexual propagation, but by the actions of the individuals. The distinction between free and bound energy is represented by the outcome of the actions, and when the outcome is expressed in economic terms, the free energy is measured by individual income and wealth, while the bound energy is measured by the common infrastructure.

Now, in our view the actions are determined by the information available to individuals at the time of action, and that information is of two types: the information received through interaction at the time, and the information stored in our memories, with the action resulting from the evaluation of the former on the basis of the latter. The stored information is acquired though interaction with other individuals, and so the evolution of society is characterised by the strength of the interaction within society, with strength being a society-level composite measure including intensity, quality, complexity, and reach (discussed in Chap. 5). The common infrastructure is a function of this strength, but it is not a simple function, and it is time-dependent. The relationship between interaction strength, common infrastructure, and the evolution of society—i.e., the appearance of new species—is illustrated in a highly simplified and stylised manner in Fig. 10.26, with the species of society and the timescale reflecting European history, and it is in the sense of this illustration that the common infrastructure can be considered a measure of the evolution of society.

Two comments on Fig. 10.26: The slowing down of the increase in the interaction strength in most recent times is intended to show the reduction in the quality of the interaction (as noted in the previous section), and the increase in the common infrastructure indicates that the species succeeding the nation is starting to emerge. What this new species might be is the subject of Chap. 12, but if the emergence of the earlier species is any indication, the potential for significant upheavals and violence cannot be discounted.

The LEI and the AAF are both indications of the state of society, i.e., of how far it has progressed in its evolution. The other measures—PEI and PEF—as well as the

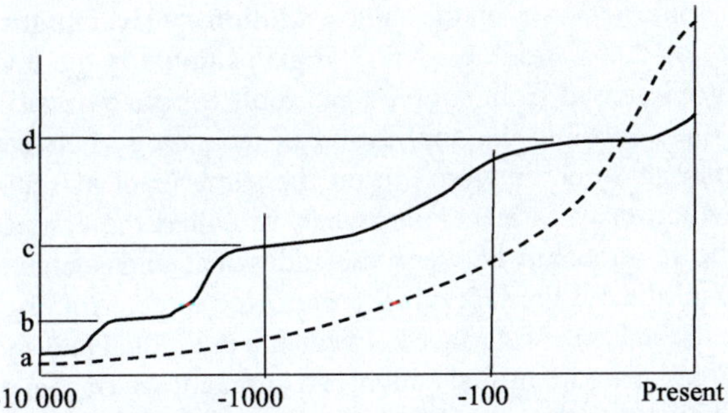

Fig. 10.26 A highly simplified illustration of the relationship between interaction strength, common infrastructure, and evolution of society over the last 10,000 years. The full curve is the common infrastructure as a proportion of GDP, i.e., the LEI; the dashed curve is the interaction strength, and the species of society are: a = clan, b = fiefdom, c = kingdom, and d = nation

two measures introduced in Chap. 6—PAL and WCI—characterise the dynamics of the evolution, which in our view of the evolution of society is determined by the operation of the collective intelligence. If we would form an average, weighted by GDP, of these four measures, then the greater the value, the more active the collective intelligence is. There is no definite scale on which this value could be measured, but as a function of time it shows in which direction and how fast a nation is changing. Because wealth, or capital, is so transnational in nature, a national version of the WCI is of little significance, but for the world, a large value is a danger signal.

10.3 Other Measures

10.3.1 Taking a Step Back

The economic data accessed in the previous section is just a miniscule fraction of the economic data available, and economic data is but one of many classes of data used to provide a quantitative description of society. They include such diverse measures as the number and types of cars, the number of homeless people, the number of patents issued, the number of books published, and the number of people with particular diseases, just to mention a few. They all say *something* about society, but what exactly do they say; what is their significance with regard to understanding the evolution of society? This brings us back to the question posed at the start of this chapter: What makes two societies different in the sense of being at different stages in the evolution of society, rather than just different versions—smaller/larger, or richer/poorer, or more powerful/less powerful—versions of the same society? Many measures are what we might call *indicators* of evolution, in the sense that they have clearly increased or decreased in value as society evolved over the last

10,000 years, while societies on the same evolutionary level might still show greatly varying values of these variables. An obvious example is the level of technology: it has certainly increased from simple hand tools to space travel, but there are still considerable differences in the application of technology between today's nation-states, even though some of them are on the same level of social evolution. The measure of evolution we are seeking, what we called the essence of society, is a measure of the relationship between the individual and society, of the extent to which the capabilities of the individual are utilised in determining the evolution. In Sect. 3.4.2 we looked at the interface between the individual and society in the action view, and in Fig. 3.6 we identified a high-level breakdown of society in terms of five components. If we use this same breakdown to describe the interaction between the individual and society as regards the individual's role in changing society, the interface to government is realised in the political system, as described in Chap. 7 and taken up again in Chap. 12. The interface to the component labelled "production" is the interface between the individual and work, and the changing nature of this interface is a further important aspect of the essence of society.

10.3.2 Work and the Individual

In the 2019 issue of the World Development Report by the World Bank, entitled *The Changing Nature of Work*, the changes taking place in the nature of work are described and analysed under the perspectives of *The changing nature of firms, Building human capital, Lifelong learning, Returns to work, Strengthening social protection*, and *Ideas for social inclusion.* The report documents the changes, both in recent times and as they are likely to develop in the near future, and their effects on society, and advocates for initiatives from government and industry that will enhance the positive side of these effects. Advances in technology are clearly the drivers of the changes, but the report points out that the issue is not, at least so far, a loss of jobs, as the number of jobs lost to the application of technology (through automation) has been less than the number of jobs created by the introduction of new technology; the problem lies in handling the transition, in the form of reskilling and social support.

In the context of the present investigation, what the report illustrates, particularly in the chapter on building human capital, is that not only is work changing from manual labour to the operation of increasingly sophisticated machines, but that the interaction of the individual with work has changed from just the particular manual activity to increasingly include the social environment of this activity.

A very simple example can illustrate this. Consider a manual labourer whose job is to dig trenches. Originally his tools are a spade and a pickaxe, and every day he is told where to dig and how wide and deep the trench needs to be. He is not told why this trench is needed, nor why it is needed at this particular time; he simply spends his eight hours digging and then goes home without any further involvement with the work. At some stage an excavator arrives, and the worker is given training in how to operate it. The productivity is now greatly increased, but this introduction of technology has not changed the labourer's involvement with the work. But then the complexity of the sites on which the work is to be done increases, and the labourer is now expected to understand how the work interacts with others working on the site and to coordinate his work with them, so that the overall operation runs smoothly. He may also have to handle complaint (and abuse) from people who are inconvenienced by the work, and some of all these social issues may stay with him when he goes home. The nature of the involvement with the work has changed from a limited physical involvement to include an intellectual involvement that is both significant and basically unbounded.

The World Development Report defines human capital as follows:

> Human capital consists of the knowledge, skills, and health that people accumulate over their lives, enabling them to realize their potential as productive members of society. It has large payoffs for individuals, societies, and countries. This was true in the 1700s when the Scottish economist Adam Smith wrote, "The acquisition of . . . talents during . . . education, study or apprenticeship, costs a real expense, which [is] capital in [a] person. Those talents [are] part of his fortune [and] likewise that of society." This is still true in 2018.

The report focuses on the return of human capital in the form of the future productivity of a child born in 2018 and defines this productivity in terms of three components: Childhood survival (probability of survival to age 5); schooling (quality-adjusted years of schooling); and health (fraction of children not stunted, and adult survival rate). The measure is as a fraction of worker productivity relative to a benchmark corresponding to 100% survival rate, complete education, and full health, determined from the extensive micro-econometric literature on estimating returns to education and health. For the nations used as representative in Sect. 10.2, the report provides the values shown in Table 10.10.

However, the report also recognises that human capital matters for society, and illustrates this by the following examples:

Over the longer term, human capital matters for societies. In the mid-1970s, Nigeria introduced universal primary education, sending a large cohort of children through primary school who otherwise would not have gone. Years later, the members

Table 10.10 Productivity, as a fraction of the benchmark, and the ranking within the total of 157 nations evaluated, for the nations considered in Sect. 10.2 (*The Changing Nature of Work*, 2019 World Development Report, World Bank)

Nation	Productivity fraction	Ranking (out of 157)
Australia	0.80	7
Chile	0.67	45
China	0.67	46
India	0.44	115
Indonesia	0.53	87
Nigeria	0.34	152
Sweden	0.80	8
United Kingdom	0.78	15
United States	0.76	24

of that cohort were found to be more engaged in political life. They paid closer attention to the news, spoke to their peers about politics, attended community meetings, and voted more often than those who did not go to primary school. Young participants in the National Volunteer Service Program in Lebanon, an intercommunity soft skills training program, display higher levels of overall tolerance. As the scientist Marie Curie once said, "You cannot hope to build a better world without improving the individuals."—Human capital also fosters social capital. Surveys typically find that more educated people are more trusting of others. And, conversely, failing to protect human capital undermines social cohesion.

The report could be the beginning of a valuable contribution to describing the evolution of society, but its focus on productivity and the associated ranking makes it more applicable as an input to the operating conditions of the collective intelligence, as treated in Chap. 6. Our concern in this section was with how the changes to work can be used as a measure of changes to the individual's social involvement.

Chapter 11
Tension and Instability

11.1 Introduction to This Chapter

The purpose of this chapter is to investigate what we in Sect. 3.3.1 introduced as *fluctuations* in the evolution of society. Fluctuations are relatively rapid deviations of the state of society from its ideal path of evolution, defined only as the path that would result from the unrestrained operation of the transformation process. We cannot say what this path will look like in the future, but we can say something about how far the operating conditions of the transformation process are from their ideal values. There are other deviations, such as the currently important global warming, that have a lesser or greatly variable rate of development; the reason for focusing on fluctuations is the assertion that we are currently at a critical point in the evolution of society, as argued in Sect. 1.2, that could lead to a major fluctuation.

In earlier chapters we developed a view of society as an information-processing system at the highest level of abstraction, where it is represented as a system of identical processors interacting by exchanging information via two paths: the collective memory and the circulating information. In addition, each individual can generate new information by discovery and reflection. At this level, the system is uniform or unstructured, and in the model of the individual presented in Chap. 4, this is reflected in the uniformity of the interaction and the resulting dynamic of the identity, as expressed by the identity partition, α. That is, this model is essentially a description of the ideal path in the information view, and it is not able to describe deviations from this path. All we could do, in Chaps. 6–9, was to identify factors that reduce the operating conditions of this model from their ideal values, such as inequalities in wealth and income. Therefore, in Chap. 10, we essentially bypassed the model and sought more direct and measurable parameters characterising the evolution of society, relying above all on the vast quantity of economic data available.

Due to its high level of abstraction, and above all its requirement for uniformity, the model and the view it represents hide issues creating tensions in today's society, issues arising from inequalities between different *parts* of society, so our first step in expanding the model is to relax the requirement for uniformity and introduce

E. W. Aslaksen, *The Evolution of Society*,
https://doi.org/10.1007/978-3-031-23054-7_11

an embryonic *structure* in the information domain and in the physical domain as a means of developing the concept of *tension*, both in describing the state of society and as a driver of the evolution of society. This is the subject of Sect. 11.2.

In the uniform model, the operation of the collective intelligence is a consensus-building process that maintains a single, albeit evolving collective identity. And this drive to uniformity and consensus does reflect a major feature of real society; we try to make others believe what we believe and behave as we behave, mainly to allay our own doubts. However, once we allow society to display a structure, a new concept arises, that of *acceptance*. That is, once we accept a certain difference between parts of the structure, it ceases to be a source of tension. Acceptance is an essential feature of society; it applies equally to structure in the information view as in the action view and is the subject of Sect. 11.3.

Section 11.4 deals with the concept of power in the context of society and how it is the most significant characterisation of structure, and finally, Sect. 11.5 considers the implications of the foregoing for the stability of society and its evolution. However, before we embark on this expansion of our model of society, just two brief notes: one on the use of analogies, and the other on equality.

Analogies are often useful in inferring features in a new domain, the *target domain*, by recognising similarities with a well-known domain, the *source domain*; the similarities can be between components or between relationships. In most cases the similarities appear at a certain level of abstraction, and this determines the level of detail, or accuracy, of the inference. In this book so far, we have introduced several analogies between physics, as the source domain, and society, as the target domain. The first one involved the abstraction of a generalised energy and its two forms—free and bound, by which evolution could be seen in both domains as the conversion of free energy into bound energy. This analogy of evolution in the two domains led to an initial understanding of integration and complexity in society. A further analogy with physics as the source domain was that between a collection of water molecules and society, illustrating the central role of the interaction and how its strength determined the structure in both cases.

We then extended the source domain to include self-sustaining processes in the form of organisms, where the concept of temperature takes on an added significance. In an organism temperature is a measure of the level of activity of the processes that maintain the operating conditions of the rest of the organism; in a society it is a measure of a deficit in the effort expended by its members to maintain its viability and thereby provide the basis for survival of its members. In both cases these metabolic processes convert energy taken from the environment into the work required to maintain the organism and society, respectively. And in both cases these processes and their representation as temperature must remain within a relatively narrow range, so that the analogy reflects the concept of homeostasis onto society.

The present chapter is dedicated to investigating the causes of fluctuation in the evolution of society, as we defined them in Sect. 3.3.1. In terms of our analogy with the human body we want to investigate society's health or, conversely, the fluctuations as the pathology of society, and as a framework for this investigation we shall employ a further analogy between the human and society. In the case of the human, health has two aspects: *physical health*, as in the absence of disease, and *mental health*, as in the absence of anxiety, paranoia, and hallucinations, with an understanding of the close coupling between the two and that the physical body provides the medium in which mental processes take place. Analogously, in Chap. 3 we defined two views of society—the *information view* and the *action view*—with the former considered as the primary view, in that actions are determined by information, but recognising that the action view describes the operating conditions for the processes in the information view. In the following we want to keep this analogy between the two aspects of human health and the two views of society in mind, and to use our knowledge of the pathology of the human to develop our understanding of fluctuations in the development of society, where appropriate.

Equality is one of those words that are used frequently, just like freedom, as we considered it in Sect. 6.6.2. As with freedom, there is an implicit understanding that equality is something good, without any further examination of the context, and equality has also always been an important feature in the evolution of society, as in the battle-cry of the French Revolution, "Liberté, égalité, fraternité". In reality, equality is a contested concept, but both because a concise definition and assessment is provided in the Stanford Encyclopedia of Philosophy (www.plato.stanford.edu/entries/equality/, accessed 20.05.202) and because our use of the concept will be quite specific, we need not consider the nature of the concept here, except to extract the following: "'Equality' (or 'equal') signifies correspondence between a group of different objects, persons, processes or circumstances that have the same qualities in at least one respect, i.e., regarding one specific feature, with differences in other features." As was clear in both Chaps. 6 and 10, our interest is not so much with equality as with its converse—inequality, and so the definition can be inverted to read "'Inequality' signifies correspondence between a group of different objects, persons, processes, or circumstances that have different qualities in at least one respect, i.e., regarding one specific feature, ignoring differences in other features."

These definitions show the importance of qualifying the use of equality and inequality by the feature to which they apply; there is no such thing a 'equality in general'. In the French motto, equality obviously does not include skin colour; it is not about *being* equal, it is about *being treated* equally. But the unqualified appearance of both freedom and equality in the national motto might perhaps give a hint as to a cause of the elevated degree of social unrest in France.

11.2 Structure

11.2.1 Structures in Society

As with so many of the concepts we have encountered, the meaning of 'structure' is dependent on the context in which it is used. In civil engineering a structure is a built object such as a building or a bridge, a structural element is a load-bearing part of

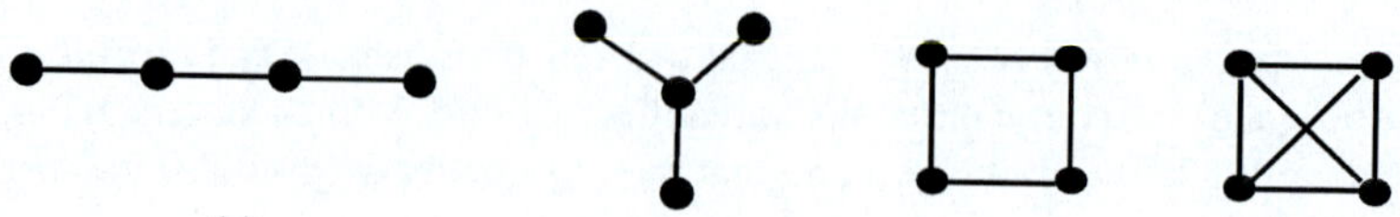

Fig. 11.1 Four structures of a system with four elements: **a** linear, **b** central, **c** ring, **d** maximally connected

this object, and structural engineering is concerned with the integrity and strength of the object. In systems engineering 'structure' refers to the pattern of the relationship between the system elements, and four different patterns are illustrated in Fig. 11.1.

In sociology, one definition is as the distinctive, stable arrangement of institutions whereby human beings in a society interact and live together (https://britannica.com/topic/social-structure, accessed 20.06.2022); another is as the patterned arrangements that are both emergent from and determinant of the actions of individuals (https://en.wikipedia.org/wiki/Social_structure, accessed 20.06.2022). We shall adopt a related definition, but one more aligned with our understanding of a system:

> A social structure is a system of institutions, with an institution defined as a system of individuals interacting through a set of rules.

As a consequence of this, we do not talk of *the* structure of a society; in any society we will generally be able to identify several different structures, each defined by the characteristic that identifies the relationship between the individuals. Examples of characteristics are consanguinity (by recognized birth) or affinity (by marriage, adoption, or other relationship), religion, and nationality; defining corresponding institutions—family, church, nation, and the behaviour of the individuals within each of these are governed by a set of rules or laws.

As a social structure is a system, it is reflected in both views of society and, depending on the characteristic defining the relationship between the individuals, one view may be more useful and descriptive than the other. If the characteristic relates to what the individual is or believes, as religion or ideology, a description in the information domain will be preferred; if it is related to what the individual does, a description in the physical domain will be preferred. In the latter case, the interactions can be in the form of an exchange of services (including products) or of information, or both, but in any case, a structure involves a process, with the individuals defined by their participation in this process. If the institution is characterised primarily by this process, it is a functional institution; and the social system constituted by these institutions is characterised by a *division of labour*. Such a structure of society into interconnected elements, each performing a necessary part of society as a process, much as the parts of an organism, was introduced by Durkheim (1893) (see also (Pope, 1975)) and further progressed by various authors, including Talcott Parsons, who developed functional analysis as a theoretical tool (Parsons, 1971).

Perhaps the best-known case where the system is defined mainly by what its institutions are is the description of the industrial society by Marx (1976), where the elements are *classes* defined by their relations to the means of production.

There are several reasons why a society might separate into groups and form a structure. One is a geographical one; the terrain on which the society is located makes interaction across the boundary more difficult that within the groups; this could be due to a wide river or high mountain ridge. Another reason is the distribution of natural resources and the benefit the grouping brings in competition for these resources. And thirdly, and of particular interest in the present context, is the emergence of differences in beliefs and customs, as such differences tend to have a self-reinforcing aspect. In all cases, the separation gives rise to a difference in the interaction between individuals; the interaction between the individuals within a group is stronger than the interaction between individuals in different groups.

Before leaving this very brief introduction to structure in society it is appropriate to note the important contribution of Norbert Elias, a German sociologist that lived in the period 1897–1990. He identified the relative autonomy but co-dependence of individuals and society, and argued that structural transformation could be explained in terms of the shifting relations between individuals and society over time. He referred to this process as figuration, with figurations as networks of individuals that form organisations and institutions (Elias, 1978), and we see how this resonates with our concept of the transformation process.

11.2.2 Structure in the Information Domain

11.2.2.1 Model of a Two-Component Structure

The simplest non-uniform structure is a society consisting of two groups of individuals, but with uniformity within each group.

Consider the following thought-experiment ("Gedankenexperiment"): An initial society is separated into two identical societies, each with the same number of individuals, *n*, with initially no interaction between them, and both experiencing the same unchanged operating condition for the collective intelligence. As time goes by, these two societies will develop different collective identities, but for simplicity we may assume that the persuasiveness, *p*, of the new identity items generated is the same in both societies, and that the strength of the assertions, ν, is the same in the two identities. Let the two societies be identified as A and B. The identity of an individual in A will now be subdivided into three subsets:

$\alpha_0\Theta$: the set of identity items common to individuals in both societies;

$\alpha_a\Theta$: the set of identity items common to all individuals in A; and.

$(1\text{-}\alpha_a)\Theta$: the set of identity elements particular to an individual in A.

And similarly for the identity of an individual in B.

Intuitively we might consider the ratio $(\alpha_a + \alpha_b)/\alpha_0$ to be a measure of the difference between the belief systems, or collective identities, of the two groups. But the *effect* of this difference depends on the strength of the interaction between the two groups; if there is no interaction, they do not know that they are different, and the difference has no effect We must consider the exchange of identity items between the two groups.

When the interaction between the two societies is turned on, there will be an influx of identity items from one society to the other, and for the moment, assume that the flow is equal in either direction and is the result of random sampling of the originating circulating information, and denote it by μ_x.

Prior to interaction, the behaviour of either society is governed by the dynamics developed in Sect. 4.3 (Eqs. 4.1 and 4.3):

$$w\frac{d\beta}{dt} = \mu_1 \cdot (1 - \alpha) - \mu_2 \cdot \beta; \tag{11.1}$$

$$z\frac{d\alpha}{dt} = \mu_2 \cdot \frac{p}{\nu} \cdot \beta - \mu_0 \cdot \alpha; \tag{11.2}$$

and the steady-state solutions for α and β:

$$\alpha = \frac{\pi}{\pi + \omega}; \tag{11.3}$$

$$\beta = \frac{\omega}{\pi + \omega}\frac{\mu_1}{\mu_2}; \tag{11.4}$$

where $\pi = p/\nu$ and $\omega = \mu_0/\mu_1$.

Once the interaction is active, society A experiences an inflow of identity items (i.e., class 3 information items), of which the major part, $(1 - \alpha_0)$, are in conflict with the identity of individuals in society A and need to be processed by their Process B. The expression for $d\beta_a/dt$ is now:

$$w\frac{d\beta_a}{dt} = \mu_1 \cdot (1 - \alpha_a) - \mu_2 \cdot \beta_a + \frac{\mu_x}{n}(1 - \alpha_0); \tag{11.5}$$

and correspondingly for $d\beta_b/dt$.

The equation for $d\alpha_a/dt$ remains unchanged:

$$z\frac{d\alpha_a}{dt} = \mu_2 \cdot \frac{p}{\nu} \cdot \beta_a - \mu_0 \cdot \alpha_a; \tag{11.6}$$

and correspondingly for $d\alpha_b/dt$.

The steady state of the two societies is now determined by four equations, but due to our assumptions regarding the societies' similarities, they can be separated into two sets of two, and for each society we obtain

$$\beta = \frac{\omega}{\pi + \omega}\frac{\mu_1}{\mu_2}\left[1 + \frac{\mu_x(1 - \alpha_0)}{n\mu_1}\right]. \tag{11.7}$$

The second term in the bracket in Eq. 11.7 is the *tension* resulting from the interaction between the two societies; it manifests itself as an increase in the stress experienced by the individuals. The factor $\mu_x/n\mu_x$, which is a measure of the *interaction strength*, expresses the fraction of identity items processed by the individual due to the interaction, and the $(1 - \alpha_0)$ factor simply expresses that the more the two societies have in common, the less the stress of their interaction. However, it is important not to forget the high level of abstraction involved in this model, both as regards the fictitious "average" individual interacting with a fictitious circulating information, and the purely quantitative characterisation of information in the form of items. In particular, throughout the whole development of the information processing model we have neglected the issue of *repetition*, and the effect this has on persuasion. In the stream of information items arriving in society A from society B, the items from the α-part of society B's identity are repeated *n* times compared to the items from the $(1 - \alpha)$ part, which means that a society that has a strong belief system will have a greater influence for the same interaction strength.

The stress that arises from the flow of conflicting information reflects a fundamental feature of human beings, a feature that can be characterised as *insecurity*. It arises as a result of our intellectual ability to process the observations of our environment and form an understanding of that environment and our place in it—we construct our understanding of reality. It is an energy-consuming process, and as we complete it, it gives us a feeling of peace and certainty. But there remains a certain niggling feeling that perhaps our view is not quite correct or complete, a certain insecurity, and as the flow of conflicting information increases, so does our uncertainty and the effort required to handle it. If we were one hundred percent certain about our understanding of reality, this uncertainty would not arise. But reality, as we are able to perceive it, is much too complex to ever have complete certainty about our construction of it, and the structuring of society is in part due to our desire to limit the uncertainty and associated stress. As the interaction between society's members increased, structuring allowed that interaction to be subdivided so that the interaction within an element contained less conflicting information, and the interaction with members in other elements was reduced. This was raised already in Sect. 5.4 and demonstrated by the study into the limits of settlement growth.

Thus, the level of interaction in a society and the complexity of its structure are closely connected; structure can be seen as the result of a human defence mechanism and increasing the level of interaction leads to increasingly complex structure. And, as we shall discuss in Chap. 13 regarding the world society, if this structuring and construction of new institutions is not carried out in a timely manner, the stress can take on levels that result in violent confrontations. This is another example of what was illustrated for the case of new technology in Fig. 2.2.

11.2.2.2 Informal Structures

In the view and model of society we developed in the first seven chapters, the members of society were represented by a set of identical individuals with identical identities that interacted through the public discourse by means of the circulating information present in the public sphere. In the previous section we relaxed the requirement for the identities to be identical and demonstrated how in that simplest of cases this introduced a change in the dynamics of the public discourse and a certain stress. Generalising this picture to a society with several structural elements, the individuals in each element have identities that are distinguishable from those of individuals in the other elements; we could say they bring a different point of view to the public discourse. But the concept and the operation of the public discourse, and thereby of the collective intelligence, remain basically the same.

A different type of structure arises when subsets of the individuals form their own interaction spheres, or *spherules*, for the purpose of discussing specific topics, with little or no interest in linking their discussions to the public discourse. When such individuals participate in the public discourse, a part of their identities is excluded from the process, and their participation is in any case reduced by the amount of time spent interacting within their own spherules. The subject of this stratification of the public sphere into types of domains of various sizes and with varying degrees of formality and of being "public" is considered in some detail in (Bruns, 2019). He notes that whereas the public sphere is an explicitly idealist concept, as is also true of our circulating information, and therefore not easily accessible for empirical validation, "by contrast, especially at the lower levels ranging from personal publics through issue publics to public spherules, some of the newer contributions to the field also seem to hold a much greater potential for empirical validation. This is in part also because they—and public communication more generally—have come to rely considerably more prominently on digital media, channels, and platforms for their operation, enabling researchers to draw on methodological advancements in the field of Internet studies, and to apply the computational methods emerging in this field to the 'big social data' provided by contemporary digital platforms". He then goes on to explore how such methods might be applied to data extracted from one leading social media platform, Twitter, to shed new light on the various elements constituting 'the' contemporary public sphere in Australia. That line of research is at a level of much greater detail than presented in this book, but it is obvious that it must be of great significance to developing our information view of society further.

It is possible to identify other types of structure of the public sphere, such as one determined by the power of the main actors. Habermas distinguishes four categories of power: political power, social power, economic power, and media power (Habermas, 2006, p. 418). The last of these was the treated in Sect. 6.5, and power in general is the subject of Sect. 11.5 but our main interest in the structure of the information domain will be as it applies to the structure of the world society in terms of nations—the subject of Chap. 13. It is different to the structure of the public sphere within a nation, or even a world society of individuals, where the structure is in terms of individuals, as in Habermas (2006). In a world society of nations, the structure is predominantly in terms of nations, with individuals, with few exceptions, playing a secondary role as representatives of nations.

11.2.3 Structure in the Physical Domain

11.2.3.1 Life and Its Government

Our point of departure is the observation that any society, of any form and size—a family, a sports club, a company, a political party, a nation—includes some form of *management*. Consequently, we can always partition a society into two elements according to their functions: The element whose function is to manage the society, and the element that comprises all the other functions, and we shall call these two elements *Government* and *Life*. And while there is nothing in this view that limits it to a nation as the society, our model is aimed at nations.

The first thing to note about these two elements is that they are functional elements. That is, it is not a partitioning of the people in society into two groups, but of the activities taking place in society. The people performing the Government functions also perform functions in the Life element, such as the functions of parents. It is also possible to characterise the two elements by the types of organisations in which the functions are performed, such as the judiciary and the legislature in the case of Government, and family and corporation in the case of Life.

Our definition of the element identified as Government is narrowly focused, as follows:

(a) Interacting with the people of the society to determine their will with regard to the development of society, both internally and with regard to its membership of the world society;
(b) formulating laws that ensure that the development of society is in accordance with the will of the people, and that it progresses in a continuous and stable manner; and
(c) administering the operation of, and enforcing compliance with, these laws.

The meaning of "laws" is to be taken as including such words as acts, regulations, and rulings, and these laws concern much more than the most common issues, such as individual behaviour in society, as enforced by the police. In particular, they include laws that define the structure of society, in the form of wealth distribution (taxation), power relationships (employer vs. employee, corporate structure and responsibility), and, indeed, the role and operation of Government itself. But even so, this definition of Government is much narrower than our understanding of what functions "government" normally includes, such as education, health care, research, and defence. There are two reasons for this: One, it provides the partitioning of society most suited to our purpose; that is, a view of society that exposes its dynamic stability. Two, it avoids the issue of public versus private ownership, and the extent to which government should be involved in commercial activities (an issue we shall take up in Sect. 12.3.2). The element identified as Life then encompasses all the other activities people are engaged in; they are so manyfold and varied that they do not fit into any single taxonomy or ordering, but there are some characteristics that can be useful in thinking about these activities, as was noted in Sect. 1.3.2.

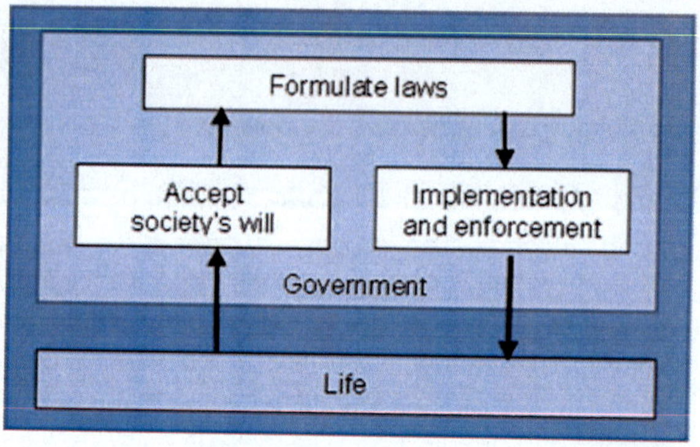

Fig. 11.2 The three main functions of Government, and the interaction of Government with the other element of society, life

Our two-element model is illustrated in Fig. 11.2, and we see that the relationship between these two elements has the form of a feed-back loop. Life evaluates the state of society and decides on appropriate changes by exercising its collective intelligence within the life processes, Government accepts these changes, formulates laws to express the changes as changes to the framework within which the Life processes take place, and implements and enforces these changes. Life then progresses within this new framework, and the whole process repeats. Note that the two arrows between Life and Government represent flows of information; Life informs Government of desired changes, and Life receives information *about* the Government's implementation (not necessarily only *from* Government).

In the process shown in Fig. 11.2 it is important to recognise that, in our model, the decisions about changes take place within the Life processes; that is the location of the collective intelligence, it is not within Government. However, for the process to work well, the conditions that promote the collective intelligence, which was the subject of Chap. 6, need to be realised to a significant extent. We might think of these conditions as expressing the *quality* of the Life processes, and a measure of this quality was proposed in Sect. 6.3 Because raising the level of this quality requires a great effort over an extended time period, it has been expedient to compensate for a low level of this quality by (temporarily) shifting the decision-making to the Government; a current example of this is China. But with the quality level present in most Western democracies, there is little excuse for this. On the contrary, a major responsibility for Government should be to ensure that the quality level and the effectiveness of the collective intelligence are monotonically increasing with time.

In order to make the relationship between the political process and the collective intelligence explicit, we recast Fig. 11.2 in the form shown in Fig. 11.3, where the political process is shown as it actually is—partly within Government and partly within Life. Within Life, we also show both collective intelligence and business as two distinct components. This view should be compared with the one presented in Sect. 7.3 and illustrated in Fig. 7.3, where we focused on the political process.

In Fig. 11.3, the path from the collective intelligence to the political system is meant to represent the democratic process, i.e., the process in which the whole

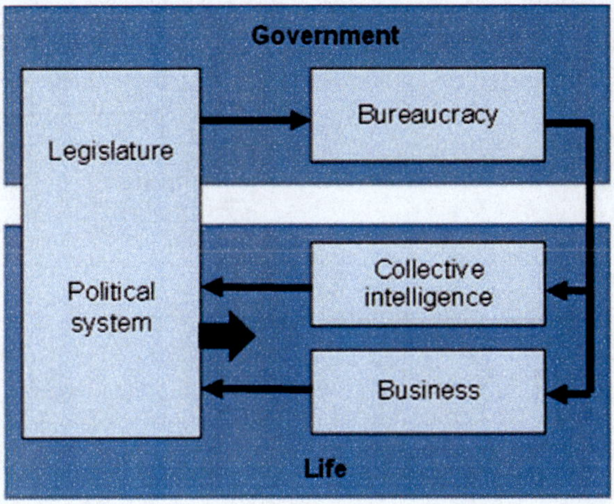

Fig. 11.3 A particular view of society as consisting of two first-level subsystems, with the four second-level subsystems relevant to the discussion of the political system

population of eligible voters—the electorate—can participate, which is the process of the secret ballot, and which is a main adaptive action of the collective intelligence. This process is normally organised by a branch of the bureaucracy (in Australia it is the Australian Electoral Commission), without any direct involvement of political parties. However, the parties are heavily involved in choosing and promoting the candidates, which is the subject of the next subsection. What is not shown in the figure is the "non-democratic" interface from the electorate to the parties, in the form of party congresses and caucuses of party members. Theoretically, this interface is not "non-democratic", as party membership is open to anyone, but in reality, party discipline ensures that this interface has relatively little effect as an expression of the collective intelligence.

The block labelled as "Business" in Fig. 11.3 represents a wide range of organisations, including corporations, NGOs, industry and labour associations, and a host of organisations promoting special interests, and the interface to the political system is partly to the representatives and partly to the party organisations, through such activities as lobbying and political donations, and, in the latter case, through exchange of personnel.

The block labelled "Bureaucracy" contains the Executive and Judiciary components of Government. As noted in Sect. 7.3, the separation between the legislature and the bureaucracy, as well as the detailed composition of these two components themselves, is somewhat different in different countries, but in all cases one can observe that the route from the people's will to its implementation, in the sense of an ideal democracy, is dependent on both components.

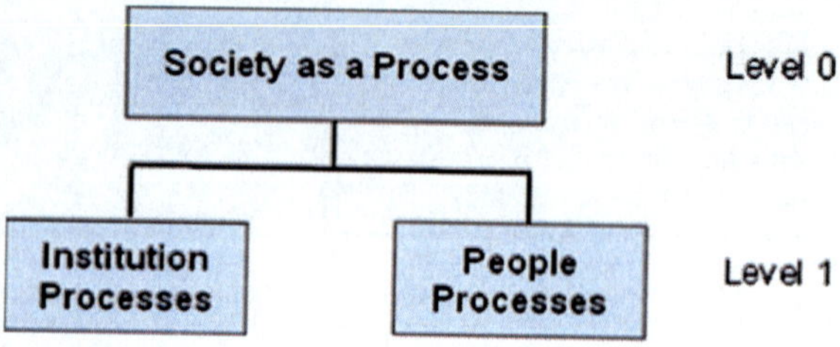

Fig. 11.4 The first level of partitioning society as a process, into institution and people processes

11.2.3.2 Economic Processes

Another type of structure is one where the elements are *processes*. There are a vast number of processes taking place within society, and following our system approach we start, at level 0, by representing this complex collection as a single entity, hiding all the complexity. There are many ways of partitioning this entity, and as the first level we partition the processes into two groups: institution processes and people processes, as illustrated in Fig. 11.4. The first are processes that produce goods and services, as well as remuneration in various forms; the second are processes directly related to people, such as work and consumption. This is not a simple partitioning, as the example of a subsistence level farmer demonstrates, but even here it is possible to think of the farm as the institution which produces products, and the farmer receiving remuneration in the form of produce and also consuming these. Obviously, this very high-level model cannot account for anything but a stationary process, where the value of the goods and services must equal the remuneration.

In the next level of partitioning, we can represent each group by two processes, as shown in Fig. 11.5. If we for a moment disregard the group of Capital processes, then the Figure represents the economy in a very early and primitive society, a society in which the raw materials used for production, such as land, timber, minerals, etc. were *free* to be exploited; there was no *ownership* of the resources. The closed loop (b,c) represents a fundamental aspect of any society, starting with a single family. The husband would hunt, and the wife would collect roots and berries, each one providing for the other. If the society expanded to encompass two families, and one was skilled in making stone tools, then the other family had to be skilled in something else, for example, making pottery, and so they could exchange products; that is the function of the income process. This mutual economic dependence is a basic characteristic of any society, and one of the major "binding forces" in our thermodynamic analogy, but it tends to become obscured by the complexity of modern society.

The major change in the development of society came with the introduction of *ownership*. This ownership has two very distinct components. One is the ownership of something the owner has created, such as a dwelling or a tool; it is created as part of the Production processes. The other is the ownership of resources; that is, ownership of something the owner has had no part in creating, and the combined value of these assets is what we call *capital*. This introduces the group of Capital processes in Fig. 11.5; they are the processes that convert capital, C, as the measure

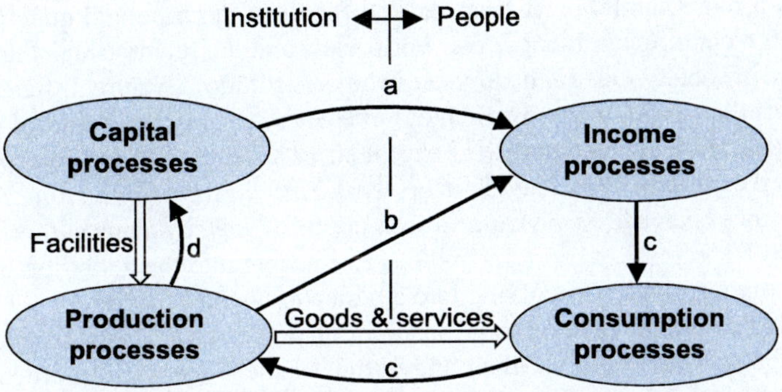

Fig. 11.5 The second level breakdown of society into four groups of steady-state processes, two institution processes and two people processes. The four arrows labelled a, b, c, and d, are monetary flows

of ownership, into production facilities and allow the owners to extract an income stream, a.

The input to the Capital processes, shown as d in Fig. 11.5, is the profit from the Production processes and includes both the cost of creating the production facilities and the interest on the capital these facilities represent. The production process then uses people to utilise the facilities to produce goods and services, for which the consumption process returns a revenue stream, *c*. It is important to recognise that, in this picture, the material input to the production processes (raw materials, construction elements, etc.) play no part; the inputs of one operation are just the outputs of another one. The revenue represents the value *added* by the application of capital and labour.

For their role in the production process (which includes maintenance), the people receive an income stream, *b*; this is part of the income process. At any one point in time there is the requirement that $c = a + b$, which is now the expression of the mutual economic dependence mentioned above, and the only interesting feature of this "steady-state" model is the fraction a/c. In a technically primitive society, this ratio will be very small, as a is almost zero, just the return on the value of some simple tools and structures. In a technically advanced society, the ratio will be much greater, as a represents the return on a very sophisticated infrastructure and, most importantly, the in-ground value of the raw materials (which in this generalised sense includes land), which in this society are no longer freely available, but tied to owners. In the limit, one could conceive of a society where the production process is fully automated and labour has been eliminated and replaced by robots, and ownership is the only parameter defining the individual lifestyle (via the consumption process). A capitalist utopia, but a human nightmare. Society is turned into a giant game of Monopoly!

What we can discern already at this high level of description is the significant influence technology has on the development of society. The obvious influences are labour-saving devices, increased mobility through mechanised transport, etc., but there is a more subtle, or at least more difficult to describe and quantify, manner in which technology influences our world view and the relationships that define the structure of society. One such influence is on our attitude to Nature. Instead of valuing Nature as the environment in which our existence unfolds, the power of technology leads us to view it as something to be exploited, as a commodity. This was perhaps first expressed explicitly by Heidegger in his work *The Question Concerning Technology*, but has since been addressed from various points of view by many authors, including Ellul and Habermas. In short, technology is entering into the definition of what it is to be human, and we are evolving into a biotechnological species.

In this picture, the Income and Consumption processes are instantaneous, or zero-sum, as is reflected by both input and output of the Consumption processes being equal to c, and also by the equation $a + b = c$. The Production processes are, of course, neither instantaneous nor zero-sum; it takes time to develop new products and to construct new production facilities, and, as discussed above, $d > c - b$. But if we include these activities in the Production processes, we can simplify the dynamics by having only one "reservoir" of value: the capital C, which is converted into the facilities required by the Production processes, and for which they pay at the rate d, consisting of depreciation, a return on investment, and the profit (or surplus) of the Production processes. For example, in the case of a copper mine, it is convenient to think of the life of the mine as consisting of two phases. In the first phase, capital is invested to buy the mining lease and to develop the mine. This includes buying equipment and paying salaries and is simply a part of the production process of the construction company. At the end of this first phase, a fully operational mine is ready to start production, and the invested capital is C_0. The second phase then starts, with expenses consisting of salaries and interest payments, and revenue, c, from the sale of the copper. However, in this case the revenue is greater than the salaries and interest payments by an additional amount, e, which is the value of the copper "in the ground", and in the revised diagram in Fig. 11.6 that is shown as flowing directly into the Capital processes, where it increases C and becomes available for investing in another project, just as the capital C_0 was generated by a previous project. That is, the simple picture presented in Fig. 11.6 indicates the value extracted from the environment in the form of non-renewable resources, such as land for agriculture, oil, gas, coal, and minerals, as an injection, e, into the capital processes that is independent of the Production processes.

This picture needs a correction and a comment regarding the system dynamics. Because the capital is represented by (invested in) facilities, it is subject to deterioration, or depreciation, at a yearly rate of, say, q. That is, the monetary flow $q \cdot C$ is simply lost within the Capital processes; it does not go anywhere and is therefore not shown explicitly in Fig. 11.6. Furthermore, it is normal to separate the rent paid for the facilities, i.e. the interest on the capital C, from what we identified as the profit, so if we denote the interest by $p \cdot C$, the profit, s, is just the difference, $s = d - p \cdot C - q \cdot C$, and the Capital processes are as shown in Fig. 11.7.

Consequently, the rate of change of C is given by

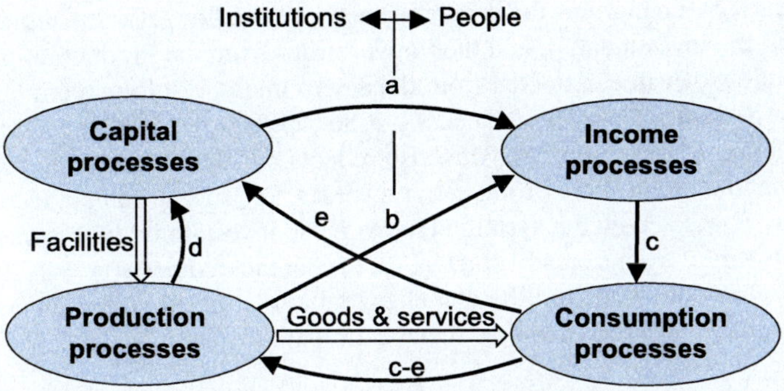

Fig. 11.6 A slightly expanded version of Fig. 11.5, showing the value extracted from non-renewable resources as the flow *e*

Fig. 11.7 The inputs to and outputs from the Capital processes. Here q is the depreciation rate, C is the capital, p is the interest rate (cost of capital), s is the profit, e is the capital injected by exploiting non-renewable resources, and a is the income derived from the capital

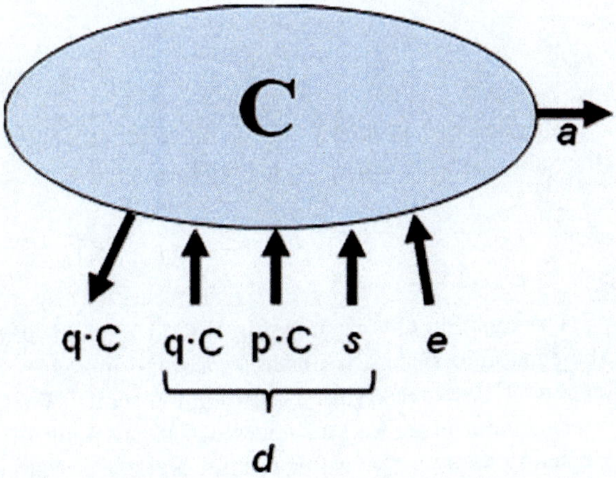

$$\frac{dC}{dt} = p \cdot C + s + e - a. \tag{11.8}$$

Here $\mathrm{p} \cdot \mathrm{C} + s + e$ is the capital generated per unit time (e.g. per year) by society and a is the amount of that capital flow that is extracted and used for consumption; the difference is invested and leads to growth of the economy.

We therefore see that, if a and e are constants, the steady state value of the capital is given by

$$C = \frac{a - s - e}{p} \stackrel{\text{def}}{=\!=} \frac{u}{p}, \tag{11.9}$$

which is only valid for $u > 0$, or $a > s + e$. That is, in order for there to be a steady state, the income derived from the capital must exceed the sum of the profit and the capital injected by exploiting the non-renewable resources.

The new parameter u is the difference between the flow of value representing the return on the investment, a, and the flow of profit from the Production processes, s, plus the flow of value extracted from the environment, resulting from the depletion of non-renewable resources, e. If $a = s + e$, no capital is required; the return is simply the depletion of the resources. This also reflects our definition of the capital, C, as the capital required for the Production processes, i.e., for the value-added processes.

Let us now see how the system behaves if the investors in the Income processes decide to vary u from a steady-state value, say, in the manner $u(t) = u_0 + \delta$ for $t > 0$ and $u(t) = u_0$ for $t \leq 0$, and thereby change the relationship between consumption and investment. We then have the linear differential equation

$$\frac{dC(t)}{dt} = p \cdot C(t) - \delta - u_0 \tag{11.10}$$

and, for $t > 0$, its solution,

$$C(t) = C_0 e^{pt} + \frac{\delta}{p} + \frac{u_0}{p}. \tag{11.11}$$

The factor C_0 is determined by the requirement that, as $t \rightarrow 0$, C(t) must approach the value given by Eq. (11.9), so that

$$C = \frac{\delta}{p}(1 - e^{pt}) + \frac{u_0}{p}. \tag{11.12}$$

The function C(t) is shown in Fig. 11.8, and it illustrates that the steady state of the economic system is unstable; a small increase (or decrease) in u results in an exponential decrease (or increase) in the capital, C. Changing C requires a dynamic management of u. And, of course, if at any point in the evolution we want to stop depleting the non-renewable resources, i.e. $e = 0$, then the equilibrium condition is $a = p \cdot C + s$.

To relate our model to available data, we identify c as the Gross Domestic Product (GDP) of the society in question, and introduce what is known as the Gross Capital Formation, γ, which in our model is represented by dC/dt/c. Furthermore, we introduce a function g, defined as C/c, which we might consider expressing a capitalisation ratio. Inserting this into Eq. 11.12 yields the following expression for g,

$$g(\alpha, \gamma, \omega; p) = \frac{\gamma - \omega + \alpha}{p}, \tag{11.13}$$

where we introduced the normalised variables $\alpha = a/c$ and $\omega = \varepsilon + \sigma$, with $\varepsilon = e/c$, and $\sigma = s/c$. (Note that the use of Greek letters and their meanings are peculiar to this section and must not be confused with the meaning of these symbols in our model of society.) The quantity $\varepsilon + \sigma$, which we for convenience denoted by ω, is the rate at which the Capital processes generate a surplus, i.e., a value beyond the normal

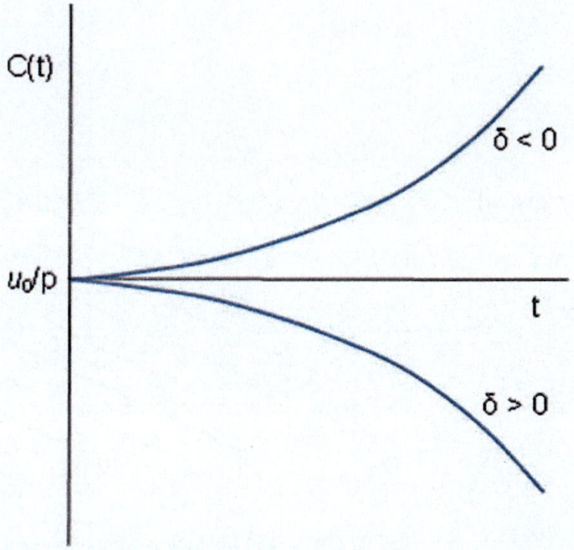

Fig. 11.8 Illustrating how the steady state of the capital C, with its value of u_0/p, is unstable; a small change, δ, in the parameter *u* leads to an exponential change in C(t)

interest on the capital, p. Furthermore, we shall make the reasonable assumption that the relative change in C over one year, ΔC/C, is approximately equal to the relative change in c, Δ*c*/*c*. That is, the growth of the economy can be expressed equally in terms of the capitalisation or in terms of the GDP.

To get a feel for what our model expresses, let us look at an actual case: the world economy. A typical value for p is 0.03 (remember that we are dealing in constant monetary values, i.e., as if there were no inflation). For the year 2013, the World Bank gives a value of Gross Capital Formation of 22% of GDP, or $\gamma = 0.22$, a GDP of about 75 trillion $US, and a value for Δ*c*/*c* of 0.02. With these values, we find that $g = 22/2 = 11$, and that C = 825 trillion $US.

Accepting these values for p and γ, the function *g* becomes a function of α and ω, and it is shown in Fig. 11.9.

For these parameter values, we obtain the relation

$$\alpha = 0.11 + \omega, \tag{11.14}$$

which just expresses the coordinates for the intersection of the g = 11 line with the ω lines in Fig. 11.7. We could also ask what it would take to stabilise the capital at its present value, and Eq. 2.2 then provides the expected answer,

$$\alpha = 0.33 + \omega, \tag{11.15}$$

as the difference between Eqs. 11.14 and 11.15 is exactly the increase in C/c.

The parameter α expresses the fraction of GDP that arises from ownership of capital, whether in the form of infrastructure or non-renewable resources; the rest

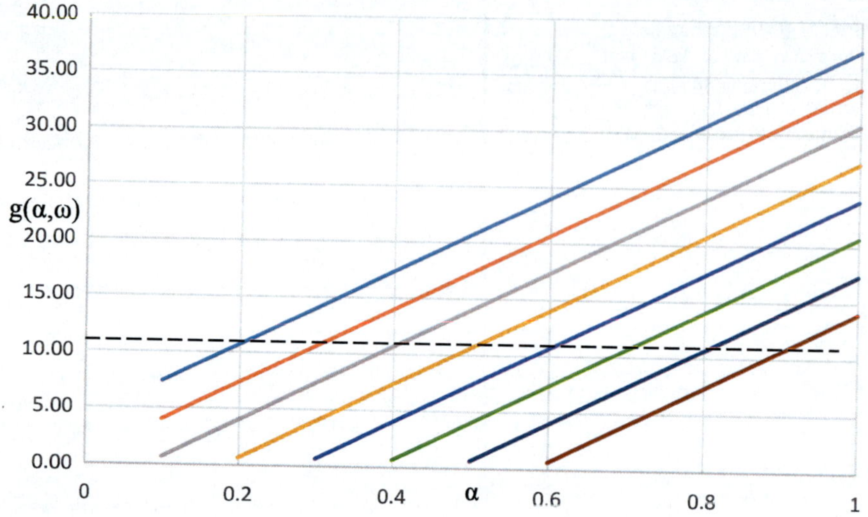

Fig. 11.9 The function $g(\alpha, \omega; \gamma, p)$, which is the capitalisation ratio, for $\gamma = 0.22$ and $p = 0.03$. The lines are for ω values 0.1, 0.2, 0.3, 0.4, 0.5, 0.6, 0.7, and 0.8, from top to bottom

of GDP is provided by labour, represented by the value flow b in Fig. 11.6. In an underdeveloped society, the value of α will be small, but as the society develops and the capital creation, represented by the parameter ω, increases, the value of α increases, and the importance of labour in the economic process decreases. The situation here is that, as Eqs. 11.14 and 11.15 show, in order to reduce or even stop this increase in capital over labour, the expenditure on consumption should increase. But the problem is, of course, that the part of the population that is investing is the part that already is able to spend more or less as much as it likes. This is illustrated, in the case of Australia, in Fig. 11.10, and Australia is a reasonably egalitarian society.

An even more striking illustration of this situation is provided by the data presented in Table 11.1, which shows the distribution of the net amount of income-producing

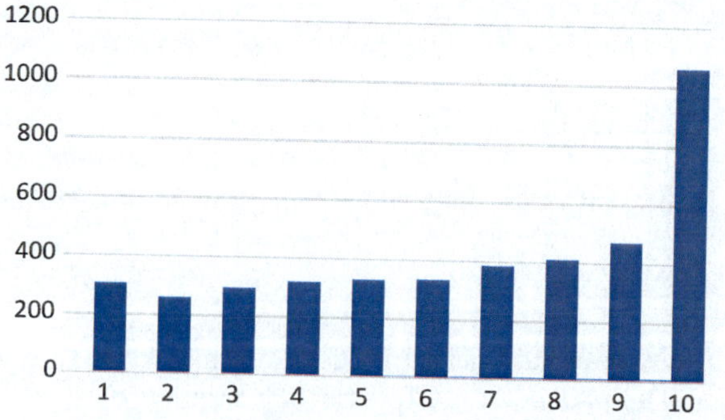

Fig. 11.10 Per capita worth, displayed against income deciles, for Australia (adapted from Australian Statistical Yearbook 2012). Values are in thousand Australian dollars

Table 11.1 The net income-producing capital, per capita, by quintiles of disposable income, for Australia in 2012 (adapted from Australian Statistical Yearbook 2012)

Quintiles, by disposable income				
1	2	3	4	5
−4	−45	8	134	1274

Net income-producing capital, in AU$'000

capital in Australia in 2012. That is, the value of owner-occupied dwellings and such items as private cars and boats is not included.

This issue of the changing balance between labour and capital, its accompanying increase in the inequality in the distribution of wealth, and how it is driven by the increasing application of technology, is a major issue in the recent evolution of society. It is clear that the current mode of capitalism cannot continue indefinitely; a contribution to solving this problem is outlined in Sect. 12.3.4.

We should note that $\omega = \varepsilon + \sigma$, and it is only σ that reflects the application of technology, whereas ε arises simply from the exploitation of non-renewable resources. So, the level of economic development of a society should be measured in terms of σ rather than ω or α. For example, Saudi Arabia would have a small value of the ratio σ/ω, whereas Switzerland would have a value close to 1. Also, a high value of ε can disguise a low value of σ, as has been the case e.g., in Australia.

From this simple view of society in terms of economic processes, we should take the following points with us as we continue to develop our understanding of the evolution of society:

1. Society, as an economic process, does not have a stable steady state. It is a process that needs to be actively managed in order to go through a desirable evolution.
2. As a society develops, there is a shift in the relative importance of capital and labour, and so we would expect a similar shift in the considerations of social equality and stability from the remuneration of labour (e.g., minimum wage) to the distribution of capital.
3. A major characteristic of this shift is the increasing presence of technology and its applications in the economic process.
4. There is a significant difference between the labour processes and the capital processes in that the latter contain a reservoir in the form of the capital, C. That is, the return on capital is in the form of a rate, $p \cdot C$, whereas the return on labour is immediate. Consequently, we expect the dynamic behaviour of the system to change with the shift in relative importance of capital and labour, reflecting the general feature of systems with regard to external interactions we highlighted in the beginning of this section.

11.3 Acceptance, Compromise, and Consensus

The meaning of 'acceptance' and of 'to accept' both depend on the context in which these words are used, as we can see by considering the following examples: If I say "I accept these conditions (e.g., of use or in a contract)", it does not necessarily mean that I believe them to be true, or that they have any value beyond allowing me to progress whatever it is I want to do. If I say "I accept that building A is the tallest building in Sydney", it does not mean that I know that it is the tallest building. And if someone says "I accept Einstein's theory of relativity", it does not mean they have any understanding of the foundations of this theory, or that they would be able to give any argument for why it is true; it is just something that is "generally accepted".

In terms of our model, an *acceptance* appears as an information item within the individual's memory that has the form of an assertion without an argument, and as such its main characteristic is that it is less susceptible to change by persuasion than are other identity items, as persuasion is about removing and providing arguments. When information items arrive as input to the individual in the information flow μ_a, they are evaluated against the items contained in the memory and, depending on the nature of these arriving items, the evaluation may require items from general memory, from the identity, of from both. In both places we find acceptances, and if the evaluation of an item involves an acceptance, the evaluation is immediate: either acceptance or rejection without any further consideration. But this rejection is different to the rejection of a such a statement as "the moon is a huge ball of yellow cheese"; that rejection is done on the basis of *knowledge*. An acceptance is not knowledge; it is somewhere between fact and belief, and it is probably best placed within what we consider *ideology*. Using an acceptance as the basis for decision-making is a part of the fast processing identified by Kahneman and Tversky and mentioned briefly in Sect. 4.1.3 (Kahneman, 2002).

We can expand our understanding of acceptance by briefly examining its relationship to three other concepts: tolerance, enframing, and prejudice. Tolerance is when we allow the existence of something we would rather be without, in the sense that getting rid of it is more effort than it is worth. Acceptance embodies some of the same resignation, in the sense of not wanting to expand more effort on determining or challenging the validity of what we are accepting, but what we are accepting does not have to be something we would rather not have accepted. On the contrary, we mostly see acceptance as beneficial; it gives us certainty.

The concept of enframing is perhaps best known in its application by Heidegger to explain how technology determines our view of the world (Heidegger, 1977); a more general use is in explaining how the acceptances inherent in our culture influence our view of another culture, as discussed by Hill (1988). Our acceptances create a frame, or a framework, that allows us to form an opinion of certain observations based solely on their relationship to our acceptances. As a result, we see reality through a lens created, in part, by our acceptances, and some of these acceptances are so deeply embedded in our consciousness that it requires a great deal of critical thinking to deconstruct our experience into an objective view of reality.

This leads us to the concept of *prejudice*. Prejudice, as a preconceived opinion that is not based on reason or actual experience, is related to acceptance in that it may be the result of accepting something that has no basis in fact, or acceptance regarding the consequences (usually negative) of membership of a category (e.g., racial prejudice). It is important to distinguish the two, as prejudice usually has a negative connotation, whereas acceptance can have either a positive or negative effect, depending on the particular acceptance and on the circumstances in which it appears. The positive effect arises from eliminating the need to involve Process B in deciding whether a class 3 information item should be accepted or not, thus lowering the individual's stress level, β. The negative effect manifests itself as inflexibility when confronted with changing circumstances. A certain degree of acceptance is essential for a smooth functioning of society—it is the oil on the gears of society—and is a means of defusing arguments and of lowering tensions in society, as will be taken up in Sect. 11.5 and again in Sect. 13.3.3.

Focusing on the acceptances of particular interest to us—those of class 3 and located in the individual's identity—we shall make the simplifying assumption that these acceptances are all located in the α-part of the identity; that is, they are shared by all the individuals in the society. The justification for this assumption lies in the nature of an acceptance—it comes to us from outside, from our interaction with other individuals, and we accept it simply because so many other people say so. An identity element that is created by the individual's own efforts, in the form of discovery or reflection, will have arguments attached to it resulting from those efforts and is therefore a normal identity element—a belief. Hence, the presence of acceptances is a characteristic of the society, and the set of acceptances form the most immediate characterisation of a society and of the difference between societies.

Acceptance may be between entities that are identical as far as their positions in society are concerned; for example, if I accept that my friend is a fan of music that I find pretty dreadful. Another example is two soccer clubs that respect each other and enjoy competing and playing against each other, even though the cultures of the clubs are quite different; the one being a free-wheeling association of friends of various backgrounds, whereas the other has strict discipline and requirements for membership. The difference is simply accepted and does not interfere with the game. But acceptance may also be between entities that are on what we might consider different levels of society, as when a parent accepts certain responsibilities for a child, or when we accept the right of the police to breath-test us when we are driving. Both types of acceptance need to be present if we are to have a world society of nations, as will be considered in Chap. 13. Nations will have to accept that they are different in many respects and work together for their mutual benefit despite these differences, and they will have to accept a degree of authority by an entity managing their relationship.

In Sect. 7.1.2 we stated that the common infrastructure can be seen as the realisation of society's belief system. We immediately think of that infrastructure as the entities that provide services, such as health, education, defence, and justice, but within this infrastructure are buried the realisation of acceptances in the form of certain 'rules', sometimes stated explicitly, as in defamation laws, but sometimes implicitly in the way the services are provided. Hence, by observing how a society operates, one can draw conclusions about the acceptances (or lack thereof) imbedded in the society's belief system.

11.4 Ideology

A few of the many definitions of 'ideology' are the following:

Britannica: a form of social or political philosophy in which practical elements are as prominent as theoretical ones. It is a system of ideas that aspires both to explain the world and to change it. (https://www.britannica.com/topic/ideology-society, accessed 20.06.2022.)

Cambridge English dictionary: a set of beliefs or principles, especially one on which political system, party, or organization is based. (https://dictionary.cambridge.org/dictionary/english/ideology, accessed 20.06.2022.)

Merriam Webster: **a**: a manner or the content of thinking characteristic of an individual, group, or culture; **b**: the integrated assertions, theories and aims that constitute a sociopolitical program; **c**: a systematic body of concepts especially about human life or culture. (https://www.merriam-webster.com/dictionary/ideology, accessed 20.06.2022).

Vocabulary.com: An *ideology* is a set of opinions or beliefs of a group or an individual. Very often *ideology* refers to a set of political beliefs or a set of ideas that characterize a particular culture. (https://www.vocabulary.com/dictionary/ideology, accessed 20.06.2022).

Consideration of ideology have appeared in numerous context and studies, such as in the Marxist formulation of ideology as an instrument of social reproduction by politically confusing the alienated groups of society via a false consciousness, an idea that was used by Jürgen Habermas in his discussion of how ideology is used to distort reality in communications, as noted in Sect. 5.3.5. The Marxist view of ideology was further developed and generalised by Karl Mannheim, and his sociology of knowledge provides a link between ideology and the public discourse (Mannheim, 1936). A particularly relevant contribution is that of Antonio Gramsci in his *Prison Notebooks* (Gramsci, 1971), where ideology is described as supporting a cultural hegemony used to maintain the capitalist order (or simply the ruling class). Gramsci's concept of hegemony is reviewed in Bates (1975).

In the past ideologies have often been presented in 'monolithic' form, as either left/right, or socialist/conservative, or autocratic/democratic and used as 'sledgehammers' in political argumentation. In the last twenty years or so a more sophisticated view of ideology has gained prominence, recognising that an ideology consists of several components, each representing a particular understanding of the underlying concept. A useful introduction to this developing view of ideology and its use in political theory is presented in Freedman (2006).

This section is concerned with developing an argument for the proposition that the culture of what we today call 'the West' is permeated by an ideology that has evolved over centuries, providing initially the justification and support for the rise of the West to its dominant position in the world, but now presenting an obstacle to the changes that the West needs to implement in view of the rapidly evolving world society in which it is embedded. To understand a culture or an ideology, a good approach is to understand its origin and the history of its development. The appearance of the Western ideology may be taken to be around 1450 AD, a time at which the world still consisted of a number of segments separated by distance and culture—Europe, Africa, the Middle East, India, China, South-East Asia, Australia, Japan, and North and South America. There was certainly significant movement and interaction within these segments, both by sea and by land, and there was also limited interaction between segments in the form of trade, exchange of information, and also through border skirmishes, raids, and piracy. Technological development varied greatly between them, and at that point in time two stood out as clearly most advanced—Europe and China. A brief review and comparison of the histories of these two segments in the four centuries following 1450 AD puts the development of the West into perspective and can act as a starting point for the more detailed examination of the developments over the last 150 years in Chap. 13.

In China, the greater part of the period of interest is covered by the Ming dynasty (1368–1644), which arose out of the collapse of the Mongol Emperors and the emergence of the founder of the dynasty, Chu Yüan-chang. When Marco Polo returned from China in 1295, the civilisation he made known to Europe was far superior in every respect to that of his native land, and despite internal tensions between the north, with the capital Beijing, and the south, with the major cities Nanking and Canton, and a disastrous campaign against the Mongols, China was prosperous and well administered during this period, and as a civilisation it was not inferior to Europe in 1450 (Fitzgerald, 1954, p. 466). At this time, agricultural production was steadily increasing, both due to new lands, including shallow lakes, being brought under cultivation and due to improved cultivation methods. Commerce and industry also expanded, with such industries as porcelain and paper manufacturing flourishing, as did the printing and selling of books (Eberhard, 1960, pp. 244–251).

Although China had an extensive coastline, its focus was inward, defending its borders against the Mongols in the north and strengthening its hold on the southern provinces by establishing military colonies. The South China Sea was dominated by mostly Japanese pirates, and the only attempt by China to project any naval force was the well-known excursions of Zheng He and his fleet, which sailed past Indochina and reached all the way to the coast of Arabia. The activities of raiders and pirates along the coast led the Chinese to the belief that basically all foreigners who came by ships were "barbarians"; when toward the end of the Ming epoch the Japanese were replaced by Europeans who did not behave much differently and were also pirate-merchants, the nations of Western Europe, too, were regarded as "barbarians" and were looked upon with great suspicion (*ibid*, p. 246).

This inward-looking was not restricted to the physical realm, it was also prominent in the intellectual realm. The Chinese did not seek to gain knowledge and insights from abroad; at best it was bound to be inferior, and at worst it could lead to a contamination of the stable and proven social and political system based on the teachings of Kungzi. It was a system that had endured for more than a thousand years, and it would endure largely intact until around 1800.

Europe was at this time experiencing a great awakening from an intellectual slumber induced mainly by the Catholic Church, and it is possible to identify at least two important sources or initiators of this Renaissance. One was the rise of the Italian city-states, including Florence, Venice, Milan, Siena, Genoa, and many others, as centres of commerce, trade, industry, and finance. They were the forerunners of modern capitalism, developing double-entry bookkeeping and the banking system, and the wealth generated by them supported a high degree of literacy and numeracy, and financed works of art, literature, engineering, and science. Another was the rediscovery of Greek science, literature, and philosophy, in part via the efforts of Islamic scholars, in part through the westward migration of Greek scholars under the pressure of the expansion of the Ottoman empire, but also as a result of the increased interaction through trade in the Mediterranean. At this time, perhaps not so much as an initiator as a response or an enabler, came the Gutenberg printing press with movable type, which, with its lower printing cost, allowed a much broader market to be addressed and greatly promoted the use of the vernacular.

And then, as this extraordinary blossoming of intellectual activity and its associated products reaches a pitch and is looking for further application, what opens up before it but several new worlds, with riches there for the taking? The Americas (Columbus 1492), India (Vasco da Gama 1498), South-East Asia (Magellan 1522). How could anyone doubt that the Europeans were God's chosen people and that the Christian God was the only true God? Least of all, of course, the Europeans themselves?

The wealth generated by access to and domination over these new lands provided a potent catalyst for an exponentially increasing level of economic activity and an underlying attitude of "anything is possible" and anyone could find their fortune, not only in the newly discovered lands, but also with the help of the wealth flowing in to Europe. This turbulence in the social order, for which the Church was the guarantor, had to be countered, and so it was no coincidence that the Reformation appeared. The influence of the Reformation on the political economy of the West has been described and analysed in considerable detail in the seminal book *The Protestant Ethic and the Spirit of Capitalism*, by Weber (1930). For Luther, the objective order of things in which the individual has been placed by God becomes a direct manifestation of divine will. "The individual should remain once and for all in the station and calling in which God had placed him and should restrain his worldly activity within the limits imposed by his established station in life." (*op. cit.*, p. 85). That is, the emphasis was on obedience to God's will rather than on any ethical considerations.

Of greater importance to Weber's analysis was the emergence of a number of variants to the reformation of the Church that he identified as ascetic Protestantism and that emphasized that the task set by God was the work in the calling. Not only was work the purpose for which God had placed us on Earth, but the performance of the work should increase the glory of God, "the valuation of fulfilment of duty in worldly affairs as the highest form which the moral activity of the individual could assume" (*op. cit.*, p. 80). Weber notes that "asceticism looked upon the pursuit of wealth as an end in itself as highly reprehensible, but the attainment of it as a fruit of labour in a calling was a sign of God's blessing" (*op. cit.*, p. 172), and he quotes Richard Baxter as stating "If God show you a way in which you may lawfully get more than in another way (without wrong to your soul of to any other), if you refuse this, and chose the less gainful way, you cross one of the ends of your calling and you refuse to be God's steward, and to accept His gifts and use them for Him when He requireth it: you may labour to be rich for God, though not for the flesh and sin." (*op. cit.*, p. 162).

Over the centuries the religious aspect of this attitude to work, as well as the prohibition against consumption beyond that necessary to sustain life, became less important and had by the beginning of the nineteenth century all but disappeared. But the work ethic, with its emphasis on profit, remained and became the core of modern capitalism; with some cynicism, the modern version can be abbreviated as "greed is good". The attitude to consumption was in any case untenable; the statement that the ethical ideal was not to consume but to earn (*op. cit.*, p. 68) ignores that without consumption there are no earnings and reflects the Protestant belief that there would always be two classes of people: the saved and the damned. As Weber says of ascetic Protestantism: "By founding its ethic in the doctrine of predestination, it substituted for the spiritual aristocracy of monks outside and above the world the spiritual aristocracy of the predestined saints of God within the world. It was an aristocracy which, with its *character indelebilis*, was divided from the eternally damned remainder of humanity by a more impassable and in its invisibility more terrifying gulf that separated the monk of the Middle Ages from the rest of the world about him, a gulf which penetrated all social relations with its sharp brutality. This consciousness of divine grace of the elect and holy was accompanied by an attitude toward the sin of one's neighbour, not of sympathetic understanding based on consciousness of one's own weakness, but of hatred and contempt for him as an enemy of God bearing the signs of eternal damnation," (*op.cit.*, p. 122).

This twofold influence of ascetic Protestantism spread out over Europe and those colonies that together became what today is the West as a drop of ink in a bowl of water. It became diluted and toned down in its expression, and the legitimacy originally provided by religion became transferred onto what Karl Mannheim called "the plane of mythical transcendence", with God replaced by history and tradition, and divine right replaced by manifest destiny (Mannheim, 1936). The latter has been, and still is to some extent, a significant feature of US culture, as exemplified in President Woodrow Wilson's 1920 State of the Union statement: "This is the time of all others when Democracy should prove its purity and its spiritual power to prevail. It is surely the manifest destiny of the United States to lead in the attempt to make this spirit prevail."

This belief in our superiority lingers on in a never stated but always present form, as an inherent ideology that distinguishes the West from the rest of the world. Mannheim called such an ideology a total ideology and described it as a quest to influence the public interpretation of reality and pointed out how the radically different premises and assumptions involved make people with different total ideologies simply talk past each other (*op. cit.*). Mannheim also introduced the concept of a *thought style* as the thinking based on a total ideology, and he pointed out that different thought styles not only deliver different answers to issues; each thought style also specifies a certain realm of reality as prescriptive for valid thinking (Mannheim, 1986, also Lundberg, 2013). This is exactly what is characterising so much of the public discourse, such as it is, in the West today, and it reminds us of Marcuse's concern in *One-Dimensional Man*, as referred to in Sect. 3.4.1. And just as it is so difficult to extract the ink from the bowl of water, it is very difficult to change or diminish this ideology. It colours, if even often just ever so slightly, all our dealings with the rest of the world, and in Chap. 13 it will be argued that it is one of the main obstacles to forming a world society.

A closely related ideology, and one that is more openly pronounced in public discussions of politics and political systems, is concerned with a belief in what is the "best" political system. In Western societies it is generally assumed that our definition of 'best' has until now most often been synonymous with the concept of a *liberal democracy*; a concept that arose out of the Enlightenment ideas of liberty and equality, and which, according to Wikipedia, is characterised by the following principles:

(a) elections between multiple distinct political parties;
(b) a separation of powers into different branches of government;
(c) the rule of law in everyday life as part of an open society;
(d) a market economy with private property; and
(e) the equal protection of human rights, civil rights, civil liberties, and political freedoms for all people.

A society realising these principles, taken at face value, appears to be a reasonable (but not the only) candidate for the "best" society. However, it is the application of these principles—their realisation in the everyday operation of the society—that is relevant to any assessment of "best". The electoral process is ineffective if the parties differ only in detail; both the separation of powers and the rule of law only become issues in a dynamic society; market economy and private property mean nothing until their extent and limitations are defined; and the items in e. are open to a wide range of interpretations. The overriding characterisation of Western societies is in terms of an *ideology* that determines the operation of all the principles in the above list; it is an ideology that has emerged as a result of two historical developments: Christianity and the expansion of European power and the creation of what we now call the West, as was discussed above, and hence the close relationship of these two ideologies. Or, perhaps they are better considered as two aspects of the same ideology, and common to both is the important role of Christianity and the Church. By promising everlasting life after death, and making belief the determining factor for achieving it, Christianity made inequality and exploitation in this life relatively irrelevant. Indeed, without inequality, how could there be any Christian charity? Suffering and death in this life, as was the fate of millions of natives, was a small price to pay for eternal life; better to be a slave as a Christian than free as a heathen. Through this relationship between faith and the qualities of daily life, the administration of the faith—the Christian church—arose as a powerful organisation in its own right and as both a collaborator with and competitor to the State, as illustrated by the power struggle between Pope and Emperor during the Investiture Controversy around 1100 AD. In the end, control over social life was shared, with the Church controlling two fundamental aspects of every person's life—sex and death—while the State controlled the economic aspects. The Church was given favourable treatment in economic matters, and the State was given more or less automatic absolution for its sins. And for the Church to be an effective partner, the original emphasis on faith was augmented and partially replaced by an emphasis on the individual's duty to contribute to the economy through diligent work and the accumulation of wealth; exemplified by what is termed "the protestant work ethic" (not that Catholics are any more adverse to accumulating wealth). It is this particular amalgamation of religious dogma with the extraordinary economic circumstances in the Western world in the last five hundred years that has created the neoconservative ideology.

When referring to societies embracing this ideology it is usually not as neoconservative or capitalist (the latter has a bad taste, like communism), but as liberal democracies. This terminology cloaks what is essentially an economic ideology with adoration of private wealth and conspicuous consumption as something noble, as the realisation of an intrinsic human aspiration, like motherhood. Conversely, labelling a society or its behaviour as undemocratic immediately characterises it as deplorable, without the need for any further elaboration or analysis. The meaning of democracy is literally "rule by the people", and Wikipedia states "Democracy is a system of government where the citizens exercise power by voting.". There is an implicit assumption that the power exercised by voting is the only power, and that power is therefore distributed equally over all citizens. In reality, in all democracies there are numerous other avenues to exercising power, and if we could measure the power exercised by each person and create a Lorenz curve, as one does in determining the Gini coefficient, I would expect the curve for power to be very similar to the curve for wealth, and the power Gini coefficient to be correspondingly large.

But besides the many questions about practical aspects of this definition, we must first ask: What is the purpose of government? Only then can we judge the efficacy of democracy as a system of government. And to answer that question, we must recognise the special nature of society as a system of interacting individuals. In all other systems, the system has a purpose defined by its relationship to something outside itself. In the case of a telephone system, the purpose is to provide a communications service to its users; in the case of a car, the purpose is to provide certain services (transport, prestige, etc.) to the owners; in the case of the education system, the purpose is to provide knowledge and understanding to the students (in the widest sense); and so on. In the case of society, the purpose can only be defined in terms of the service it provides to its members, i.e., to the very elements that make up the system. It is as if the purpose of a car would be to make the engine run better.

Therefore, to answer the question, we first must define the purpose of society. One approach is to see the purpose in terms of supporting the individual in satisfying its needs, and with reference to Maslow's hierarchy of needs (Maslow, 1943), we might define a measure of how well a society meets its purpose as the degree of self-actualisation reached by the individual, averaged over all the individuals in society.

The extent to which the lower needs have to be satisfied in order for effort to be allocated to self-actualisation; i.e., the relative strengths of the needs, will vary from person to person, from society to society, and with time within each society, and hence it is unlikely that a single form of government will be appropriate for achieving the most effective distribution of society's resources over these five levels for all societies. And there does not seem to be any rational argument for why Western liberal democracy is an optimal form of government for any society; it is an ideology that has emerged as a result of the special circumstances experienced by the West in recent times.

The dominance of the neoconservative ideology in the West reached its peak in the 1980s, personified by Ronald Reagan and Margaret Thatcher, and claimed its ultimate vindication in the collapse of the Soviet Union in 1991. In the words of Francis Fukuyama: "That is, the end point of mankind's ideological evolution and the universalization of Western liberal democracy as the final form of human government." (Fukuyama, 1992). However, this exuberance glossed over many of the problems arising from different cultures and religions. In the book *Spectres of Marx*, Jacques Derrida criticised the Fukuyama view, provided a list of the problems in the form of "10 plagues of the capitalist system", and pointed out some of their effects:

> For it must be cried out, at a time when some have the audacity to neo-evangelize in the name of the ideal of a liberal democracy that has finally realized itself as the ideal of human history: never have violence, inequality, exclusion, famine, and thus economic oppression affected as many human beings in the history of the earth and of humanity. Instead of singing the advent of the ideal of liberal democracy and of the capitalist market in the euphoria of the end of history, instead of celebrating the 'end of ideologies' and the end of the great emancipatory discourses, let us never neglect this obvious macroscopic fact, made up of innumerable singular sites of suffering: no degree of progress allows one to ignore that never before, in absolute figures, have so many men, women and children been subjugated, starved or exterminated on the earth. (Derrida, 1994)

In the context of this book, the central feature of this ideology is its emphasis on a person as an individual relative to a person as an element of society. It is the individual's faith that is the determinant in achieving salvation and eternal life; the only relationship is between the individual and God; and a person's achievement is expressed predominantly in the accumulation and individual ownership of wealth. Society, as the entity created by individuals through their interactions, does not feature in this narrative. The opposing ideology, socialism, emphasizes the importance of society and the mutual dependence of society and individual—the Social Contract—and recognises the importance of the state as the realisation of the structure arising out of the interaction between individuals. And as the interaction increases, it is acknowledged that society becomes increasingly dynamic and increases the demands on its control. One way to look at this is that, as the level of interaction increases, the number of possible configurations increases and, as an expression of the Law of Requisite Variety (Ashby, 1956) that we mentioned in Sect. 3.6, for society to remain stable, the control system, i.e., government, needs to develop accordingly and have an increasing number of options at its disposal. The neoconservative war cry of "small government" reflects an ideology removed from reality; government needs to be complex, dynamic, effective, and efficient; size is not, in itself, a significant measure.

The question that remains is how rational, well-educated, and highly successful businessmen (and women) give allegiance to an ideology that to us seems to be so obviously at odds with reality? An interesting and insightful contribution to answering this question—although not providing a definitive answer—is provided by the book *Democracy in Chains* by MacLean (2017). The core of the ideology is the idealisation of the individual, free in its actions and in the disposal of the results of its actions in the form of property. Any interference of government in the activity and property of the individual in the form of taxes and restrictions on individual liberty must be kept to a minimum. It is an ideology completely at odds with democracy and its ability for a majority to force a minority to act against its interests and to levy taxes to support a common infrastructure beyond a minimum for defence and upholding law and order.

When reading MacLean's book, it is difficult to discern whether the behaviour of the far-right is simply the tactics of the very rich to hold on to and increase their wealth, or if there is a true belief in its righteousness and as the way forward for the evolution of society. Some of the not-so-rich adherents to the ideology may simply be hangers-on hoping that some beneficial crumbs may fall their way, but for someone like the key figure of James McGill Buchanan that does not seem to be the case. His dedication to finding a strategy to subvert democracy brought him no riches, and his Nobel Prize came from solid academic research. There are some people who have an unusually strong objection to being told to do something they do not want to do and refuse to see its relationship to society rather than just to themselves. And such people, who might be considered extreme individualists, are found at all levels of income and wealth, it is just that the rich ones, like the Koch brothers, can do more about it.

However, irrespective of any belief or adherence to an ideology, Buchanan's strategy for circumventing democracy, expressed succinctly as changing not *who* rules, but changing *the* rules, is alive and well. In Australia it can be observed at all levels of government, even at the lowest level which should be closest to the people, where, as an example, the ability for people to object to development plans has been removed and replaced by certification.

As a final comment in this section on ideology, where we have focused on the West and on neoconservative ideology and its characterisation as being on the right-hand end of the political spectrum, we should recognise the importance of right wing, or even far right ideologies that are not related to capitalism, and in some cases directly opposed to its US-based realisation. One such ideology, and one which is relevant with regard to the situation in Ukraine, is the *Eurasian idea* promoted by the Russian philosopher Aleksandr Dugin. It is a revolt against the Western liberal capitalist society, with its impersonal Internet and lack of piety. Dugin believes that the world order currently promoted by the West—the globalised unipolar structure—is headed for an apocalyptic collapse; the only possibility of avoiding this outcome would be the emergence of a multipolar structure with non-competing components, of which

Eurasia—the western half of the Asiatic continent—would form one component. This would not be a nationalistic component, or one based on race; it would be based on a common culture, a partly romanticised and idealised Russian culture that has survived the ravages of modernism and capitalism, with mystical overtones. A simpler society with closer and more meaningful interpersonal relations, reviving the importance of the Church, and led by a new aristocracy. President Putin is well aware of these ideas, and we might wonder to what extent they are influencing his behaviour vis-à-vis Ukraine.

11.5 Power, Tension, and Stability

11.5.1 Types of Power

'Power' is, just like 'freedom', a word with many meanings, depending on the context in which it is used, a context that ranges over various *domains*, such as politics, economy, military, physics, and mathematics. Within the domain of social science, a possible first partitioning, and one that is useful for our purpose, is into the application of power in two different levels of society. One is the basic level, where the individuals in the definition of society are persons; this is the level contemplated throughout this book so far. The other is the level where the 'individuals' are nations, and the system of interacting nations provides a description of the world society. Within both these levels there is, with suitable interpretation of 'person' and 'institution', a further partitioning into two types of power: *interpersonal* power, and *institutional* power. There are many definition of interpersonal power, for example, "the means and resources available to enforce compliance with one's requirements", or "the capacity or ability to direct or influence the behaviour of others or the course of events", or "A has power over B to the extent that A can get B to do something that B would not otherwise do", the last one due to Dahl (1957). The concept of interpersonal power can be further decomposed, and of the many such decompositions the best known one is perhaps the one due to French and Raven (1959, 1965) in relation to the power of a person in society—*social power*, consisting of six types:

Personal in Fig. 11.11 means that the power types in this group are characteristics of the person itself. *Expert* power is the power that arises from specialised knowledge, although it should be noted that this might also be due to a *perception* that the person has such knowledge. *Referent* power is power due to the person's reputation in another field, such as a famous actress having power of persuasion in an ad for toothpaste. The power types in the group *Situational* (sometimes also called *Positional*) depend on the person's situation in society. *Legitimate* power arises from occupying an accepted position of authority, such as a president or a queen; it is the power persons acquire through their position in a stable and integrated system of social rules. (As the position is normally defined within an institution of some sort, this type of power is sometimes also called institutional power, e.g., in Stahl (2011)). *Reward* power is

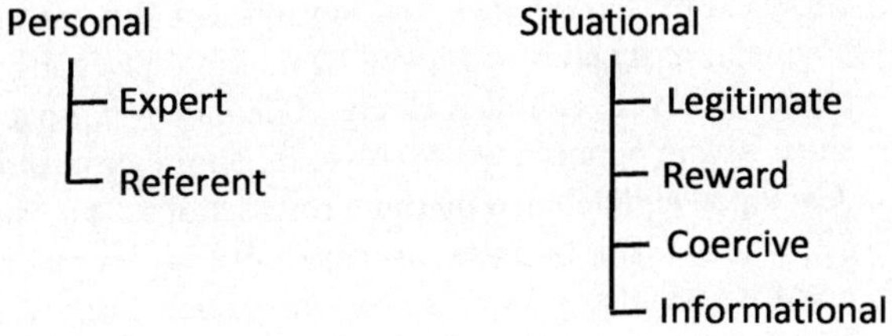

Fig. 11.11 The six types of power wielded by a person, according to French and Raven

the power to offer others rewards (positive or negative) for what is wanted of them. *Coercive* power is the ability of a person to threaten or force others to comply with his or her demands, and *Informational* power is the ability to use information to bring about change, but also the power that comes from ownership of information desired by others, such as contained in patents and military secrets.

While this taxonomy was developed for the basic level, there is no reason why it cannot be translated to also apply to the world society level.

There are various definitions of 'institution'; one is: "An institution is a type of collective activity based on a set of accepted rules and in which some form of evaluation and sanction is present". Institutional power can be exercised in various ways and, again, within each of the two levels, we can identify three ways: One is the power of an institution in a society over the individuals in that society. If, at the basic level, the society is a nation, typical examples would be the police and the taxation office; a more subtle example would be a news corporation, and most businesses exercise power through advertising. At the world society level, the institutions are such organisations as the International Monetary Fund (IMF), the World Trade Organisation (WTO), the World Health Organisation (WHO), and the many alliances, such as the North Atlantic Treaty Organisation (NATO), and these organisations can exercise a varying degree of power over nations.

A second way is the power exercised between institutions; within a nation an example would be the mutual power relationship between unions and the government; in the world society this would be rare.

The third way is also a personal aspect of institutional power, in that individuals may have the ability to set the rules, agendas, norms and customs of institutions to get others to do what they would otherwise not do. That is, this type of institutional power can be seen as a two-stage process: first, an agent influences the institution, and then the institution influences the target person or group of persons. At both levels this type of power is more often than not in the form of economic power; on the basic level often personal wealth, on the world society level the size of the national economy.

A further distinction that is often made is between *hard* and *soft* power, where hard power is the traditional understanding of power as being able to force an outcome. The concept of soft power was first promoted by Nye (1990) as an argument for the strength of the position of the US in world affairs but has now been recognised as an important component of power and incorporating many concepts that were not thought of as power, such as persuasion and attraction. If power is the ability of an entity to effect a change in its environment, this can be achieved with coercion or with payment, or it can be achieved by making the change seem attractive, by making the others involved want this change. The latter approach is soft power, and in the years since it was popularised by Nye, its application in international relations has been developed and debated by numerous scholars, and some of this will be referenced in Sect. 11.5.3. But in these discussions of power and its different types it is important not to forget that power is a *means* of achieving something, and not to become so absorbed in searching for and developing the most efficacious means that we lose sight of what we set out to achieve. Or even worse, let the development of the means determine the outcome, as reflected in the saying "when all you have is a hammer, every problem looks like a nail".

In the context of the present work, hard and soft power could be seen as the two aspects of power that arise when power is considered in the action view and in the information view, but while there is some merit in this, it is an oversimplification; any application of power involves both action and information. If we refer to Dahl's definition of power quoted above, the resulting action by B is preceded by a change in B's information (or identity). In the case of hard power, B needs to observe and assess A's action before using this information to change its existing information and then take adaptive action. The issue here is that A must either have knowledge of B's existing information or make an assumption about it in order to design the action so that B's interpretation of the action will be as intended, and the possibility of a misinterpretation is present. In the case of soft power, A provides the information it wants B to adopt, and while the issue of A's knowledge of B's identity arises here also, as discussed in Sect. 7.1.1, soft power appears as the more direct approach, perhaps contrary to what is often assumed. With the primacy of the information view soft power is subsumed in the most general type of power—the power to change the individual's identity. Reflecting this, the following two sections focus more narrowly on two applications of power that are particularly relevant to the present work, one on each level: the power of institutions over the individual arising through information technology, and the application of institutional power in international relations.

11.5.2 Influencing Society's Belief System

Until, say, five hundred years ago, the individual's by far most significant mode of exchanging information with society was face-to-face interaction, and while this mode is still important, technology has revolutionised our means of interacting, in a development outlined in Sect. 2.3.4. As far as information is concerned, the individual's interface to society is increasingly provided by various *media* based on information technology (IT), as treated in Sect. 8.5, with the result that our view of reality is to some extent determined by our use of these media, This was the reason for including, in Sect. 4.1.4, the media as an intrinsic component of the individual; the characterisation of the individual must include how it chooses to integrate the media into its daily life.

In Chap. 5, Fig. 5.7 presented a picture of the interaction of the individual with society. That picture reflects our high-level model of identical interacting individuals, where the role of the media is simply to facilitate this interaction, as in overcoming the distance, time, and cost limitations of face-to-face interaction. When we now extend this picture to include structure, it appears in two places. One is the structuring of the individuals according to their access to and use of the media, the other is the structuring of the participants in the public discourse, i.e., as contributors to the circulating information. In this book the first of these is treated only in a summary fashion as a characteristic of the operating condition of the collective intelligence on a national basis, as was done in Sect. 6.3. The elements of the second structure are *institutions* of various kinds, including government institutions, think-tanks, institutions dedicated to pursuing a single agenda, news providers, political parties, etc., and we are interested in how these institutions influence the behaviour of individuals by means of information exchanges. We can conceptually look at this process from two perspectives. One is the type of information: either information instructing the individual to do something, or information intended to change the individual's identity with the purpose of encouraging a certain behaviour. The other perspective is the relationship between the institution and the individual: either *authority* or *credibility*. The resulting two-dimensional characterisation of institutions, with a few examples, is shown in Fig. 11.12.

The main issue that arises regarding authority is the extent to which the society's belief system is well-defined or, alternatively, the degree of dissent regarding the belief realised by the institution. In terms of our high-level model, this is illustrated by the concept of commonality, in Sect. 3.3.3. In the idealised case of the rectangular approximation, the belief system is completely well-defined, every item in the belief system is also contained in every identity. In reality the situation is something like the curve in Fig. 3.5, with some dissenters to most items. An example of this are the current objections to COVID-19 vaccination, even among parliamentarians, and the, in part, violent anti-vaxing demonstrations; the principle that our freedom should not be harmful to other people is by no means universally accepted in our society.

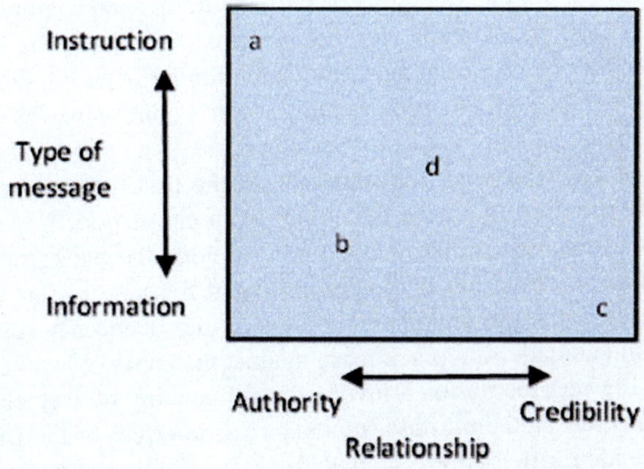

Fig. 11.12 A conceptual two-dimensional characterisation of the relationship between individuals and institutions within a nation, with a few examples: a = law enforcement institution; b = education institution; c = marketing institution; d = religious institution

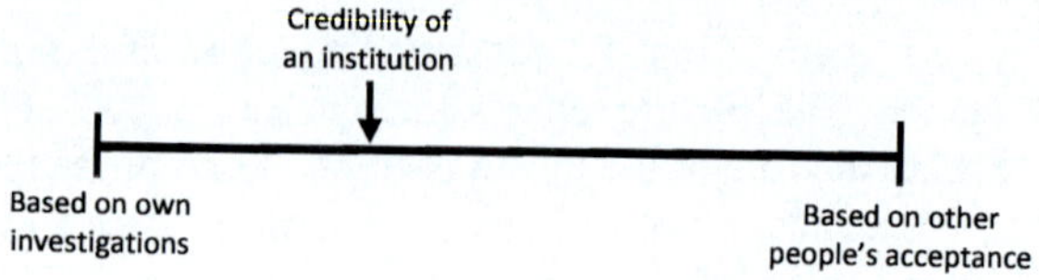

Fig. 11.13 A scale of an individual's assessment of an institution's credibility according to what it is based on

To explore credibility, we can think of the individual's assessment of the institution's credibility to be placed on a qualitative scale that goes from an assessment based entirely on the individual's own investigations of the institution's performance on the one side to an assessment based entirely on other individuals' acceptance of the information provided by the institution on the other side, as illustrated in Fig. 11.13.

The position of an individual's assessment of an institution's credibility on the scale in Fig. 11.13 is indicative of the relative importance or value of two quantities that are not directly commensurable. We could perhaps measure them indirectly by the amount of time the individual spends on own investigations and on considering and accepting other people's use of the institution's information, and just considering my own experience, which is probably not atypical, convenience will place the arrow in Fig. 11.13 closer to the right-hand side of the scale most of the time.

Credibility is often conflated with truth; an institution has credibility if the information it presents is truthful. But this is only one component of credibility; another, and at least equally important component is the extent to which the information is presented in its relevant context, as illustrated by the following example: A society (nation, state) has two political parties, one in power, the other in opposition. An opposition party politician makes the following statement in Parliament: "In the past year, we experienced 106 fatalities on our roads, including several children. This is both a tragedy and a disgrace for a developed society like ours, and I call on the Minister to take effective action without delay." This is reported verbatim in one newspaper, without any further comment, whereas a second newspaper points out that the party in power introduced an initiative to reduce traffic fatalities when it came to power three years ago, and that as a result the number of fatalities has declined significantly each year since, and that the current result places the society near the lowest per capita fatality rate of any developed society.

The statement by the politician is not untrue; the number of fatalities is a fact, most people would agree that any number of traffic fatalities is a tragedy, that it is a disgrace could be seen as the speaker's opinion on the performance of all developed societies, and the call for action expresses a desire on part of the speaker. Now, most people who read this statement in the first newspaper would come away with the impression that here is another example of poor performance by the government, and they are unlikely to go and research the background of this statement; if they did, it would lower the credibility of the newspaper. Whereas those who read the statement and the related context in the second newspaper would shake their heads and consider it a black mark against the speaker—either ignorant or deceitful.

Every institution is involved in influencing society's belief system; some explicitly as their stated purpose, others as a side-effect of their operations; sometimes unintended, often simply undisclosed. And power is not necessarily greatest when the intention is explicit, as that allows one to mobilise one's defences against it; the effect of an undisclosed intention seems like part of a "natural" evolution. Things don't just happen; every change in society is the result of human action, and while some actions performed by individuals appear unintended or random, in the sense that it is impossible to identify the information that caused them, all actions performed by institutions are caused by information located within a framework of information particular to the institution, so that we may say that every institution has its own identity.

11.5.3 Tension

The concept of *tension* appears both in physics and in psychology. In physics it is a force applied to an object, typically a string or a metal rod, but generally resulting in a distribution of *stress* (force per unit area) and associated *strain* (incremental deformation), with the two being linearly related by the modulus of elasticity. In psychology tension describes a nervous state, such as a feeling of fear or anger. It is therefore tempting to indulge in another analogy between physics and society by representing society as a sheet of metal and with stress being the analogue of tension. If we apply external forces at the edges of the sheet the result is a deformation of the sheet and an associated distribution of stress. The deformation can be seen as the metal's means of coping with the external forces without losing its intrinsic shape; if the forces are not too great, the sheet will return to this shape once they are removed.

When a society is exposed to external forces as in a change to its environment, it will adjust its operations and structure in order to respond appropriately, and this will result in tensions within the society, as was demonstrated most recently in the case of the COVID-19 pandemic. But if the change was not too great, society will revert to its normal operation and structure once the changes are removed. The society's ability to adjust to external circumstances without losing its identity—its *resilience*—is the analogue of the elasticity of the metal.

If the external forces on the metal sheet increase above a certain limit, the further deformation becomes permanent, so that the sheet does not revert to its original form when the forces are removed. And if the forces are increase even further, cracks start to appear, and the sheet may eventually tear apart. In the case of society, strong external forces may well result in such extensive changes to its operations and structure that even if the forces are completely removed the society will not revert to its original form. An example of this would be colonisation; colonies that become independent states do not revert to their identity prior to colonisation.

Metal, even if it is pure, does not have a uniform structure; the metal atoms form small crystals, called *grains*, and the boundaries between grains show a discontinuity in the orientation of the crystal structure. The boundaries are weak points in the bulk metal where tiny cracks can appear when the metal is exposed to high stress, and with repeated exposure to high stress one can observe a phenomenon known as *fatigue crack propagation*; the tiny cracks grow until at some point they become fatal to the integrity of the metal object. Such repeated exposure to high stress occurs in aircraft structures, requiring periodic inspections for the appearance of cracks, and it is a limiting factor in the lifetime of aircraft. A society also has a structure made up of numerous different groupings of its individuals—families, associations, political parties, and states within a federation, and the cohesion within groups is stronger than the cohesion across groups, so that the boundaries between groups can constitute weak points when it comes to adjusting to external forces. A small example of this were the tiny "cracks" that opened up between states and between states and the federal government in Australia under the pressure of COVID-19, with accusations of mismanagement flying in all directions.

But the forces do not have to be only external; forces can be generated internally through a non-uniform distribution of some parameter. In the case of the metal sheet, it would typically be temperature; high temperature areas expanding and creating a distribution of stress throughout the sheet. In the case of society, it is *inequality* in its various forms that causes tension; inequality of income, of wealth, of privileges, and so on. But also "inequality" in the form of qualitative differences, such as different religions, different political ideologies, or different ethnicities. Examples of societies that have come apart due to such tensions are the Soviet Union (USSR), Yugoslavia, and Sudan, and examples of societies that were significantly transformed, but did not fall apart, are France, Russia, and China.

The relationship between inequality and the resulting tension is complex and far from linear, and the problem of understanding it starts with developing any sort of *measure* of inequality. For example, just starting with the intrinsic capabilities of individuals, how can we compare someone's talent in one of the arts—painting, architecture, sculpture, literature, music, performing, and cinema—with someone's talent in another one? And what about talent in mathematics and science? How would we measure the inequality between Picasso and myself in painting? Or between Mozart and myself in music? Or between Einstein and myself in physics?

The next aspect of the problem is how to identify and measure tension in society. One possibility is through appropriate interrogations—through questionnaires and interviews—of a representative sample of society or of an element of a social structure. In such cases tension is commonly expressed as dissatisfaction or anger, but as tension is between two elements of a social structure, such interrogation often does not clearly identify the source of the dissatisfaction or anger.

Another approach is to observe behaviour and actions. Within a nation the most obvious sign of tension are public protest gatherings and/or marches; crime rates are another indicator. In the world society the signs involve the treatment of the opposing element in the public discourse, diplomatic manoeuvres such as recalling ambassadors, trade disputes, and military sabre rattling. But in both the national and the international case there is then the problem of measurement of inequality and/or tension. If not quantitatively, which is unlikely, then at least qualitatively to the extent that we can make a significant estimate of how close the tension is to the critical point—in the analogy this is the elastic limit. This can only be attempted on a case-by-case basis, and in Chap. 13 we consider the East–West tension.

11.5.4 Stability

According to the Canterbury English Dictionary, stability is "a situation in which something is not likely to move or change". Depending on the nature of the "something", the concept takes on a more specific meaning. For example, the stability of a structure, such as a building or a bridge, refers to its ability to withstand external forces, as might be experienced in an earthquake or a cyclone. In the case of a control system, it is its ability to maintain the controlled parameter close to its set-point value under varying conditions. And in materials science it is the ability of a substance to remain unchanged under changes to its environment. When we now want to consider the meaning of stability in relation to society, it can obviously not mean that society is unchanging, as by its very nature society is continuously evolving. If we view this evolution as the result of a process, then perhaps stability could mean that this process remains unchanged? In the case of an industrial production process, stability means that the characteristics of the product remain within certain limits; but this type of stability cannot apply to the process driving the evolution of society, as the changes that took place a thousand years ago were quite different to those taking place today, both with respect to the rate of change and to the substance of the changes. So, if social stability can mean neither no change nor a prescribed change, including progressing toward an end state, is the evolution of society just the result of a random process, in which case the concept of stability has no meaning?

Both our intuition and the historical record would indicate that this is not the case. From the early examples of society, say, in the form of nomadic tribes, until our current nation-states, there has been a desire to regulate change through a set of rules and customs, and what was considered normal changes at one point in time, e.g., intertribal warfare, became undesirable fluctuations at a later date, when the tribes had been integrated into a larger structure, such as a kingdom. At each stage, there has been an understanding of what is an acceptable change and what is an unacceptable change—a fluctuation away from the desired evolution of society. That understanding is based on what we introduced as the *embedded basic knowledge* in Sect. 3.3.1. The frequency and severity of such fluctuations constitute the manifestation of the instability of society.

The concept of stability arose already in the previous section in the form of the limit to the tension a society can endure without a lasting change to its essential character. In the analogy with the metal sheet it was represented by the elastic limit; another often-used analogy from mechanics is that of a ball resting on a curved surface, as illustrated in Fig. 11.14. In the stable configuration, (a), we can think of stability as the amount of energy that can be imparted to the ball before it goes over the ridge and disappears down the outer slope.

Referring to the previous section, it might appear that stability is the same, or very similar to the ability to survive tension, and while instability and tension can both result in the demise of a society, there is a significant difference, which may perhaps best be clarified by another analogy, this time not with physics, but with the relationship between two persons, as in a marriage. Tension might arise from a

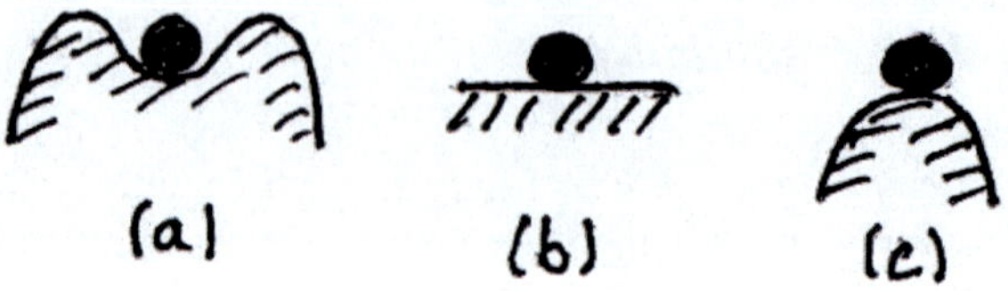

Fig. 11.14 Three states of stability: **a** stable, **b** indifferent, and **c** unstable

domineering attitude on the part of one partner, from excessive complaining, from infidelity, and from physical abuse, and at some point, this tension may reach a level where the abused partner leaves. But it may also happen that the two partners have less and less in common or, perhaps, that the reasons for forming a partnership have become less relevant, and that they have more in common with, or benefit more from interaction with, other people. There is no tension, but they still part company. In the case of society, *stability* is the ability to maintain the internal *structure* in the face of external forces, and it is distinguished from *resilience*, which is the degree to which society is able to maintain its *operation* in the face of changing operating conditions.

This little analogy and the importance of what the partners have in common should remind us of the developments around identity, identity items, commonality, society's belief system, and the identity partition, α, in Sect. 3.3. Stability now appears as a measure of the relationship between identity items in the circulating information that conflict with the identity of society's members, in the sense that a high level of stability is associated with a high value of α and a small rate of conflicting items, whereas a low level of stability is associated with a low value of α and a high rate of conflicting items. And we would expect the measure of the rate of appearance of conflicting items, high or low, to be related to the ability of the individual to process such conflicting items, resulting in a value of the stress level, β. This stress level is a parameter of the state of the individual, but if it extends to a group of individuals, and the conflicting items originate from another group of individuals, β and tension may become identical variables.

While this picture of the variables involved and their relationships is useful in a conceptual and qualitative sense, it is of little use for a quantitative development, and the reason is that while our high-level model assumes a structure-less society of identical individuals, both tension and stability refer to the structure of society. An extension of this model to the general case of a structured society is a matter for future work, but in Chap. 13 we apply these concepts to the special case of the world society and a simplified view of its current state.

In Sect. 3.5 we introduced the concept of dividing the activities taking place in society into two cycles, with all the activities we associate with daily life, and described by economics, politics, ethics, religion, and, more generally, by social science, making up the economic cycle. In our model of society as an information-processing system, there is a part of the information processing associated with the economic cycle; let us call it the *secondary* process. This process includes the research, innovation, and increase in experience that drives the evolution of the economic cycle—the increase in GDP, the new applications of technology (discussed in (Aslaksen 2017)), and even new works of art, but it also includes such changes as wars, economic crises, and the like; changes that we would consider *fluctuations* of the evolution of the economic cycle. The severity of a fluctuation is measured by such characteristics as the deterioration in the satisfaction of basic needs, loss of life, and destruction of the built and natural environments. The *stability of society* is a measure of the degree to which such fluctuations are limited in frequency and severity; it does not mean that the economic cycle is not evolving.

Underlying the actions of the economic cycle is a belief system or an ideology; we might say that the actions and evolution of the economic cycle *represent* this belief system, and the stability of society is a characteristic of this belief system. The view put forward in this book is that this belief system is itself evolving, and that there is another process—the *primary* process—that drives the changes to this belief system. The primary process evaluates proposed changes to the secondary process, and the ideal version of this primary process is what we, in Chap. 4, called the *collective intelligence*. It is the belief (and it can only be a belief) supporting the work in this essay that the operation of the collective intelligence will provide the most stable evolution of society, i.e., of society's belief system.

The core of the collective intelligence is the individual's ability to survive, to determine to what extent a new situation is relevant to survival and what adaptive action is most likely to increase the probability of survival. This ability is inherent in every living creature; its power and sophistication are appropriate to the creature's environment and to its ability to take adaptive action. It is not something that at present, if ever, can be further defined or detailed; its existence is a matter of personal interpretation of evolution. It may perhaps best be viewed as the defining feature of *life*; life is about survival. In humans, this ability is highly developed and based on a learning process, so that as society evolved in the level of integration and complexity, the understanding of what constitutes survival and the criteria for achieving it evolved; accordingly, mostly through a process of trial and error, coupled with reflection, and transmitted from one generation to the next through the process described in Sect. 9.2.2.

However, the power of the collective intelligence as the guiding process for the evolution of society comes from the interaction between individuals—the public discourse that leads to a consensus and the potential for action—and this interaction can be defined and parametrised in an indirect manner by identifying the characteristics of society that determine the operation of the collective intelligence. This is what was proposed in *The Social Bond* and outlined in Sect. 6.3, where we defined a single composite parameter that represented the restrictions society places of the operation of the collective intelligence; in the absence of any restrictions, society will evolve along a trajectory that minimises fluctuations.

Due to the activities of individuals, there will always be a large number of ideas and proposals for changes to society's belief system. If there are no restrictions, so that the collective intelligence is operating freely, the overwhelming majority of these will not have any significant impact, mainly because they are seen as beneficial only by a small part of the population, but some of these initiatives will have a more significant impact within a local part of society and a limited timeframe; they are the fluctuations associated with an *equilibrium* state of the belief system. Only a very small fraction survives the assessment and consensus-building process of the collective intelligence, and these are the changes that constitute the evolution of society—a slow, continuous change of the equilibrium state. (Note the similarity with the dynamics of equilibrium thermodynamics.) Fluctuations will be of limited duration due to the averaging effect of the collective intelligence; they are to society as infections are to the human body, and the collective intelligence plays the role of a social immune system.

This picture changes if there are restrictions on the operation of the collective intelligence, which is, of course, the case in all societies today. The primary process is then no longer identical with the collective intelligence, and while there are various reasons for this, depending on the particular society, they are basically of two types: One, arising from the lag between the problems arising from the technology-driven changes to society and the development of their solutions; this is the situation depicted in Fig. 2.2. Two, arising from the resistance to any change on the part of society benefiting most from its current state. It is this latter type of actions that are most relevant to our concerns about stability, because, while the economic cycle may be in an equilibrium state relative to the operation of the secondary process under the current belief system or ideology, the overall evolution of society, under the combined primary and secondary processes, is diverging further and further from the ideal path, with an increasing probability of a catastrophic fluctuation. This divergence is reflected in a corresponding divergence from the uniform society assumed in our model—a segmentation of the society that appears as one segment imposing a limitation on the free generation and flow of information. Now, when we speak of the collective intelligence, of its operation, of the restrictions placed on it, and of social stability, this is always in relation to a particular society. Societies range in size from a family to the world society, and the conditions under which the collective intelligence operates vary accordingly. And when we treat a society as a stand-alone entity, that is always a simplification, in that any society, with the exception of the world society, is embedded in a larger society, as an element of that society. In the last two chapters of this essay we look at two societies—the nation-state and the world society—that are of particular interest in the context of stability, not least because the relationship between them—the "embedding" of nation-states in the world society—is unique due to the fact that the world society is "the end of the line", so to speak, in the sequence of societies, as was shown in Table 1.2.

Before leaving this section on general aspects of social stability, we should note the rising visibility in the last 25 years of the central role of religion in politics and in the underlying ideology. This is, of course, nothing new; the priesthood and its organisation has been a major component of any state since the beginning of historical times. But now, perhaps in reaction to the long domination of Christianity, all the three other main religions—Islam, Hinduism, and Buddhism—are showing signs of reasserting their role in politics, as evidenced by the persecution of the Rohingya's and rising militant Buddhism in Myanmar; the increasing emphasis on Hinduism in Indian politics; the blasphemy conviction of the mayor of Jakarta and a shift towards a more strict interpretation of Islam, with a leading cleric as the running mate of Widodo in the recent elections in Indonesia; the strength of the Taliban in

Afghanistan; the call for the introduction of Sharia law in some African states; and the rise of such a movement as Islamic State. And even in the West, the influence of religion seems to be on the rise, as evidenced, for example, by the current discussions about religious freedom in Australia and about abortion in the state of New South Wales and in the US, the election of an ultra-conservative Catholic as president in Brazil, and the religious component of nationalism in Eastern Europe.

Chapter 12
Modifying the Nation

12.1 Utopia and Reality

According to the view presented in this book, the key to developing and maintaining a society is the unrestricted exchange of information among its individuals. This supports a degree of consensus—the society's belief system—which is then reflected in a common infrastructure and the services required to maintain the society. However, the individuals are an abstraction and an idealisation; in reality they are persons going through a lifecycle, as described in Chap. 9, and this constant maintenance of a body of individuals is society's metabolism. The rectangular approximation to the distribution of identity items in Fig. 3.5 is utopian, but keeping the real distribution from departing too far from it is part of the metabolic process, with education playing a main role.

If we now consider a world society with nations as its individuals, as we shall do in the next chapter, we can extend the concepts of our model, such as circulating information, collective intelligence, identity, belief system, and a common infrastructure to this level. But there is one concept that cannot be extended, and that is metabolism. Nations are not continuously created and destroyed, and they do not go through a lifecycle that includes education and transfer of knowledge; this renewal takes place *within* each nation. At the level of this world society, nations are represented by an element of their common infrastructures, the realisation of that part of a nation's belief system that relates to international relations. Therefore, a prerequisite for the establishment and maintenance of this form of a world society and, indeed, for any form of a world society, is that the operation of the transformation cycle in each nation operates as close to the ideal as possible, and two of the main obstacles to this are the political party system and the neoliberal economic paradigm. For each of these, suggestions for how they could be improved are presented in this chapter.

E. W. Aslaksen, *The Evolution of Society*,
https://doi.org/10.1007/978-3-031-23054-7_12

12.2 Politics and People Power

12.2.1 Liberal Democracy

In most of the nations making up the west, the political system is a version of what is called a liberal democracy. But what exactly is the definition of a liberal democracy, and how closely do the political systems in these nations realise this definition? Let us first examine the components of the concept, starting with 'liberal'. Oxford Languages gives the following definition of 'liberal':

adjective

1. willing to respect or accept behaviour or opinions different to one's own; open to new ideas
2. relating to or denoting a political and social philosophy that promotes individual rights, civil liberties, democracy, and free enterprise

noun

1. a supporter of policies that are socially progressive and promote social welfare
2. a supporter of a political and social philosophy that promotes individual rights, civil liberties. democracy, and free enterprise.

Already here we see two definitions that, if not exactly the opposite of each other, are significantly different. As a stand-alone concept, *liberalism* is defined as a "political doctrine that takes protecting and enhancing the freedom of the individual to be the central problem of politics" (www.britannica.com/topic/liberalism, accessed 20.06.2022).

Leaving the type of democracy and the realisation of 'liberal' aside for the moment, Oxford Languages defines democracy as "a system of government by the whole population or all the eligible members of a state, typically through elected representatives". However, Britannica gives the basic meaning as "rule by the people", and then sets out to examine this in an excellent article by Robert A. Dahl (www.britannica.com/topic/democracy, accessed 29.06.2022). There it is noted that any practical realisation of democracy, liberal or otherwise, will have to provide answers to a number of questions, and it is the third question that is particularly relevant for the present work:

> Assuming a proper association and a proper *dēmos*, how are citizens to govern? What political organizations or institutions will they need? Will these institutions differ between different kinds of associations—for example, a small town and a large country?

That is, in the view and model of society developed in this book, the ideal society is governed by what can be seen as the ideal (if utopian) democracy, from which we are currently very far removed, so the main issue in the evolution of society is how to move towards this ideal, and the answer to that question would basically also be the answer to question three above.

As the 'association' we first consider the nation; in Sect. 13.2 we then consider the world society. And of the 'organizations and institutions' we focus on the institutions involved in transforming the will of the people into directives for action, i.e., directives to government, and central to these are the *political parties.*

12.2.2 Role and Purpose of Political Parties

In rethinking the role of political parties, we should remember that societies have been governed for thousands of years, but political parties are a relatively new appearance. Real political parties emerged only after about 1850; before that time there were such structures as informal alignments among parliamentary representatives and political clubs, for example those associated with the French Revolution—Jacobin, Girondin, and others—and in England the Tories organized the Carlton Club and the Whigs the Reform Club. In NSW an early political organisation was the Australian Patriotic Association, founded by W.C. Wentworth in 1835 together with a group of influential colonists; the Adelaide City Council held its first election in 1840, and the secret ballot (with Government supplied ballot papers) was introduced in Victoria, South Australia, and Tasmania in 1856. In the US, the two components of a two-party political system emerged soon after the founding of the republic, in the form of the Federalist Party, led by Alexander Hamilton, and the Republican Party, led by Thomas Jefferson. These two components went through a number of transformative stages, during which the original Republican Party became what is currently the Democratic Party, and the current Republican Party arose out of the remnants of a Federalist-Whig Party transition in the 1850s.

Correspondingly, the theory and understanding of political parties, both as a concept and in practice, as a main part of political science, is a relatively immature science, as was pointed out by Riker (1982). One of the standard works is the book *Political Parties*, by Duverger (1964), which introduces a taxonomy of political parties, based on the manner of their creation and their internal structure, and then goes on the describe different party systems. The work is based on detailed examination of actual parties, with focus on developments in France and Britain, with lesser coverage of Germany, US, Belgium, the Scandinavian countries, and Switzerland (very little). Of particular relevance in the present context are the following statements (note, by "elector" Duverger means "eligible voter"):

> Democracy requires that parliamentary representatives should take precedence over the members of the party, since the electors constitute a larger group than party members, who are moreover included in it. In practice the opposite often takes place: in many parties there can be seen a tendency of party leaders to give orders to the parliamentary representatives in the name of the militant members. This domination of the party over its elected representatives constitutes a form of oligarchy that might be termed 'external' by comparison with the oligarchic nature of leaders within the community of party members. (*op. cit.*, p. 182)

The reference to "militant members" reflects the time in which it was written, the post-war era of strong ideological motivation for political participation. Today, as a result of increased affluence and the conversion of a large part of the "working class" into "middle class", the preoccupation of political parties is mainly with economics and with their own survival as entities within this economic view of society.

> In practice elections, like the doctrine of representation, have been greatly changed by the development of parties. There is no longer a dialogue between the elector and the representative, the nation and parliament; a third party has come between them, radically modifying the nature of their relations......If we wish to maintain the juridical theory of representation we must admit that the representative receives a double mandate: from the party and from his electors. The importance of each varies according to the country and the parties; on the whole the party mandate seems to carry more weight than that of the electors. (*op. cit.*, p. 353).

> Here the word 'representation' is applied to a sociological phenomenon and not to a legal relationship; it signifies the resemblance between the political opinions of the nation and those of parliament. Deputies represent their electors not as a mandatary represents his mandatory but as a photograph represents a landscape, a portrait its model. The fundamental problem consists in measuring the degree of accuracy of representation, that is the degree of correspondence between public opinion and its expression in parliament. (*op. cit.*, p. 372).

> The party system existing in a country is generally considered to be the result of the structure of its public opinion. But the converse is equally true: the structure of public opinion is to a large extent a consequence of the party system. (*op. cit.*, p. 372).

In these three last excerpts, Duverger seems to capture some of the main reasons why citizens would become increasingly disillusioned with and disinterested in political parties, but he does not draw the conclusion that maybe there is something wrong, or anachronistic, with the role of political parties.

Two further relevant contributions to the theory of political parties are Strom (1990) and Levy (2004). The first develops a model of party behaviour based on a consolidation of the three rational choice behaviours originally proposed in Downs (1957, 1957a)—vote-seeking, office-seeking, and policy-seeking; it is an example of the common view of the political party as an organisation with its own identity and purpose, with the voters making up the environment in which it exists. If the party is the plant, the voters constitute the soil in which it grows. The message of the second paper is essentially that the party allows the politician to represent a richer and more optimised and effective combination of policies than would be possible for stand-alone politicians. This is, to a systems engineer, fairly obvious, with the effectiveness of the combination as an emergent property, and it is also something that can be observed in that independent candidates promote themselves mainly on single issues, but it highlights an important function of the political party that needs to be preserved and further developed. And this brings me to the article by Ann-Kristin Kölln, *The value of political parties to representative democracy* (Kölln, 2014), which addresses some of the main issues in any assessment of the role of political parties. The point of departure is her statement that the core of representative democracy is the authorisation and accountability through means of regular elections; these two measures are then used to compare parties with individual representation. Individual representation is essentially what Strom, in a taxonomy of political organisation called pluralistic democracy, and as it is the organisation most closely related to party democracy in that taxonomy (Strom, 1990), Kölln focuses her comparison on these two types of political organisation. She concludes that parties have a slight edge, and this is further strengthened by a more detailed comparison of how the two systems fulfil an expanded set of requirements. In a thought-experiment, the author illustrated how she imagines individual representation might operate; in the case of e.g., Australia, that model, with multiple representatives per constituency, would apply only to Senate elections. In elections for the House of Representatives there is only one representative per electoral division.

However, for all the good points Kölln makes in her comparison about fulfilling valuable functions, there is no mention of the adverse effects of the party organisation on the democratic process; effects that basically arise as a result of the necessity for the party to maintain itself as an organisation. Viewing the actions of parties through the media here in Australia, one might easily get the impression that our political system has turned into a business, with the parties as corporations, and with a business model in which the main purpose is to maintain the viability of the party by trading favours with special interest groups, giving preferential legislative treatment in exchange for electoral and financial support, and with the politicians as executives (SMH, 2017). This mode of operation was, in the case of US political parties, noted by Lenski (1970, p. 358), already half a century ago, and it seems to now have invaded most Western democracies. An example of the conversion of the party into a business was provided recently in Australia leading up to the 2022 federal election, where the federal executive of the Liberal Party (i.e., the conservative party) decided to impose its candidates over the choices of the local party members in three electorates (SMH, 2022). A somewhat similar case of overriding local preferences was that of the Labor Party imposing Ms. Keneally on the electoral seat of Fowler, NSW (SMH, 2021).

Thus, when assessing the current status of political parties, we must differentiate between the function of the party as a component of the business world and its function as a key component of the democratic political process. If we compare these two functions of a party, or, perhaps better, simply consider the balance of their importance in society, it appears that this balance is shifting in favour of the business function, and we could imagine that in the end, society will be run as a business, with the citizens as the shareholders.

That is unlikely, because there is another trend that opposes this shift, and that is the increasing level of education and social awareness of the population. But harnessing this increase in what is, in effect, the collective intelligence, is proving to pose several challenges for the current organisation of the polity. A discussion of these challenges, together with a survey of the scholarly literature on political parties, can be found in the article by José Ramon Montero and Richard Gunther (Montero & Gunther, 2003). The authors list the following challenges: Trends toward secularisation have sapped the strength of denominational parties; increasing affluence has shrunk the electoral base of working-class parties; the greater participation of women in the professional workforce has placed new demands on the internal structure of parties; massive international migration has resulted in groups with no previous representation; and a higher level of education has given rise to greater individuality and critical assessment. And, finally, they note that various applications of information technology have had a significant influence on structure and operation of political parties; something we touched on in Sect. 8.4.3, and which will be central to the proposal put forward in the following section. However, while the authors have great faith in the continued existence of political parties, the most interesting feature of this review of the state of relevant research is that it also seems to ignore the influence of business and capital on the structure and function of political parties, and on their focus away from the citizen and toward business interests. It is these adverse effects that have prompted me to include a proposal for change.

12.2.3 A Proposal

The underlying purpose of the following proposal is to enhance the conditions for exercising society's collective intelligence, as it was developed in the previous chapters. In particular, it addresses three of the conditions that were considered as essential: The availability to all individuals of true information in a useful form, i.e., in a form adjusted to the abilities and circumstances of the individual; opportunity to participate in public discourse as part of the deliberation process; and the ability of individuals to express the outcome of their evaluations of this information as meaningful actions.

As is demonstrated e.g., in the analysis by Kölln (*op.cit.*), both the party and the individual representative bring value to the democratic political process. So, rather than choosing one or the other, why not try to combine the valuable features of both, while at the same time eliminating those features of the party that are detrimental to democracy? The key to doing this is to separate the individual representative from the operations of the party but use valuable products of the party as an important part of the interface between the representatives and their electors.

The rise of political parties was due to the enlargement of the franchise and started out as a means of gaining popular support for the election, or re-election, of candidates. Today, the key tasks of political parties are considerably more numerous and demanding. In the article *Types and functions of parties*, Richard Gunther and Larry Diamond identify a number of functions that a party should perform:

(a) candidate nomination;
(b) electoral mobilisation;
(c) issue structuring;
(d) social representation;
(e) interest aggregation;
(f) forming and sustaining government; and
(g) social integration. (Gunther & Diamond, 2001).

In addition, the following functions (not totally different from the above) have been introduced and discussed in various places in the literature:

(h) Soliciting and articulating public policy priorities and civic needs and problems as identified by members and supporters;

(i) socialising and educating voters and citizens in the functioning of the political and electoral system and the generation of general political values;
(j) balancing opposing demands and converting them into general policies;
(k) activating and mobilising citizens into participating in political decisions and transforming their opinions into viable policy options;
(l) channelling public opinion from citizens to government; and
(m) recruiting and training candidates for public office.

Based on the belief that, in a democratically constituted society, the *primary purpose* of a political system, including parties, the elected representatives, and the structure within which these representatives operate, is to ensure that the will of the people finds effective expression in the performance of government, and if we view this political system to consist of parties, electors, representatives, and the government, my proposal now proceeds as follows:

The political **party** is reborn as a *think tank*, dedicated to analysing the state of society, identifying its problems, defining its preferred policies for dealing with these, and advocating these to the public through various channels, such as the Internet, but also including public meetings and rallies. Parties promote policies, not candidates; and they are financed by private donors: the public, industry, unions, etc. Of the above functions, the party covers the following:

c, e, h, j, and k: This is the main purpose of the party: to produce a coherent, realistic, and well-argued package of costed policies and, where necessary, specific actions, all anchored in current factual information.

Regarding the **electors**, each citizen of voting age is provided with a *secure password*; this is no different to what is already provided for communication with the government in Australia via the myGov site, and most of us do our financial transactions online, using a password.

Regarding the **representatives**, any citizen of voting age and good standing (no outstanding criminal charges, etc.) can become a candidate in his or her electorate by first *registering* with the government and placing a bond of, say $10,000, and then receiving *endorsement* from a proportion of the voters in that electorate, say, on the order of a thousand "signatures" (see below). Once this level of endorsement it reached, the bond is released. *Support* for individual candidates is on a voluntary basis within each electorate, financial support is limited to say, $500 from any one citizen of voting age (i.e., owner of a password), and any other form of financial support is not allowed. Once elections have been held (see below), the successful candidate becomes a representative. The formation of the legislative arm of the government, i.e., ministers, speakers, etc., is carried out by the elected representatives through an internal election; the prime minister position might rotate on a yearly basis among the senior ministers (as in Switzerland). This will avoid the unseemly brawling and finger-pointing that goes for deliberation in the Australian Parliament, promote cooperation, and focus attention on the task of governing. People who cannot work together as a team for the benefit of the nation should not be elected, and certainly not re-elected. Of the above functions, the representatives cover the following:

a, f, and l: The legislative body formed by the representatives is responsible to the people for all aspects of the Government's operations.

Within **government** there is an organisation dedicated to the political system (in Australia this is the Australian Electoral Commission). Besides being responsible for the security of the IT infrastructure supporting the political system, this organisation maintains a web site on which all registered candidates are listed, with a CV and a statement from each candidate about which policies they will support if elected. This may be a selection of policies from different think tanks. At this web site there is also a link to an online discussion forum (e.g., blog) for each candidate, also provided by the government. Access to such a site requires the secure password, so that there can be no anonymous entries.

An important function of this organisation is to organise elections, say, every four years, and half-yearly referendums on major issues. Any citizen, or group of citizens, may call for an issue to be decided by referendum, following very much the same procedure as for candidates, except that the required bond and the number of signatures would be substantially higher. The organisation provides educational material about the political system, including to schools and to prospective candidates, and advertises elections, referendums, and all deadlines associated with them, so that, of the above functions, it covers the following:

b, d, g, i, and m: The purpose of this organisation is to administer and maintain the infrastructure supporting the political system.

The main feature of this scenario is that it restores the primacy of the individual citizen in the democratic process (both as elector and representative), eliminates a parasitic organisation (the current party), and reduces the influence of business and wealth. It also elevates the relationship between citizen and candidate to a more personal level. At present, even though people who vote, say, Liberal in different electorates vote for different candidates, they might as well just vote for the Liberal party. Unless they know the candidate socially, they cannot know what the candidate really thinks, as he or she is not allowed to deviate from the party line. In the proposed scenario, the candidates must *reveal* themselves, and not hide behind a party platform. This was what Hannah Arendt saw as an essential feature of the *vita activa* as political action, as we noted in Sect. 4.2.

The account of the deliberative democratic process given in (Landemore, 2017) can be summarised by describing the process as composed of two sub-processes: the deliberation process and the aggregation process, with the latter represented by majority voting. The epistemic competence of the process—that is, the degree to which the outcome is superior to that of any other political process—is achieved through both sub-processes, with the important features being inclusiveness and cognitive diversity, as prerequisites for the effective operation of the collective intelligence. From this perspective, our development of the social bond is seen to provide a measure of the outcome in the form of the measure of the Good Society defined in Sect. 6.3.1, and our proposal has the following advantages:

- The process of deliberation is extended from a small proportion of a small number of political parties to a significant proportion of all members of society, greatly extending both inclusiveness and cognitive diversity;
- the quality of deliberation is increased by improving the access to the information that serves as input to the deliberation; and
- the aggregation process is more closely reflective of the state of public opinion following deliberation, instead of being limited to a choice between the policies of a few political parties.

Besides all the other obvious objections to this scenario (e.g., the demise of a profitable business), the problems associated with online voting will be prominent. That is basically a red herring, and only demonstrates how the thinking on this subject is mired in the past. If our whole financial system functions online, including the international transfer of billions via the SWIFT system every day (although that now also seems to have fallen prey to the world's master hacker, the NSA, as reported in SMH, 2017), there is no reason why this relatively simple interaction with the public could be conducted online. And it is not a new issue or a new technology; there are several voting systems available (such as Google "online voting"), and a report (Second Interim Report) issued in 2014 by the Australian Parliament discusses the arguments in favour of electronic voting. There is no doubt that there are a number of currently unsolved problems (and unimplemented solutions), as detailed e.g. in the report from (Verified Voting), but given the importance of the issue, which is that a system at the centre of democracy is essentially broken, as demonstrated by the 100 million that did not vote in the 2016 US presidential election, it is difficult to believe that they could not be readily solved.

12.2.4 One-Party System

All representative democracies in the West are based on a multi-party system; the candidates we vote for are mainly representatives of one party or another. There is no rule that prevents independent candidates, and in Australia we are seeing more of them lately, but they struggle to have much impact except, perhaps, on a narrowly defined single issue. This means that for most individual citizens their input to the political process is casting their vote once every few years for one out of a few broadly defined options for the general direction of government. The individual actions taken by government is decided by the party in power and do sometimes not correspond to the preferences of the majority. An example of this in Australia is the principle of indefinite detention of asylum seekers, which is being taken to ludicrous lengths, even though opinion polls show a clear preference for abandoning it.

Instead of realising the individual's influence on the actions of government in the form of a choice between parties, there is a different approach that requires only a single party. Unfortunately, single party states have a poor image, as they are mostly autocratic and dictatorial, but here is a highly simplified and stylised model of a single-party system that allows for a measure of democracy. In this model, the legislative functions described below are separated from the executive functions, but that is more for simplicity than necessity.

For the purpose of illustration only, consider a nation with 100 million citizens eligible to vote, divided into 30 equal provinces, each of which is again divided into 100 local communities of about 33,000 voters each. In a two-yearly election cycle, each community elects two representatives to their provincial parliament. These representatives are elected for a period of six years, so a provincial parliament has 600 members, and every two years 200 leave and 200 new members arrive; re-election is not allowed. A provincial parliament sits twice a year for three weeks, and in the first sitting following an election, the members elect a standing committee of, say, 20–30 members, each responsible for all or part of a provincial portfolio (education, healthcare, public works, etc.). The members of the standing committee are paid full-time; the other members of the provincial parliament are paid for 10 weeks per year only. The standing committee is supported by a staff for research and formulating legislation, as required on the provincial level.

In the year between local elections, each provincial parliament elects three of its members to the national parliament and again, they are elected for a period of six years, so that the national parliament has 900 members. The national parliament sits twice a year for six weeks, but its members are paid full-time, as they are expected to spend time in their provinces identifying and investigating issues and projects of national importance, as well as undertaking study tours abroad or similar activities; they also each have a small staff. At its first sitting following a renewal of the membership, the parliament elects a standing committee of, typically, 30 members; they are Ministers, each responsible for a portfolio (defence, foreign affairs, treasury, etc.). The standing committee elects the President from its membership for a two-year term.

Any citizen above voting age that has resided in the community for at least two years prior to election can stand for election; registration as a candidate requires 500 signatures and the payment of a modest registration fee. Following close of registration, six months prior to election, the local electoral commission, supervised at provincial level, produces a photo, short bio, and a statement of political belief and intention for each candidate, and publishes this documentation on a website, in newspapers, on notice boards, or whatever is appropriate for the local conditions. No other form of paid publicity is allowed.

This legislative structure is mirrored in a three-level executive and administrative structure, where admission and promotion are by examination and by performance, and there will be strong and defined interaction between the two structures. However, that is not relevant here; the purpose of this model is simply to show a way of transforming the will of the people into government action by means of a one-party system; the legislative structure is the Party.

There are two main features, or characteristics, that make this one-party political system special. The first is that there is no Party in the sense of multi-party or autocratic single-party systems; the function of the party is only as a process of administration, it is a function within the common infrastructure. The Party has no relationship to the individual politicians, so there is no lobbying or means of influencing policy via the Party. Of course, this does not exclude lobbying or even corrupting of individual politicians, but this is much less cost-effective than influencing a politician within the power structure of a "real" Party.

The second feature is the facilitating and encouraging of a deliberative process at each level. At the community level, the size of 33,000 voters should make it possible to have a vigorous exchange of information throughout most of the community, and besides neighbourhood interactions ("over the garden fence"), such means of communications as letterbox drops, group meetings, locally focused newsletters in various media, and the like would have more of an impact. Of course, groups will form at this level, but they will be in support of persons, not parties. At the provincial level there is again only information exchange and discussion between individuals, and while groups will undoubtedly form and some members will be more persuasive than others, there is no structure within which power can be institutionalised (except, of course, within the Government function). And the same applies at the national level. It is my belief that this deliberative process would bring Government actions more into line with the will of the people, and it could also be augmented by a referendum process, something that is shunned by political parties as it restricts their freedom of action.

12.3 A New Economic Paradigm

12.3.1 The Need for Change

The importance of economic inequality, treated in Chap. 6, as a driver of tensions within society was discussed in Sect. 11.5.4, and it has been the focus of concern of a significant group of knowledgeable people, from Guterres (2020) through Stiglitz (2021) to Piketty (2014, 2021). There is a wide-spread consensus among this group that the economic paradigm prominent in the West—neo-liberalism or neo-conservatism—is due for replacement and, as in any campaign, social or military, the first step is to understand the opposition. With the earlier considerations regarding freedom (Sect. 6.6) and the role of information technology (Sect. 8.5), both of which play a major role in creating the public understanding of the economic system, the scene is set for reviewing the current economic paradigm, its relationship to the concept of freedom, and its problems, and demonstrating the need for a paradigm shift in both our economic system and in our understanding of freedom.

12.3.2 The Current Paradigm: Neoliberalism

The concept of a paradigm, and of a shift in paradigms, was brought to prominence by Thomas and presented in his seminal work, *The Structure of Scientific Revolutions* (Kuhn, 1996), to which we have referred several times. Kuhn showed that, for long periods of time, science works within a set of accepted truths and norms, which he called the scientific paradigm, and that it is the existence of this stable framework that, to a great extent, underpins the efficiency of scientific work. However, these periods of stability are interspersed by relatively abrupt changes to the paradigm, brought about by two factors: the number of experimental results that cannot be explained within the existing paradigm reaches a critical value, and a new theory is put forward that explains all or most of these results.

In economics or, perhaps more specifically, in political economics, it is possible to discern a similar behaviour, but with two important differences: As economics is an empirical science tied to the state of evolution of society, an economic paradigm can vary in importance over its existence, whereas this is not possible for a scientific paradigm; there is no way the Newtonian paradigm will come back to replace relativity. And economic paradigms contain a significant component of ideology, which is no longer the case with scientific paradigms. Despite these differences, Kuhn's paradigm concept can be usefully applied to economics, and we shall apply it here to capitalism.

Under the capitalist system, an increasing proportion of income is in the form of return on investment, i.e., on capital. Capital has been with us from the beginning, and over time, it has taken on various forms, such as hoards of precious metals, land, and slaves. And, increasingly, it has taken the form of applications of technology, starting with the first stone axe. With technology came the concept of *production*, the transformation of a material into a more useful form, and with it the *means of production.* For simple transformations, these means of production could be the tools or the plot of land belonging to each worker, but as technology evolved and allowed more elaborate transformations, establishing the means of production required an expenditure in excess of what the workers could afford, and so they would have to borrow money, or conversely, someone would have to lend the required capital. Assuming the enterprise was successful, the workers would be able to pay back the loan at some stage, but having had the use of the money over that period clearly had been of significant value to them, for which it was only reasonable that the lender should be reimbursed, and so money, in addition to its immediate value as a means of exchange took on an additional, time-dependent value when it was invested, in the form of an interest rate. This is, essentially, the difference between money and capital. In this way, money took on some of the nature of a product, in that it could be traded without any other, physical product being involved; the most immediate example of this is insurance.

Now, this feature of money as capital, and the process of lending and borrowing, and of paying interest, has been around for a very long time, certainly it was prominent in Rome, and was the *modus operandi* of most of the traders that travelled between

Europe and the East in the Middle Ages. It is one of the pillars of the economic system we call *capitalism*; another one is a particular extension of the concept of *ownership*. Going back to the example of the workers that borrow money to set up their production facility, they own the means of production, and they reap the full benefit of whatever success they achieve. There is, in principle, no problem with this, and it is realised in many small workshops and other businesses, and it was also realised in the early phases of many large enterprises. However, in practice there are two factors that mitigate against this: the *economy of scale* and *risk*. Most industrial applications of technology display an increase in efficiency with size due to the sharing of common overheads between individual processes, from management to maintenance, so that, unless a small company has some special advantage, such as a patent, or is operating in a small niche market, it cannot survive in open competition with larger companies. All new enterprises involve a measure of risk; and according to the US Small Business Administration, about half of all new employer firms fail within five years. Lenders therefore require some form of guarantee which, in our example, the workers would have to provide, e.g., in the form of their houses or other private assets, something that is unusual in all but small firms. In the overwhelming part of industry (e.g., by turnover or capitalisation), the capital is provided by people who can afford to take the risk, i.e., in the worst case lose the investment, which is one way of defining a capitalist, and so the owners of the means of production and the workers are separate groups of people; the workers are employed by the owners and provide the labour power required for production. The concept of ownership is extended from things that are directly related to me, such as my house, my block of land, my tools, etc. to things that are only related to me through the capital involved. The converse of this is that, in those investments that are successful, the investor can make a substantial return on the investment, or profit, and on the average, the returns on the successful investments must exceed the losses on the unsuccessful ones, otherwise the investment process would come to a halt, and so the stock of capital accumulates. This positive feed-back mechanism is the driver of wealth inequality, which we consider in the next subsection.

In all of the above, there was no assumption made as to the identity of the investor/owner; it would all apply equally to the case of private owners as in the case of government ownership, or any combination of these two. Furthermore, the two factors just discussed are closely related to the third pillar of capitalism, the *market*. Successful investment in production facilities requires that the products can be converted into money, to cover the production costs (labour and materials) and produce a profit, and the mechanism for this conversion is the market or, more specifically, the free market, although there are numerous factors that make this freedom less than absolute. It is the competition in this market that favours the larger enterprises, and it is also this mechanism that generates the risk, not only from other firms producing the same product and competing for the same group of consumers, but also from completely different products that find more favour with the consumers.

We identified three pillars of capitalism: the concept of capital (extending the nature of money from a means of exchange to a product), an extended concept of ownership (mediated by capital), and the concept of the (free) market. They form one part of the capitalist paradigm, what might be called the rational part. Another part is what might be called the ideological part; it is essentially a view of human relations that sees competition as the ideal form of these relations. The most immediate and visible support for this view is sports, where competition brings out the best performance in each individual; other cases that seem to provide support are found in the competition for promotion, for awards, for the most desirable partner, and so on; all of which are seen as expressions of Darwin's theory of evolution and expressed as "survival of the fittest" by Herbert Spencer, in the human domain. This view has several consequences, of which three important ones are:

- Competition should be encouraged in every instance, and where there are barriers to competition, e.g., in the form of government restrictions, these should be removed. And as the competition takes place in a market where the only measure is a monetary one, this means that the valuation of all human endeavour is simply what the market is willing and able to pay for it—an exchange value.
- The logical outcome of competition in a limited market is that there is a winner, and so an inherent trend in capitalism is an increase in size and a reduction in number of competitors, i.e., an oligopoly. And should this process end in a monopoly, capitalism sees no in-principle problem with this, as long as the monopoly is open to a competitive challenge in a free market.
- Almost as a corollary to this dogmatic stance is the view of private ownership; the right of every person to accumulate wealth, both in the form of things and in the form of financial capital, over which he or she can dispose and benefit from at will. It is seen not only as a fundamental component of liberty, but also as something admirable and a moral duty, in the sense that making a profit and accumulating wealth is the opposite of sloth and dissipation, as noted in Sect. 11.4. One expression of this attitude can be observed in people who go to any lengths to minimise their tax payment, but then give large amounts as philanthropy to charitable organisations. (Of course, the latter approach might earn you a medal or something; nobody ever got any thanks for paying tax.)

This has been the paradigm of political economy for much of the time since the beginning of the nineteenth century. It faced a serious challenge in the period between the two world wars, but since the last war it has regained its position as the dominant paradigm, championed by such public figures as Reagan and Thatcher, and vindicated by the collapse of the Soviet Union and by an increase in living standards, and it has now become the meaning of *neoliberalism.*

It is worth noting that a recent book by Suarez-Villa is concerned with a particular aspect of this stage shift—technocapitalism (Suarez-Villa, 2009). The hallmark of this new stage of capitalism is the commodification of innovation and creativity on a global scale by means of what Suarez-Villa terms "fast accumulation", which is supported by IT, and he sees technocapitalism as more aggressive and farsighted than any previous versions of capitalism, as it seeks to re-engineer humanity and nature itself for its own ends.

However, as with all paradigms, issues are emerging that conflict with the paradigm's claim to provide a better life for all and the economic foundations for a more just society, such as the perversion of democracy by wealthy individuals or organisations, and economic inequality, both within nations and between nations. It is becoming increasingly clear that "more of the same" is not going to work, and numerous researchers and social activists have analysed the problems and suggested ways forward; one excellent example is Shiller (2012). Here we shall only look briefly at the aspect of particular importance in the context of the social bond: the influence of IT.

Due to the strong and ubiquitous interaction between IT and society, as already introduced in Sect. 4.6, the presence of IT in the structure and functioning of society can be considered a defining characteristic of society, and one speaks of *the information society*. Garnham emphasizes the need to distinguish between two types of theories about the impact of IT on society: those that focus on the direct impact of developments in IT on politics and culture, and those that focus on the impact on the economy or mode of production (Garnham, 2001). With regard to the former, he states that "this whole strand of thought is the view that problems of social and cultural inequality can be solved and barriers to full political participation removed by technologies for the production and distribution of information. Put the processes of democratic politics on-line and full political freedom will be achieved. Put university and school courses on-line and the age of universal education will finally arrive."

Regarding the latter, Garnham refers to the work of Daniel Bell and the argument, based quite explicitly on both Marx and Weber, that post-industrial society is developing in stages. Capitalism is moving from a stage of industrial capital based on the exploitation of matter and human energy to a post-industrial stage based on upon the exploitation of what Bell called "organised knowledge". The core resource has shifted from monetary capital to knowledge. Garnham makes the observation that much of the talk of an Information Society is just the Service Economy relabelled, and that arguments about IT having caused an epochal shift in either economy or society are largely specious.

In the book *Information technology: Impact on the way of life* (Bannon et al., 1982) several authors point out that what we are witnessing is a second "industrial revolution" in which an increasing number of human activities in all fields are being taken over by robots. Robins (1982) states that the history of multinational corporations gives little promise that they will have democratic interests at heart in the future. For them, IT represents just another series of exchange values that by no means correspond to real social needs, and they have shaped it—without democratic participation and consultation—to express specific corporate values and priorities. There is no reason to believe that IT will not reinforce, and aggravate, existing inequalities at both national and international levels. He quotes Schiller (1981) as saying "contrary to the notion that capitalism has been transcended, long-prevailing imperatives of a market economy remain as determining as ever in the transformation occurring in the technological and informational spheres", a view that was formulated already in Schiller (1972), where he argued forcefully that the current role of the media in modern society, and in the US in particular, is to support the capitalist economy and suppress any meaningful debate on alternatives; a condition expressed as "packaged consciousness". Similar concerns were raised earlier, in Mills (1956).

In the same volume, Garnham presents a view that we will have to take into account in our development in the section after next, and it goes as follows (Garnham, 1982): He challenges the view that the information technology will usher in a new era of cultural freedom, diversity and abundance and shows that we are here in the presence of ideology in its pure, classical form; that is to say, a social analysis that not only misrepresents its object of analysis by focusing on its surface rather than its underlying structure and by denying its real history, but also misrepresents it in such a way as to favour the interests of the dominant class. In this case, the trick is played by concentrating upon the technical potentialities, rather than upon the social relations that will determine the form in which those potentialities are realised and by denying history by exaggerating the novelty of the process in question. We are witnessing merely the latest phase in a process integral to the capitalist mode of production; an "industrialisation of culture" or "colonisation of leisure" by which massive market interests have come to dominate an area of life which, until recently, was dominated by individuals themselves.

Garnham then discusses the growing information gap in western societies, between the information-rich and the information-poor, and how this is related to the widening gap between rich and poor. Also, the information technology is structured to reinforce this, splitting the culture into two classes. Choice, being increasingly expensive, is offered to upper-income groups, while an increasingly impoverished, homogenised service is offered to the rest. Eventually, we will have to choose between two social forms. On the one hand, we can choose a society which primarily fosters social relations, based upon the Aristotelian notion of men and women as essentially social animals, based therefore upon notions of social reciprocity and interchange, upon the public as opposed to the private as the essence of humanity. Without such a notion, politics in any true sense is unthinkable. On the other hand, we can choose (or more likely have forced upon us) a society which is merely a social structure within which atomised, privatised individuals interrelate, primarily through commodity exchange, and by so doing necessarily reproduce the dominance of the capitalist mode of production and of those who control it, namely the owners of the means of production. Such a society will necessarily subordinate the public to the private sphere and destroy politics in favour of a manipulative form of elite control if we are lucky—what Bertram Gross has dubbed "Friendly Fascism". These latter tendencies will be powerfully reinforced in the cultural sphere by the introduction of information technology under market conditions. Such introduction should therefore, as far as possible, be opposed.

12.3.3 The Road to a New Paradigm

The focus on competition as the main interaction between individuals is a limited and primitive view of human nature, discounting our perception and intelligence. It is a view appropriate to the progression of animal species, but of decreasing applicability for our progression of society. There are numerous examples of how people perform at their best without any significant competitive incentive, from nurses and missionaries to teachers and judges, and there are many organisations that function very well without being in a competitive environment, such as Médecins Sans Frontières.

There is no doubt that competition plays a significant role in the behaviour of both people and organisations, but that role should change as society evolves; from a means of survival towards a refreshing intellectual and physical challenge—it should be fun. Its interpretation in capitalism, and its expression in accumulating wealth, is going in the wrong direction, as was explained so well by Weber (1930) and noted in Sect. 11.4. Capital (i.e., the existence and use of capital) has been around for a long time, since before Antiquity, but it was always either a property of authority, administered by bureaucracy (and tax farming), as in ancient Egypt and the Roman Empire (note also the importance of slaves as capital), or then, in the Renaissance, a necessary adjunct, in the sense that making a profit was more like a sport. The change came with Protestantism, when hard work and making a profit became a moral duty; not making a profit on capital was seen as a waste and an affront to the Lord's intention for mankind, and Weber illustrates this with quotes from Benjamin Franklin. Thus, capitalism became a central part of culture and of society. As an aside, this finds an analogue in technology. In the early days, much of the development of technology, often but not always rooted in science (e.g., Hero of Alexandria, also Leonardo da Vinci), was for amusement; only with the industrial revolution and the rise of capitalism did the application of technology become a moral duty; not doing so would be a waste.

Relaxing the insistence on competition means that the importance of private ownership and its reflection in the Holy Grail of Small Government could be somewhat reduced. In particular, should we not differentiate the ownership of things from the ownership of the means of producing things? And within the latter, is there not a significant difference between the production of services seen as part of the infrastructure of our society, as services everyone has a claim to, such as education, health, justice, and safety, and the production of more differentiated and discretionary items, such as e.g., clothing, household goods, and entertainment? This would also give recognition to the fact that the rate of change of infrastructure is much less than that of consumer goods, and that this is reflected in the business strategies, with an emphasis on long-term planning rather than on agility.

In defence of the current paradigm one can bring many examples and studies that show how government owned enterprises are performing less well than comparable privately owned enterprises. However, there are also numerous counterexamples, as documented in Hanna (2013), where it is also noted that the current extent of public ownership is much greater than most people would realise. A recent book, *The Public Wealth of Nations* (Detter & Fölster, 2015) shows that poor management is not an inherent feature of being publicly owned; it is a result of the manner in which governments have traditionally managed state-owned enterprises (SOEs). The authors demonstrate, through an array of examples, that public assets can be managed just as successfully as private assets and emphasize how this frees politicians and government administrators from the pressures of vested interest groups for a share in the spoils of public wealth and allows them to concentrate on issues concerning individual citizens and the economy as a whole.

The key to this is to consolidate all SOEs and other commercial assets into a national wealth fund (NWF), a ring-fenced holding company at arm's length from short-term political influence by the government and with professional management. Quoting from the book:

> The NWF uses all the appropriate private sector tools and institutional frameworks that enable the government to consider the portfolio of commercial assets as a whole from the perspective of operating income and liabilities without any of the constraints of a public sector bureaucracy. Consolidation under a private sector vehicle allows the government to establish strategies for handling lossmaking assets, while the priority of improving performance provides greater opportunities to raise financing and choose optimal circumstances for disposals, in the same way as a private sector owner. (p. 144)

While Detter and Fölster definitely see privatisation as one of the management mechanisms, their other suggestions for how to set up and run a NWF provide much of the basis needed for a new national economic paradigm in which the economy consists of two parts: the private sector economy, with private ownership, and the public sector economy, with government ownership. The details of how to make this transition would vary considerably from nation to nation, as the current organisation of the national economy varies, with some nations having already realised elements of the transition, as reported by Detter and Fölster. But the transition does not have to be done all at once; it would be a gradual process, sector by sector, with each sector having issues peculiar to individual countries, such as labour relations, current level of privatisation, and international obligations and contractual arrangements. The sectors for immediate consideration would be those where government already has a significant role, such as education, health, justice, safety, and transport, but in the next section we shall consider how other sectors of the economy might transition from the private into the public part within this new economic paradigm.

However, the greatest difference between nations would be in the required shift in ideology, with current economic ideologies ranging from a Wild West belief in the free market and private ownership to almost complete government control and ownership. Not that the shift would be any easier in the latter case than in the former; politicians are as loath to give up the power that direct management of state resources gives them, with its possibilities for nepotism, corruption, and cronyism, as private sector entrepreneurs are to give up their ability to engage in any part of the economy. In Western nations, in particular in the US, the cultural shift, and how it could be coupled with a successive shift in business structure, is discussed by Kelly (2013). She defines a class of business structures that she calls Generative Ownership Designs, because they generate the conditions for our common life to flourish, rather that focus exclusively on maximising profit. However, it is not clear to what extent this reorientation would change the power relations in society, and the reference to "the butcher, the baker, the candlestick maker" as an idealised form of enterprise seems to neglect the role of technology and the capital that goes with it.

So, given the problems involved in this paradigm shift, why should it take place? Referring to Kuhn, it is not enough that problems with the existing paradigm become significant, but that new models, results, or insights provide convincing resolutions of the problems; only then will the paradigm shift take place. In that sense, what is outlined in this section is just a proposal for what could become a new paradigm, and what is provided in the next section are arguments for why a shift should take place. Both the development of the details as well as the proof that it results in a better society will only be provided if and when it is gradually implemented.

12.3.4 Arguments for the New Paradigm

In advocating a transition to the new paradigm, we need to answer the following four questions:

- Why is the transition to a new paradigm relevant in this point in time?
- What current problems will it solve?
- What are the obstacles to the transition?
- What is the relationship to freedom?

One answer to the first question is that as society is undergoing rapid change, measured e.g. by the intensity of interaction between individuals and the integration of individuals in the structure of society, or by the development and application of new technology, it is unrealistic to expect our understanding of one of the major aspects of society—its economy—to continue to be based on a view of the role of the individual that goes back to Antiquity. Another answer is that while in the case of paradigm shifts in science the timing is not critical, it happens when it happens, the speed of change makes the need to address emerging problems in society critical, to prevent them resulting in significant, and in the worst case catastrophic, fluctuations in the evolution of society. But let us now turn to some specific answers to the second question.

Better return on public wealth. Much of the world's public wealth is providing little or no return; by consolidating it into NWFs, under professional management, the increased return is, for the government, the equivalent of an additional tax, and offers the same options for its employment (e.g., increased investment in public infrastructure, or as input to a sovereign wealth fund).

Reduction in such management failures as corruption and cronyism. Replacing short-term political manipulation by fully transparent, professional management greatly reduces the opportunities for corruption, cronyism, and the undue influence of vested interests.

Freeing up politicians to focus on their job. For many politicians, the pressures exerted by vested interest groups for access to some of the "hidden" public wealth is an unwelcome, but unavoidable distraction. As a corollary to the previous answer, this pressure would be practically eliminated.

Control of public resources. For many countries, Australia among them, a significant portion of the public wealth is in the form of natural resources, such as oil, gas, coal, and minerals, but in many cases the public receives only a small fraction of this wealth; it is essentially given away to private industry. The new paradigm provides the framework for nationalising extractive industries, without suffering the criticisms regarding poor management previously attached to nationalised industries, and at the same time providing the means for greater compliance with environmental requirements. One example of such a SOE is the Norwegian oil and gas company, Statoil.

Reduction of economic inequality. If we allow that public wealth belongs to the people, and that the return on this wealth is distributed in a manner that meets the needs of the people, then the new paradigm will reduce inequality. Detter and Fölster estimate that the world's public wealth held by national governments is at least 75 trillion USD, probably considerably more (p. 52). This number does not include local government assets and natural resources, but from the limited data available, it is at least clear that the total value of public wealth is comparable to the total value of private wealth, which is about 250 trillion USD (Credit Suisse, 2015). Hence, the public wealth would provide a significant counterpart to the private wealth, and under the new paradigm the two components of the economy would be comparable. This would not eliminate the inequality inherent in the private economy, nor should it, as a degree of inequality reflects the inherent difference in individual capabilities and motivates innovation. But it would eliminate the excessive inequality manifested as poverty and lack of basic opportunities, and it would reduce the pressure exerted by capital accumulation and introduce a stabilising effect on the inherent instability of the free market.

Regarding the third question, there are several reasons for why a transition is resisted; most of which are related to the very nature of a paradigm. As Kuhn explained, a process becomes a paradigm because it works successfully over an extended period of time, and through this success is able to repel any challenges from divergent processes. A wealth of experience is built up, together with significant investments in whatever infrastructure is required to make the process work, and any problems that appear are always solved *within* the paradigm, simply because it is unthinkable to go outside it. This is the issue of *enframing* we encountered earlier. We have become so conditioned to the concepts and language of the capitalist society in its neoliberal version that we have difficulty thinking outside this framework and question its foundations.

Another reason is the fact that much of the success of capitalism, in the form of wealth and a high standard of living, is concentrated in the West, i.e., Europe and America, whereas most of the problems—inequality and poverty—are concentrated in other parts of the world. So, for the majority of us in the West, the problems seem fairly remote and not sufficient to require any change to our comfortable lifestyle, even though it is from us that the impetus for change should come.

A further reason is the current extension of the paradigm to the international level, in the form of clauses in trade agreements that limit the ability of national governments to implement changes that would affect private business interests. It might remind us of the regulations imposed by colonial powers on their colonies, making national governments seem like colonies of the TNC.

The answer to the fourth question is that the issue of freedom is inherent in any structural part of society, including its economic structure, and that underlying the two economic paradigms we have considered here are two considerably different views of freedom. The current neoliberal paradigm sees freedom as the freedom of the individual economic agent, and the role of the societal framework to be to allow the greatest possible expression of this freedom. The restrictions of the framework should only be restrictions on actions that would limit this freedom; it is a concept of freedom focused on the individual, with little concern for its effects on society as an entity in its own right, with its own "individuality". Indeed, to quote Thatcher (1987), "There is no such thing as society: there are individual men and women, and there are families."

The concept of freedom associated with the new paradigm is one of being able to use our abilities to contribute to society; it is a concept that only makes sense when the individual is viewed as an integral element of society. It involves an explicit recognition of the fact that almost every feature of our lives is dependent on us being embedded in a society, and that the concepts describing our existence, including that of freedom, need to reflect the complex structure of society. There is something simplistic, almost primitive about the view of freedom within the neoliberal paradigm, with its focus on personal wealth and ownership, when what is needed is a *reorientation*, from the economic to the political aspects of freedom, and from independence to participation. This is not a reduction in freedom; on the contrary, it is an increase by extending its applicability and relevance, and the shift to the new economic paradigm would be a significant and very visible step in this reorientation.

Chapter 13
The World Society

13.1 Recent History

Our view of society, what is good about it and what is bad about it, and what we should do to improve it, is a very personal matter, whether the society is the local community, the nation, or the world. This view is based in part on facts, or our observations, but also, in varying degree, on our beliefs or ideology. For example, one person believes that society is simply the sum of its members, and if each person only strives to work hard for himself and his family and makes the best out of every opportunity that arises, the best society will result. This was the belief of Thatcher (1978). Another person believes that the best society arises out of the harmonious and efficient cooperation of all its members and by giving each member the means to realise its potential. And similar differences can be reflected onto the world society, with nations as its members. The existence of different views is important; it is the dialectic processing of these views in a vigorous and free public discourse that drives the evolution of society.

But every view, no matter how ideologically pure, is based to a significant extent on past experience—on history, and our public discourse would be so much more effective if it were based on a common knowledge of history, instead of the present use of selective views of history as weapons in the discourse. The following very brief review of history over the last two centuries is focused on providing the reader with a convenient reminder of the main events and developments relevant for understanding the current international power relations and associated tensions. Setting the horizon at two centuries is somewhat arbitrary, but I believe it is possible to define and explain the current situation to a reasonable degree of accuracy and completeness in terms of changes that took place in this period.

E. W. Aslaksen, *The Evolution of Society*,
https://doi.org/10.1007/978-3-031-23054-7_13

13.1.1 The Starting Point—Colonies and Imperium

The structure of the world in 1850 in terms of nations, colonies, and zones of influence developed in several stages following the discoveries around 1500. In some cases, such as the Spanish and Portuguese discoveries and incursion in the Americas, and the establishment of Australia as a penal colony under the British Crown, the remote territories were defined and administered as colonies from the very start, but in most other cases the initial exploration and settlement was by private enterprise in search of new business and profits and/or, in some cases, to escape from religious persecution, and the designation as colonies under the sovereignty and administration of the mother country came only later.

The Portuguese had started exploring the west coast of Africa from early 1400 onwards, pushing ever further south and establishing trading posts along the way, some of which later developed into colonies, such as Angola, Guinea, and Cape Verde, until, in 1498, Vasco da Gama continued around the southern point of Africa and reached the south-west coast of India. Portuguese traders took over the spice trade from the Italians, replacing Venice with Lisbon as the prime European trading port, and then expanded their reach into South-East Asia and to the coast of China. However, the arrival of the Dutch and British, and the establishment of their two large government-sponsored trading companies gradually reduced the presence of the Portuguese in India and in the East. The focus of Portuguese overseas interests changed from Asia to South America and resulted in establishing Brazil as a colony of the Portuguese Crown.

As the Spanish conquistadores advanced in the Americas, Spain developed settlements in Mexico and from there northwards up the west coast of North America, and in Peru and from there southwards along the west coast of South America. Sailing westwards, the explorer Miguel López de Legazpi arrived in the Philippines and established it as a Spanish province.

The Dutch expanded their trading activities in and around what is today Indonesia, with headquarters in Jakarta (Batavia), and when the fortunes of the Dutch East India Company declined in the latter half of the eighteenth century, the company was nationalised, and its holdings turned into a colony of the Dutch republic. Dutch explorers and traders also established a trading post at the mouth of the Hudson River that in 1624 became a Dutch colony—New Netherland, with the city of New Amsterdam on the island of Manhattan. In 1664 the Dutch possessions were ceded to the British in exchange for Dutch sovereignty over Surinam.

The British advance in the footsteps of the Portuguese into India and East Asia was also fuelled by commercial interests, and the British East India Company expanded rapidly both in economic activity, accounting at one stage for almost one half of the world's trade, and in the area under its control, including large parts of India and settlements in South Africa, Hong Kong, and Canton. However, in all these areas conflicts with the native population arose that went beyond what the Company could handle, resulting in the British government stepping in with military force and turning the holdings into colonies under the British Crown. The involvement with North America started out with government-sponsored piracy of Spanish ships transporting treasure from the Caribbean to Spain; the first settlement was established by the Virginia Company of London at Jamestown in 1607, soon followed by the Plymouth Colony in Massachusetts, and from then on, a rapid expansion all along the east coast and also penetrating ever further west. This expansion brought British settlers into conflict with French settlers, both inland and in Canada, and at the settlement of the Seven Years' War in 1763, France ceded all its possessions in North America to Britain.

French explorers were active in all parts of the world in the seventeenth century, and significant settlements were established in India and in North America—in Canada and in Louisiana along the Mississippi River. But by 1800 all had been lost, including the holdings in India, mainly as a result of the Seven Years' War with Britain, as noted above.

If the period 1500–1775 was one of exploration, conquest, and the establishment of colonies, the period of 1775–1825 was one of revolution. Inspired by the French Revolution, the purpose was to sort out the tensions that had arisen in the relationships between colonies and mother countries. The American War of Independence saw the creation of the United States of America, and, in the first decades of the nineteenth century, the revolutions against Spanish rule in South America led eventually to the independent nations there today. Thus, by 1850 there had been a shift in colonial power from Latin to Anglo and from Catholic to Protestant, and the by far most significant colonial power was Britain, with Canada as a Dominion, control of most of India through the British East India Company, and significant interests in China.

The following period, up until the start of the First World War in 1914, was dominated by several interrelated issues. One, and perhaps the primary one, was the rapidly increasing industrialisation of the European powers, resulting in an accumulation of capital looking for investment opportunities, and a need for raw materials and markets for products that went beyond what the individual nations could provide. This led to what has been described as the New Imperialism; a scramble for colonies in the two areas of the world still available—Africa and East Asia. In what has been termed the Rape of Africa, most of Africa came under the rule of Britain, France, Germany, Portugal, Italy, Belgium, and Spain, as set out in the terms of the Berlin Conference of 1884–1885. In South-East Asia France acquired Cambodia, Laos, the Chinese territory of Guangzhouwan, and Vietnam. China was a very lucrative trading partner for the European powers and made more so by using military force to subject the Chinese to humiliating defeats and highly discriminatory trading conditions. But by and large China maintained its territorial integrity, including its resistance to Russian incursions in Central Asia and French attempts to gain a foothold on Taiwan and in Tonkin, until the Japanese started their campaign to create an empire.

Another central feature of this period was the entry of three new players in the game of imperial Monopoly—Germany, Japan, and the United States. After defeating the French in 1870, followed by the unification of the German states under the leadership of Prussia, the new nation of Germany set out to cement its status as a major power, both on land and on sea, causing increasing angst and tension both with France and Russia, but above all with Britain, who saw their control of the maritime trading routes threatened. In the Far East, the rapid modernisation of Japan following the Meiji restoration and its defeat of both Russian and Chinese forces led to Taiwan becoming the first Japanese colony, followed by Korea, and to extensive rights in Manchuria.

The most aggressive and successful practitioner of imperialism was the United States. In a series of wars and acquisitions it increased its territory manyfold, starting with the Louisiana Purchase in 1803, followed by the acquisition of Oregon Country and Spanish Florida from Spain in 1821. The annexation of Texas in 1845 led to the Mexican–American War, which ended with the US obtaining the northern half of Mexico. In 1867 the US purchased Alaska from Russia, and after defeating Spain in the Spanish-American war in 1898, Guam and Puerto Rico became unincorporated territories of the US. In the same year the US annexed Hawaii, having overthrown its government some years earlier.

A feature of this period that should not be ignored, in particular in view of its current role, was the expanding armaments industry in the major powers. Not only did the industry promote the increased military expenditure within the country but fanned the flames of conflict in other parts of the world. In the end, the military establishment in each country became a capability looking for an application.

A final notable feature, and one that is again very relevant today, is the process of forming alliances; generally for the purpose of mutual defence, but often also with a significant commercial component. Now, there was nothing new, around 1900, about forming alliances, and a substantive list of military alliances going back to antiquity can be found at https://en.wikipedia.org/wiki/List_of_military_alliances, accessed 20.06.2022. What distinguished the Allied Alliance from most earlier alliances was its ability, due mainly to British diplomatic skill honed over centuries of forming alliances with the purpose of destabilising its competitors on the Continent, to bring a great number of additional nations into the Allied alliance and achieve a *polarisation* on a world scale. "If you are not with us, you are against us".

The root cause of the conflict was the rise of the newly established German Empire and its obvious chafing at the bit to become a major power on a world scale. On the one hand, this frightened its direct neighbours, France and Russia, on the other hand it made Britain concerned for its far-flung empire. The alliance of Germany with Austria-Hungary in 1879 did nothing to assuage these fears, so first France and Russia formed an alliance in 1892, and then Britain joined through its Entente Cordiale with France and Anglo-Russian Entente to form the Triple Entente.

Thus, the decades leading up to 1914 were characterised by a peculiar dichotomy: On one side, the West was experiencing a period of local peace, stability, and prosperity that supported a flourishing of science, technology, and all areas of the arts. The French call it *La Belle Époque*, and its hallmark was the opulent lifestyle of the *nuveau-riches*. On the other side, and largely unseen or ignored by the general public, was a constant jockeying by the governments, including the heads of state, for position and influence on the international scene, forming relationships and building up real and imagined intentions and a tension which it only took a single shot to ignite.

13.1.2 The World Wars

The First World War did not start with an attack on a member of either of the two primary alliances—the Triple Alliance and the Triple Entente, but with the invasion of Serbia by Austria-Hungary in response to the assassination of the heir to the Austro-Hungarian throne. One of the numerous secondary alliances obliged Russia to support Serbia, and from there the whole edifice of alliances unfolded until, towards the end of the war, the Allied Powers included more than fifteen members.

From its start in July 1914 until its end in November 1918 the war resulted in 15 million fatalities and millions of wounded and crippled, and a combined cost of US $5.5 trillion in today's purchasing power. And at the end, four empires—Austro-Hungarian, German, Ottoman, and Russian—were gone, Europe had lost its leadership role and was no longer a cultural ideal, and the United States had greatly enhanced its international role, although it still harboured strong isolationist views.

The aftermath of the war, starting with the Paris Peace Conference and the reorganisation of the European and Middle Eastern national landscapes was, if anything, as disastrous as the war itself, as it did little to reduce the ethnic and religious tensions in the Balkans and the Middle East, and exacerbated the pre-war tension between Germany and its neighbours through casting all the blame for the war on Germany and through punitive reparation demands. In short, through either a complete lack of understanding or disregard of the forces operating both within and between nations and of the resulting dynamics, it made WW2 more or less inevitable.

It is interesting to note that although China was at first neutral, Japan saw the war as an opportunity to strengthen its position on the Chinese mainland, first by taking the German concessions on the Shandong Peninsula, including the city of Qingdao, and then issuing the Twenty-One Demands (later reduced to Thirteen Demands) on China in 1915, which the local warlord, Yuan Shikai, had no option but to acquiesce to. China ended its neutrality in 2017 and entered the war on the side of the Allied Powers, but at the Paris Peace Conference the German concessions were not handed back to China, but remained in Japanese hands, something which infuriated the Chinese and made their delegates leave the conference without signing the peace treaty. However, under pressure from the US, China gained control over these ex-concessions in 1922.

The Second World War had two aggressors, Germany and Japan, both driven by the need for new markets and access to raw materials, in addition to imperialistic ambitions. After successfully furthering its aim by acquiring Austria and Czechoslovakia without any serious opposition from other European powers, Germany invaded Poland on 1 September 1939 despite France and the United Kingdom having guaranteed their support for Polish independence, and the European part of WW2 was underway. In Asia, Japan started its campaign by attacking the Republic of China in 1937, followed by the attack on Pearl Harbour and offensive actions in Southeast Asia in late 1941.

The war ended in Europe when Germany surrendered in May 1945, and in Asia when Japan surrendered in August/September 1945. The total cost of the war in human lives was in the range 70–85 million, with the Soviet Union and China by far the worst hit, and the economic cost ran into many trillions in today's $US. [Accurate and complete data on the cost of WW2 is difficult to obtain, see e.g., (Shirras, 1949) and (Harrison, 1998).] Germany lost a quarter of its 1937 territory, mostly in its eastern provinces, and Japan lost Karafuto Prefecture and the Kuril Island to the Soviet Union, and Taiwan and the Spratly Islands to China. But the immediate aftermath of the war was very different in these two nations, mainly because of the US perception of what the future geopolitical situation would be. In Europe it was important that Western Europe could provide a first line of defence against any Soviet expansion and serve as an example to Eastern Europe of the benefits of capitalism. To this end, the Western European nations would have to work together, which required Germany to acknowledge its guilt for the war and relevant atrocities, to punish the guilty, as in the Nürnberg Trials, and remove Nazi party members from official positions. Also, numerous locations were dedicated to commemorating Nazi war crimes, such as Auschwitz and concentration camps and museums in many countries, and to this day, even in Australia, a country that is far away from Germany and was never occupied by the Germans, there are weekly reminders in the mass media of the war and of the Holocaust.

In Asia the situation was quite different. By the end of the war, there was already some doubt in the US administration about the chances of the Kuomintang, under Chiang-Kai Shek, overcoming or coming to an agreement with the communist opposition, and by 1947 it was becoming clear that the communists were going to win, as described in the memoirs of the then Secretary of State, Acheson (1969, pp. 302–307). As a result, the US now saw Japan as its first-line defence against two communist opponents, Russia and China, and there was now no need to reconcile Japan with its former enemy, China, in order to form a united front against Russia. On the contrary, there were now two good reasons to minimise Japan's war crimes: firstly, to reduce any antagonism against the US, and secondly, to make any reconciliation or rapprochement between Japan and China as difficult as possible. Consequently, the American occupying forces, under General MacArthur, carried off the incredible feat of absolving the Emperor, Hirohito, and the imperial family of all guilt, to reduce the number of death sentences to a minimum, and not to object when most of the prison sentences were either not carried out or commuted after a fraction of their original duration and some of these went on to occupy high positions in government. They even provided the head of the infamous Unit 731, Colonel Shiro Ishii, the Japanese equivalent of Germany's Josef Mengele, with immunity from prosecution and a plum job, a puzzling apparent difference in the valuation of Jews and Chinese. Also, on US insistence, China, as the main victim of Japanese aggression, was not invited to the San Francisco Pease Conference in 1951. And with Australia more or less an extension of the US there is hardly ever any mention of the Chinese part of WW2 or of Japanese war crimes in Australian mass media.

So, the US considering China as an adversary is nothing new, they just got caught out by the suddenness of the transition from potential to actual.

13.1.3 Superpowers and Hegemony

By the end of WW2, the world had undergone a major transformation, to some extent completing the changes that had been initiated by WW1 but adding some new features. The most significant change was the polarisation of the geopolitical landscape around two superpowers, the US and the Soviet Union, changing what had been ideological adversaries into adversaries in a global power struggle backed by military strength. But these two superpowers were very different in both nature and genesis. The US rose to superpower status as the only major power to be spared the material destruction caused by the war. With all its competitors more or less devastated by the war, and with its own industrial capacity greatly enhanced by the demands of the war, the US emerged as the major economic, technological, and military power, with the financial capacity to shape the West in its image. The Soviet Union experienced the war very differently; taking the brunt of defeating the German army, it ended up with several of its cities, including Stalingrad, in ruins, its industry and infrastructure severely damaged, and the loss of some twenty million of its citizens. It was therefore in no position to advance its sphere of influence by offering financial support; on the contrary, backed by its army, which at the end of the war exceeded ten million soldiers, it stripped occupied Eastern Europe of machinery and equipment, and strengthened its grip on the governments of those countries.

In the following decades, Russia and the US formed the cores of two opposing groups of nations—the USSR and the West, both with their military infrastructure—the Warsaw Pact and NATO—and both using the fear of an attack by the other as an excuse for excessive expenditure on development and maintenance of a military infrastructure and for the sabre-rattling known as the Cold War. For Russia, military strength was a prerequisite for its hold on power within the USSR and in its sphere of influence, as demonstrated, for example, in quelling the Hungarian and Czechoslovak uprisings. For the US, military strength served to keep the Russians in place, but it was also an important driver of commercial development through the arms industry and the commercialisation of military technology.

But to be a superpower requires more than just military power; it is about economic power in general and the ability to project this power through trade, support and aid, effective diplomatic activity, and a culture and lifestyle that others would admire and aspire to, and despite all the hype, Russia was never a credible rival to the US, as was demonstrated when the USSR imploded around 1990. The tragedy was that the US became so focused on 'communism' as the epitome of all things evil, conflating it with Marxism, Stalinism, socialism, anti-colonialism and basically anything that was perceived to threaten capitalist consumerism, that it missed out on many opportunities to support other people's desire for self-determination and a better life; sometimes opting for the easy solution of installing or supporting a repressive regime, as, for example, in Indonesia, Taiwan, and several cases in Middle and South America, sometimes simply causing enormous and unnecessary suffering on all sides, as in Vietnam.

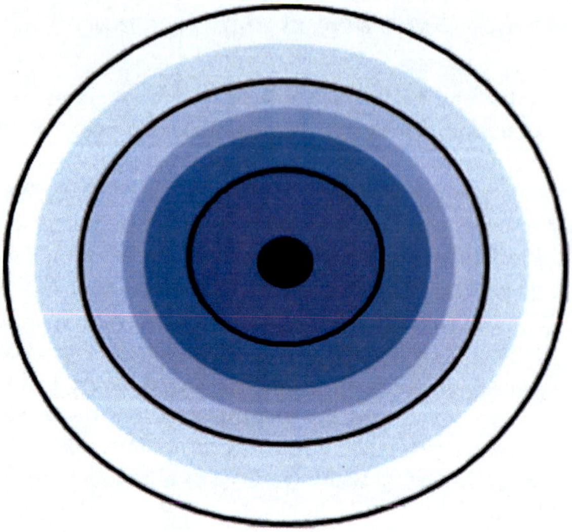

Fig. 13.1 US superpower and hegemony. The US is in the centre, with the three rings of nations subject to its superpower. The shading is intended to show the degree of hegemony over these nations, illustrating that while the superiority in physical power is unlimited, hegemony requires the acceptance of a non-material component of power

After the demise of the USSR, the US was the world's only superpower, and the geopolitical landscape now became one centred on the US surrounded by an inner ring of closely allied and more or less like-minded developed nations—what we identify as the West, a middle ring of developing nations hoping to share some of the West's riches by adopting and adhering to most of the West's rules, and an outer ring of nations opposed, in varying extent, to the US, either on ideological or political grounds, or both, as illustrated in Fig. 13.1.

In this landscape, there is not one single nation that could mount a successful military attack on the US. On the other hand, the US has been less than successful in using its military capability to extend its dominance in the outer ring, and this brings us to the concept of *hegemony*. Hegemony and superpower are closely related and often used indiscriminately, but in addition to coercive power, be it military or economic, hegemony should include a non-material component, an acceptance of all or a part of the hegemon's belief structure. This connection with belief or ideology was raised briefly in Sect. 11.4 in the form of the work of Antonio Gramsci and his concept of cultural hegemony—how the ruling class uses cultural institutions to promote its values and norms so that they become the "common sense" of all, very much as taken up by Herbert Marcuse in *One-Dimensional Man*. There are many views of exactly what constitutes hegemony, and a short and readable review of the concept and its application to the US is the article *The debate on American hegemony*, by Schmidt (2019). He presents several different variants of hegemony, with reference to their major proponents, showing how the different components of hegemony—power and coercion, consent, consensus, rules, and institutions—are given different weight and interpretation in each variant. However, there are two aspects of hegemony that get no mention in this review. Firstly, if we take a step back and consider hegemony in its most general form, simply as influence, then we should differentiate the people being influenced into two groups: the political and business leaders—the elite or Gramsci's "ruling class", and the rest of the population, the common people. For the elite, US hegemony provided order and stability through various international institutions and agreements and through the US version of liberal democracy; that is, the market as the regulatory force, the pre-eminence of private over public, the superiority of representative democracy, and the sovereignty of states, subject, of course, to not stepping out of line. So, for the elite, the benefits of succumbing to US hegemony were mainly material benefits.

For the common people, it was a matter of admiration for the American way of life, for all the visible manifestations of the popular culture—the music, the show biz, the flashy cars, the fashion, from clothing to haircuts, the big houses full of wonderful gadgets, and, yes, Coke and McDonalds, just to mention a few. A view of the Promised Land, which could be realised by following the American lead. I am personally very aware of this "cultural hegemony" as I was only six years old when the German occupation of Norway ended, and we had an influx of Allied service personnel. As my mother was American, we had significant contact with American military and diplomatic personnel, and as I could speak quite a bit of English, I became good friends with the son of a diplomat who joined my primary school class, Stephen Baldwin. I visited the US three times as a teenager—in 1952, 1956, and 1958—and I was thoroughly impressed by Americans and things American. When I graduated as an electrical engineer in 1962, I was in no doubt about where I wanted to go—to the Mecca of electronics—and I spent seven years at Bell Labs. It was during this time that my view of the US started to change; not because of the people I worked with or met otherwise, but because of the Vietnam War. The length to which the government was prepared to go in deceiving its own people was an eye-opener and started me wondering to what extent my own view of the US was based on a carefully crafted fiction—soft power *par excellence*.

The second aspect is the extent to which US hegemony relies on *fear*; we all need to stick together and be vigilant under US leadership to avert the terrible fate that would otherwise await us at the hands of our *enemy*. This justifies a large defence budget and also interfering in the operation of any nation that shows an inclination, real or invented, to have any relations with the enemy. So, it is essential to identify an enemy, and following WW2 it was not too difficult to paint the USSR under Stalin's leadership as the enemy, and NATO was the response. But once the USSR collapsed one might have thought that Russia could become a competitor rather than an enemy, but no, there was just no one else around at this time that could take the mantle of enemy. To ensure that the tone of Russia's interaction with the West could be interpreted as threatening, the border of NATO was pushed steadily eastwards. Now another nation, China, is being painted as a second and more formidable enemy, despite having engaged in no aggressive actions towards the West, leading to the dangerous situation that is the subject of Sect. 13.3.4.

13.1.4 The Rise of China

Until recently, the history of the world over the last two centuries, as taught and studied in the West, was focused on the West, on the rivalries in Europe, on colonialism, the two World Wars, the Cold War, and the emergence of the US as the global hegemon. As for the East, Japan received some attention, as did the nations arising out of colonial status after WW2, but China was somehow not treated as a 'real' nation. The European powers and the US tended to treat China mainly as a market subject to their rules, as demonstrated, for example, by the "Open Door" policy promoted by the US. That this was the very thing the Chinese authorities did not want is noted in a brief comment, "American Notions About China", in the previously referenced memoirs of Acheson (1969, pp. 7, 8). In the same comment he also makes reference to the importance of the US missionary activity in China on influencing public opinion in the US about China as an exotic land of heathens, whose minds needed education and whose souls needed salvation. So, to put the current view of China in perspective, the following is mainly my severely abridged and edited version of the information provided in the second edition of *A History of China*, by Eberhard (1960), as well as some other material, as referenced.

Leading up to the end of the eighteenth century, China had been developing into an increasingly prosperous and well-managed nation-state under nominal Manchu rule, but in reality administered by a Chines bureaucracy. External relations were very limited; external trade was only a few percent of the internal commerce, and the activities of foreign traders were severely restricted. The social structure was composed of the ruling Manchu, with their emperor and military garrisons, an administrative bureaucracy, a class of landholding gentry, a small and weak middle class of traders and manufacturers, and then the greater part of the population (> 80%) as tenant farmers. But the time around 1800 became a turning point in the development of the Chinese society and the start of an accelerating decline. There were several contributing factors, including a rapid rise in population, with a consequent decrease in the per capita arable land, a smouldering antagonism against the ruling Manchu that had to be sustained by the Chinese, and, not least, an incompetent and corrupt administration that was being sidelined by local elites and wealthy traders. But increasingly it was the aggressive interference by foreign powers taking advantage of the government's weakness that brought China to its knees during the course of the nineteenth century and beyond, until its resurrection by the Communist Party in 1949. Led by Great Britain, with its commanding naval force, but joined to the extent allowed by their means by other European nations and then, increasingly, by the United States and Japan, China was forced into the one unwinnable war after the other, each time resulting in harsh terms of surrender. Major milestones on this downward path were the Treaty of Nanking (1842, loss of Hong Kong); Treaty of Tientsin (1860, loss of Kowloon), Japan's annexation of the Ryukyu Islands (1874) and penetration into Korea (1885), which until then had been a Chinese protectorate; the Chefoo Convention (1876); the Treaty of Shimonoseki (1895, loss of Taiwan and of the protectorate over Korea); and the defeat of the Boxer Uprising (1900).

These all involved war indemnities, foreigners' exemption from Chinese courts, the reopening of the opium trade (which the Chinese government had attempted to stop), and, notably, the promotion of missionary activity throughout China. This linking of the Christian church to foreign oppression was bound to leave its mark.

However, these events, which had a devastating effect on the Chinese economy, also had their effect on Chinese society. Firstly, in the century leading up to 1912, there were numerous internal revolts and uprisings, some regional, but a few encompassing large parts of China. Some were inspired by foreign ideas or beliefs, some were encouraged by the disrespect shown by foreign powers for the Chinese government, and some were simply a result of the declining standard of living. Not only did they directly weaken the authority of the central government and tax its already stretched resources, but the campaigns to defeat them gave rise to several powerful generals that pursued their own political objectives, which further reduced the national cohesion. Secondly, through the interaction with the Europeans in the treaty ports, a small middle class of Chinese emerged that adopted European methods and became familiar with European thought and political and philosophical concepts. This led to an increasing conflict between a progressive and revolutionary south and a traditional north, based on its conservative landholding gentry and with its anchor point in the imperial government in Beijing.

In 1912 the Manchu emperor abdicated, and China was proclaimed as a republic, albeit with the strongest of the generals, Yuan Shikai, as president. He was in reality a conservative and a supporter of the gentry, and with ambitions to reinstate the monarchy with himself as emperor. However, he died before he was due to become emperor, and the years that followed, until 1927, were marked by infighting between a succession of generals and a collapse of the political power of the Beijing government. Several provinces declared themselves independent, and in the south, the revolutionary party—the People's Party—based in Guangzhou under the leadership of Sun Yat-sen, was steadily gaining in strength. After the death of Sun Yat-sen in 1925, the new leader, general Chiang Kai-shek, embarked on a campaign against the north which, after much fighting and many compromises, was successful in establishing a "united" China under the dictatorial leadership of Chiang Kai-shek. It was the nature of these compromises—essentially a conservative turn with persecution of socialists and communists, such as the Shanghai massacre in April 1927—that resulted in the separation of Chinese society into two irreconcilable camps—the Chinese Communist Party (CCP) and the Nationalists, the Kuomintang (KMT)—and the inevitability of a civil war *à outrance*.

This was interrupted only by their common stand against the Japanese invaders. In this, China was supported by the US, but the support went manly to the KMT, and once the Japanese were defeated by the US and the civil war resumed, the US sided with the KMT and provided significant material and logistic support. Despite this support, the poor military leadership, and the corruption within the KMT, led to an increasing loss of territory, and by the end of 1949 the whole of mainland China was under communist control. Chiang Kai-shek and the KMT government with their families, and about half a million troops, fled to the Chinese island of Taiwan, taking with them the Chinese gold reserves of around 115 tons and various national treasures, as well as the Air Force and Navy. The KMT were not particularly welcome and instituted a repressive government operating under emergency decree for some 38 years, but they stayed under US protection.

In the seventy years that have gone by since the founding of the People's Republic of China (PRC), the country has undergone a remarkable, but also very turbulent rise, from a downtrodden backwater at the mercy of the West and with an economy in ruins to the second-largest global economy and a major participant in international politics. This despite being hampered by the legacy of its revolutionary birth and the hero-worship of its leader, and when compared to how Russia has managed a similar transition, the Chinese transformation has been outstanding. But it would be unreasonable to expect that the experience of the last two hundred years should now be forgotten; today's China is justifiably wary of the West's intentions and is under no obligation to fit into a global system (euphemistically called "the rules-based order") based on a political and economic ideology created by the West.

China's internal development has been, and is today, one of unprecedented scale and dynamism, and requires a management system tailored to meet the challenges of this development. The management is a difficult balancing act that involves constant vigilance and resolute action whenever required; quite the opposite to the *laissez faire* so admired by neoconservatives. Today, China is a major member of, and contributor to, the international community, and should be welcomed as such. It has not used any military force in attaining this position, and if its commercial practices are at odds with Western norms, this is a matter for negotiation, possibly within an existing framework, for example, the WTO. The present hysteria surrounding the rise of China is unfounded in fact, but the longer it is promoted, the more likely it is to become reality. At present there is considerable tension between China and the West, as will be examined in Sect. 13.3.4; the following section examines the historical background of some of the main issues of contention.

13.1.5 Background to the Main Issues of Contention

13.1.5.1 Economic Power

(All data in this item are from the World Bank database, https://data.worldbank.org/indicator/).

The rise in China's economic performance, as measured e.g., by GDP, has, if anything, outperformed that of the earlier rise of the "Asian Tigers"—Singapore, Taiwan, South Korea, and Hong Kong, areas that benefited from foreign trade advantages that set them apart from other countries, most significantly economic support from the United States. If we start by looking at the GDP measured in current US$, as shown in Fig. 13.2, then the US is still clearly ahead of China, and Russia is not going anywhere. Measuring the GDP in PPP current international dollars, as shown in Fig. 13.3, the picture looks somewhat different, in that China has already overtaken the US, but this picture is less important in the context of power in international

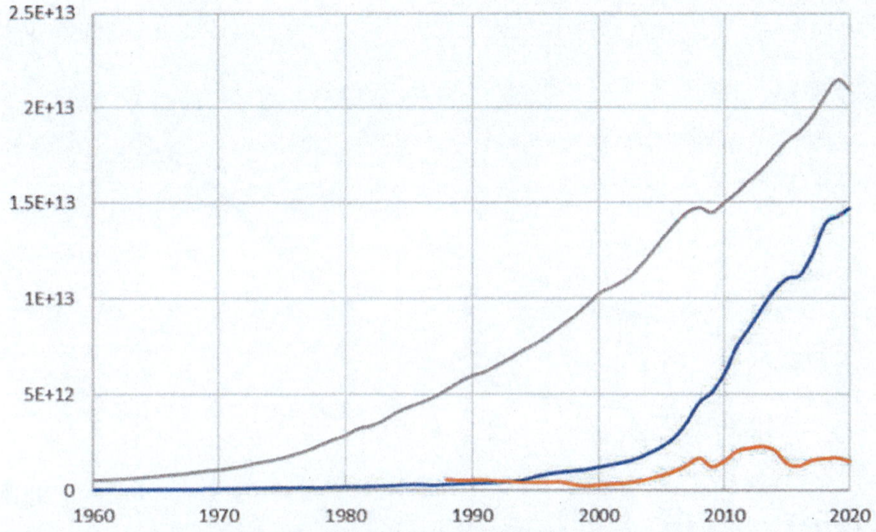

Fig. 13.2 GDP in current US$ for, from top to bottom, the US, China, and Russia

relations. What is most significant is the rapid increase in the Chinese GDP starting around 2004, and the reason for this can be seen if we look at the export of goods and services minus the import of goods and services for China, measured in current US$, Fig. 13.4. What happened in 2004 was that the developed nations, as a logical consequence of neoliberal globalisation, started to outsource more and more of their manufacturing to China, lowering manufacturing costs and increasing profits, but at the same time becoming more dependent on China and losing lower-skilled jobs at home.

This outsourcing of jobs did get some attention in the West, but for a while the concern, as reported in the media, seemed to be mostly with substandard working conditions and outright exploitation in countries other than China. There were some complaints in the mid-1990s about China keeping the value of their currency—the renminbi—artificially low in relation to the dollar, but the issue of China's growing economic power and strength as a trading partner only became explosive with President Donald Trump's initiation of a trade war by placing a 25% tariff on some 34 billion dollars of Chinese goods and accusing China of "stealing" US jobs and manipulating the currency. In the second half of 2020 a moral component was added to these trade sanctions by linking them to the treatment of Uighurs in Xinjiang province. This trade war has spilled over onto other nations, including Australia, and is still ongoing, but it should be seen as just one manifestation of the basic issue—China's rise as a major economic power. If we look at the GDP per capita in PPP dollars, shown in Fig. 13.5, which is the most reasonable use of GDP as a measure of living standards, then we can see that China still has a long way to go to catch up with the US and the West. Why do we believe the Chinese should continue to have a lower living standard than we have?

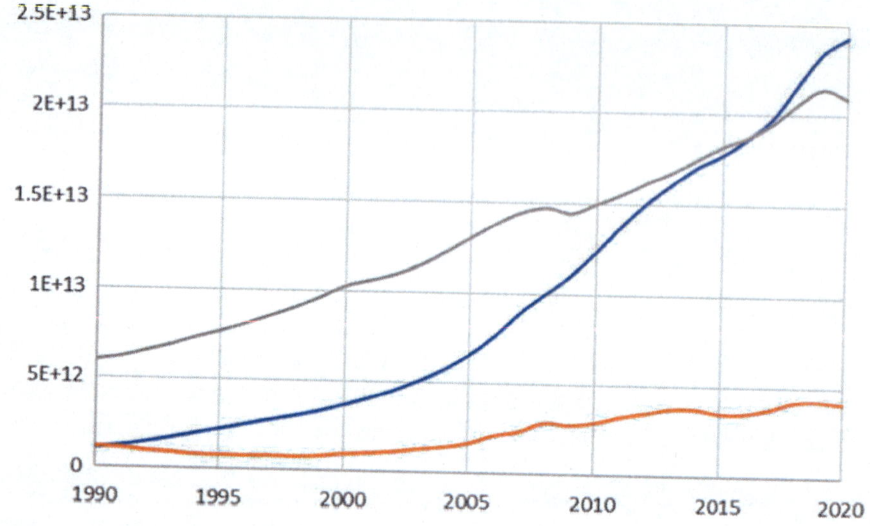

Fig. 13.3 GDP in PPP current international dollars, showing China overtaking the US in 2017

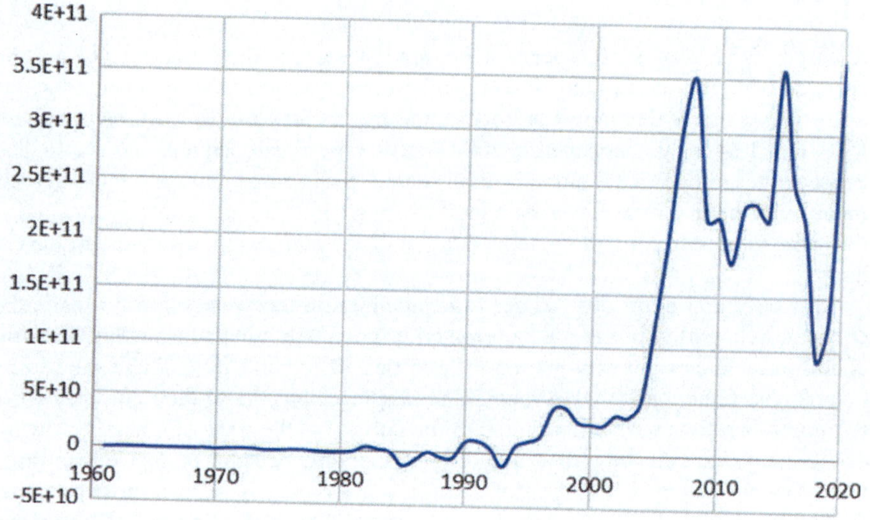

Fig. 13.4 China's export of goods and services, minus its import of goods and services, in current international dollars

13.1.5.2 Military Power

Military power is related to economic power, but not necessarily proportional. Figure 13.6 shows the military expenditure in the US, China, and Russia, and it is obvious that the US has increased its spending significantly since the beginning of

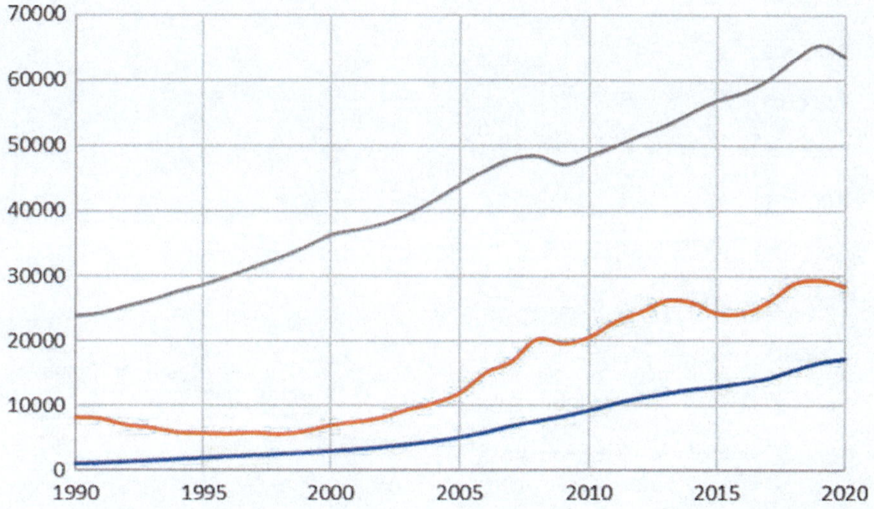

Fig. 13.5 GDP pro capita in PPP international dollars for, from top to bottom, the US, Russia, and China (World Bank)

the millennium, partly in response to a significant increase in China. In comparing expenditure in the US and China, it is important to recognise three main differences. The first is that the US maintains troops and installations in numerous places over much of the world, which is disproportionally costly. The second is that the data for China are estimates by the Stockholm International Peace Research Institute (SIPRI), so has a certain degree of uncertainty, and the third is that the values are in US$ at current prices, million US$, which makes the Chinese expenditure less representative of actual force strength.

If we look at force strength in terms of military personnel, then, in 2018 the numbers were US: 1,379,800, China: 2,695,000, and Russia: 1,454,000 (International Institute for Strategic Studies, quoted in the World Bank database), which clearly demonstrates the much higher cost of maintaining US personnel, and also probably the greater sophistication of the hardware (so far). A major concern in the West is the increasing assertiveness of China in the South China Sea and the extent to which the domination of the US over the Western Pacific can still be maintained. In the number of ships, it would appear that China is ahead, 360 versus 297, as reported in Table 2 of *China Naval Modernization: Implications for U.S. Navy Capabilities—Background and Issues for Congress*, Congressional Research Service, December 2, 2021, but this is a very simplistic comparison, as is discussed in some detail in the same report, and the outcome of any confrontation would be uncertain and very dependent on the details of the severity and geographical extent of the confrontation.

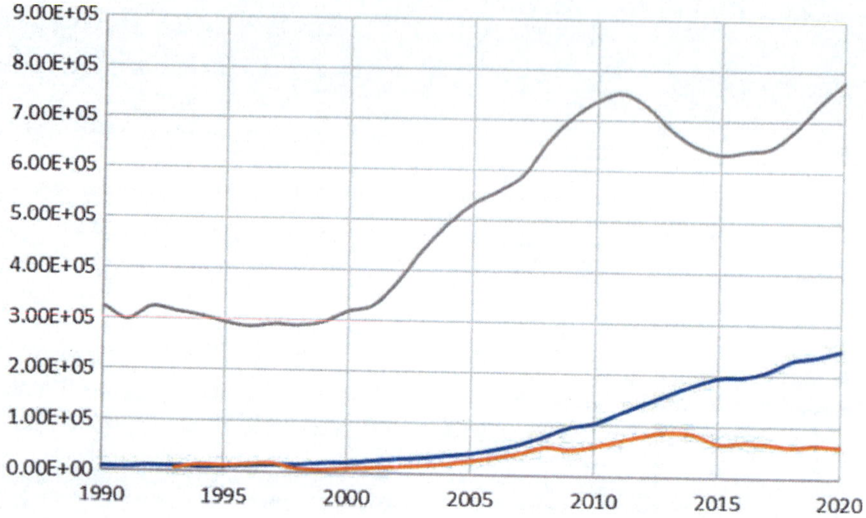

Fig. 13.6 Military expenditure, in million US$ at current prices, for, from top to bottom, the US, China, and Russia (https://sipri.org/databases/milex)

13.1.5.3 Science and Technology

Science and technology are drivers of growth in both economic and military power and are also important in a nation's image as a leader in humanity's quest for ever greater applications of and domination over Nature, including in medical, electronics, nuclear, and space science and technology. Until somewhere around the start of the new millennium the US was considered by all to be the undisputed leader, not least because it was able to make its research and development visible to everyone as products that were sought after and readily available. Many other countries in Europe and Asia were by this time ramping up, both in research and the manufacturing of advanced products, and China was no exception. The jewel in the crown was the electronics firm Huawei; it is the world's biggest producer of telecoms equipment, and it also ranks second in global smartphone sales, behind Samsung and ahead of Apple. Within the new generation of mobile technology—5G—it was clearly the world leader. This was a serious blow to Western supremacy and so, not wanting to blatantly go against the neoliberal ideal of the free market, Australia, under some pressure from the US, took the lead in banning Huawei from any 5G network rollout on the basis of it being a risk to national security, and this was then followed by many of the Western nations. In addition, President Trump put a ban on the sale of certain critical semiconductor components to China, and we can expect this approach to slowing China's technology development to become a major feature of the US–China relationship.

13.1.5.4 Taiwan

The status of Taiwan is a major and potentially dangerous contentious issue between China and the West. From 1683 onward, Taiwan became increasingly integrated into the Qing empire, but following the war with Japan in 1894/95, and under pressure of the overwhelming military strength of Japan and the lack of any international support, the island was reluctantly ceded to the Empire of Japan as part of the Treaty of Shimonoseki. So, there is no doubt that until this point Taiwan was considered a part of China, otherwise it would not have been able to cede it. As part of the conditions of surrender by Japan on 2 September 1945, Taiwan was returned to China, at that point considered to be represented by the government of Chiang Kai-Shek. This is supported by a chain of documents starting with the Japanese Instrument of Surrender, signed on 2 September 1945, referring to the Potsdam Declaration of 26 July 1945, which referred to the Cairo Declaration of 27 November 1943, which reads as follows:

(https://en.wikipedia.org/wiki/1943_Cairo_Declaration#Text, accessed 20.06.2022).

> The several military missions have agreed upon future military operations against Japan. The Three Great Allies expressed their resolve to bring unrelenting pressure against their brutal enemies by sea, land, and air. This pressure is already rising.
>
> The Three Great Allies are fighting this war to restrain and punish the aggression of Japan. They covet no gain for themselves and have no thought of territorial expansion. It is their purpose that Japan shall be stripped of all the islands in the Pacific which she has seized or occupied since the beginning of the first World War in 1914, and that all the territories Japan has stolen from the Chinese, such as Manchuria, Formosa, and The Pescadores, shall be restored to the Republic of China. Japan will also be expelled from all other territories which she has taken by violence and greed. The aforesaid three great powers, mindful of the enslavement of the people of Korea, are determined that in due course Korea shall become free and independent.
>
> With these objects in view the three Allies, in harmony with those of the United Nations at war with Japan, will continue to persevere in the serious and prolonged operations necessary to procure the unconditional surrender of Japan.

Again, the identification of the three territories "stolen" by Japan as being part of China is quite clear, and it is a shame that claims by the Taiwanese administration to the contrary ("Taiwan was never Chinese") are reported in the Australian press without comment. (It is worthwhile to note, as an example of propaganda, the following information about Manchuria being a part of China):

> In view of the pre-war Japanese propaganda which treated the Manchurian region, renamed "Manchukuo" as a territory distinct from, and never having formed any part of China, it is worth recording that the Ming province of Shantung included not only the peninsula of that name, but also the whole of the Manchurian region on the north side of the Gulf of Chihli, from the Great Wall up to Mukden, and as far east as the Korean border, then as now defined by the Yalu River. This territory was already fully occupied by settlers of Chinese stock. (Fitzgerald, 1954, p. 459)

The Paracel Islands also became contested following WW2; this time by the French, who had annexed the islands in 1933 as part of French Indochina, and subsequently by South Vietnam, but the issue was settled by the Battle of the Paracel Islands against South Vietnam in 1974.

The issue of the status of Taiwan is a result of US intervention in the Chinese civil war, first trying to avert a defeat of the Nationalist forces on the mainland, and when that was not successful, providing support for Chiang Kai-Shek to establish his regime on Taiwan. The Chinese did not, at that point in time, have the capability of pursuing the Nationalists, as was demonstrated by the ill-fated attempt to take the island of Kinmen in 1949, and later, US protection made it impossible, as described in the following passage:

> It is probable that if American protection for the regime of Chiang Kai-shek had been withdrawn, or merely allowed to lapse, an invasion assisted by aircraft supplied by Russia would have followed. The American declaration neutralizing Formosa was made by President Truman, without the concurrence of other member states of the United Nations, at the outbreak of the Korean War, but several months before Chinese intervention in that conflict. It was stated that the neutralization of Formosa was a measure of security for the American forces engaged in Korea and would endure only as long as the war in Korea continued. When the war was brought to an end by a permanent cease-fire, it was not easy to see what justification could then be produced for continued American intervention in what remained in law a Chinese civil war. It can be taken as certain that no government ruling in China, whether the present Communist regime or any other, will abandon the claim to Formosa (Fitzgerald, 1964, p. 235).

13.1.5.5 Hong Kong

Hong Kong Island was ceded to the British Empire following the defeat of the Qing Empire on the First Opium War in 1841/42. The colony expanded in 1860, after the second Opium War, to include the Kowloon Peninsula, and again in 1898 when Britain obtained a 99-year lease on the New Territories. All was returned to China in 1997, to be governed under *The Basic Law of the Hong Kong Special Administrative Region ofThe People's Republic of China* (https://www.basiclaw.gov.hk/en/basiclaw/index.html, accessed 20.06.2022). Given Hong Kong's close ties to the West and the great differential in prosperity between the mainland and Hong Kong (GDP per capita in 2020: Hong Kong US$ 46,300, mainland China US$ 10,400, https://data.worldbank.org/indicator/NY.GDP.PCAP.CD?locations=CN-HK, accessed 20.06.2022), this was always going to be a very difficult situation, and one might wonder if this arrangement was not a Western Trojan horse or poison pill. That is certainly how it has worked out, with protesters in Hong Kong forming a pro-independence party and waving US flags, the Hong Kong authorities responding with mass arrests and repressive laws, and the West having a field day in attacking China, both in words and in deeds (e.g., US sanctions under the Hong Kong Human Rights and Democracy Act accessed 20.06.2022).

13.1.5.6 The Uighurs and Human Rights

China has been conducting what appears to be a massive, forced re-education and indoctrination program for Uighurs and other Muslim ethnic minorities in Xinjiang province, interning them in what are essentially concentration camps at the rate of well over a million per year over the years 2014 to 2019. The stated purpose is to counter extremism and terrorism, but this would apply to only a very small number of activists; the real purpose is likely to be to suppress any resistance to the absolute power of the CCP and any thoughts of separatism and the breakaway of an independent Islamic state. There have also been allegations of human rights abuses, including mistreatment, rape, and torture, but these are more difficult to verify.

Human rights are a thorny issue for China in its international relations and image and provides an obvious and easy point of attack for the West. Human rights abuses have occurred, and still do, in every nation, and the intensity depends on many factors, of which one is the stage of development of the society. The British Empire in the process of building its power in India, in the Caribbean, and in South Africa, the US through genocide of the indigenous population and slavery, Australia in its treatment of the Aboriginals and currently in its treatment of asylum seekers, and so on; plenty of human rights abuse to point at. China is in the process of creating a new society out of the shambles in which it found itself in 1949; a delicate balance between transformation and stability, and not made any easier by the West's constant sniping in the hope of destabilising the whole process and so remove the niggling doubt about the supremacy of the Western ideology.

13.2 Representations of the World Society

13.2.1 Structure and Infrastructure

In the view presented in this book, a society consists of a set of interacting individuals that share a common belief structure and a common infrastructure that reflects this belief structure and realises its benefits. Without a common infrastructure there can be no benefits arising from the common belief, and without benefits there is no incentive to maintain society's belief structure (excepting, perhaps, in the case of benefits that accrue in a life after this). In the view of the evolution of society put forward in Chap. 2, as the current stage of a general process and with the sequence of species as suggested in Table 2.3, we would expect the next stage to be the emergence of a world society with nations as its members. The elements of the nation-societies are e.g., states (as in the US and Australia); the elements of the state-societies are communities (counties, local councils), the elements of the community-societies are families, and the individual human beings are the ultimate foundation of the whole edifice. At each level, each society has a common infrastructure that reflects a part of that society's belief structure, and the benefits of that infrastructure accrue to the

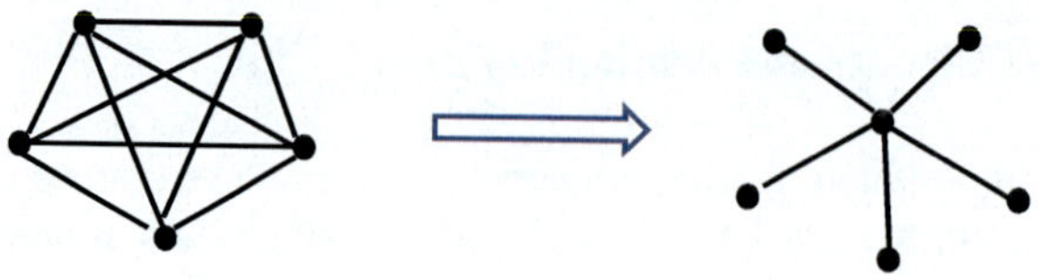

Fig. 13.7 The change in structure when the interactions between all members are replaced by interactions between the members and the common infrastructure

members of the society. At each level, starting with the family, the society and its infrastructure came into being because of the benefits it brought to its members were greater than their evaluation of the partial loss of autonomy. So, if we now want to create a world society, we must look for common beliefs that, when implemented in a common infrastructure, will result in a net benefit to every nation. Not necessarily the same benefit to each nation, but at least some benefit to every nation in that all get their common belief implemented.

The central feature of what we designate as "common infrastructure" is that it replaces one or more functions that were previously performed by each member, and that required a substantial amount of interaction between members. The benefit arises from the elimination of duplication (e.g., of research and data management) and from the rationalisation of the interactions; effectively replacing a network structure with a star structure, as illustrated in Fig. 13.7.

In a society, the change in structure shown in Fig. 13.7 applies only to a subset of the interactions; there will, of course, still be interactions directly between members. But for the interactions where the change applies, and with reference to the measure of structural complexity in Sect. 10.1.4, the complexity is reduced from N-1 to 2. The change means that the members relinquish a component of their sovereignty; individual members cannot interfere with the operation of the common infrastructure. In a nation, the administration of justice is (or should be) an example of this.

In a recent publication, *On Building Peace*, by von der Schulenburg (2017), the author emphasizes the importance of sovereign nations (he uses the term *nation-states*). He states that collapsing or failing nations are increasingly becoming the dominant risk to global peace and stability and illustrates this by numerous examples of what happens when nations fail. His vision of a world society is one where equal and sovereign nations cooperate and compete under mutually agreed rules and supported by mutually agreed instrumentalities. The main such instrumentality would be a suitably modified and strengthened United Nations; in particular, taking account of the fact that the majority of hostilities since WW2 have been intranational rather than international (even though, as he also notes, many have been—and are—proxy wars advancing international interests). The author emphasizes the difference between peacekeeping and peacebuilding; the former a retroactive activity following the termination of hostilities, the latter a proactive prevention of hostilities in the first place, and it is the UN's ability to practice peacebuilding within nations that needs to be strengthened.

However, such modifications will not turn the UN into a common infrastructure for a world society, and will therefore, under our perspective, not provide a basis for a world society. The high ideals and professions of a common belief in the Charter and the Universal Declaration of Human Rights are mainly good words expressing aspirational intentions, and the structure of the organisation is designed to ensure that the interests of the main players—the permanent members of the Security Council—are not interfered with. The many funds, programs, and specialised agencies of the UN, which are doing a good job within their means, are essentially international charities, much as the charities within nations that rely on donations from members of society (persons and businesses).

13.2.2 Representation as a Society of Individuals

Today, the world's population is contained in around two hundred mostly stable nations, as listed in the Appendix, and it is immediately obvious that they differ greatly in all four of the indicators listed. A further demonstration of this is the range of variation of some of the characteristics of nations, as shown in Table 13.1.

They also vary greatly in the strength of their internal cohesion and in the extent of their common infrastructure, something that is, in part, reflected in their ranking in the Fragile States Index (http://fragilestatesindex.org/, accessed 20.06.2022); however, all of this variation is no greater than that found within nations, and so does not constitute an a priori reason why nations could not form a world society.

So, if we, almost as a digression, want to consider the world as a society of some seven billion people, how do we apply our framework— the two views and our model of the collective intelligence—to this society? If we adopt the straightforward approach and retain the concept of the individual as a nominal representative of a person in the world, then the list of nations shows that this is a much greater approximation or idealisation that in the case of the society being a single nation. Not only do we have the inequalities within each nation, but also the inequalities between nations; not just as indicated above, but in culture, including history and language. With the rapid growth in international information exchange the concept

Table 13.1 Variation of some national characteristics (2019 data)

Parameter	Smallest value	Greatest value
Population [,000]	Tokelau 1.367	China 1,440,000
Size [km^2]	Monaco 1	Russia 16,376,870
Agricultural land [ha/capita]	Various < 0.01	Australia 1.90
Wealth [$US/adult]	Sudan 534	Switzerland 564,653
Median household income [$US]	Togo 571	Luxembourg 52,493

of the collective intelligence still has its significance on the global level, but the operating conditions are more complex and further from the ideal of complete equality. However, if we recall the concept of commonality and that of the identity partition, α, in Sect. 3.3.3, then we can extend these concepts to the global society. In Eq. 3.4 the number of individuals, n, becomes the number of individuals in the world, and the global identity partition, $\aleph$, is defined as in Fig. 3.5, but with i_{max} now the global number of assertions. And, as before, the assertions in the $\aleph$-part of the global set of assertions constitute the *global belief system*. The assertions in this set are those held by every individual, and so an equivalent definition of this set is as the intersection of the α-part of Θ for all individuals in the global society.

In the case of the society being a nation there was a means of converting society's belief system into corresponding actions—the adaptive actions—so that, eventually and ideally—society in its physical realisation would reflect the belief system. The means of performing that conversion was the *common infrastructure*; effectively, the various institutions making up the government. At present there is no such common infrastructure on the level of the world society of individuals, and therefore no institutional means of converting the global belief system into physical reality. However, there are organisations that are concerned with developing and strengthening the global belief system, rather than the many organisations engaged in promoting a particular belief on the international level, such as religious organisations. One such organisation is Humanists International, an umbrella organisation founded in Amsterdam in 1952 and made up of more than 160 secular humanist, atheist, rationalist, skeptic, freethought, and Ethical Culture organisations from over 80 countries (https://humanists.international/, accessed 20.06.2022). In addition to a presence at the United Nations Human Rights Council in Geneva, the United Nations General Assembly in New York, and the Council of Europe in Strasbourg, among other international institutions, the organisation provides, through the communication with its member organisations, a small component of what might be considered the global circulating information.

There are certainly many organisations that consider matters of global importance and interest, such as a number of the UN System organisations (see Sect. 13.2.4), but also more commercially-oriented organisations where their operations have a global impact—mining and energy companies are typical examples. And environmental concerns are finding their way into school curricula. But, while this is all good, and does slowly build a global consensus regarding such particular issues, it is far from constituting a global public discourse about what a global society should look like. Of course, one might also be of the opinion that the world's population is not ready for such a discourse.

13.2.3 *Representation as a Society of Nations—The Information View*

13.2.3.1 The Relationship Between Nations and the World

A different approach is to consider the nations to be the elements of the global society, which requires us to recast most of our concepts to reflect the characteristics of these new elements. What we do want to retain are the two views and a model of the information processing within the nations, and, as before, we will initially examine the level of least detail (or greatest abstraction) and consider the nations to be identical as information-processing elements. We now recast our model to apply to this society of nations, concept by concept, starting with the purpose of the interaction between nations.

> The purpose of a nation's interaction with other nations is to increase its relative position of power, with power as defined in Sect. 11.5.3.

This becomes the decision criterion when it comes to nations taking adaptive action. But how does this fit in with our assertion, in Sect. 2.3.1, that the basic criterion in the evolution of society is survival, and that the ideal operation of the collective intelligence will provide the greatest probability of survival, as stated in Sect. 2.3.1. The explanation is that, while the decision criterion for international actions by a nation is to improve its relative position of power, the information on which the decision is based—the *national intelligence* applied in the formulation of options—is not only information relating to this and previous actions, but also the nation's belief system, defined previously by Def. 3.7, which is developed by the nation's collective intelligence and therefore reflects the survival criterion. This influence of the nation's belief system on the international order is illustrated in Fig. 13.8.

The box labelled "International order" is a set of agreements, accords, treaties, etc., that regulate, to some extent, the interactions and relationships between nations. This is not the same as the world society's belief system; in this representation by means of nations there is no global belief system. However, as was the case in the nation as a society, the international order is reflected in an international infrastructure consisting of intergovernmental organisations, of which the United Nations (UN) is the major one and the only one encompassing all nations. These organisations are the subject of Sect. 13.2.4, but as they are not, in themselves, significant powers, we will not consider them for the time being.

In the information view of a society consisting of individuals we found it useful to group information items into three classes. Is there a similar grouping of items in the information exchanged by nations? Before we can attempt to answer that, we must first describe what an item of information consists of in this case. If an individual in nation A interacts with an individual in nation B in relation to private or business matters, this is not what we understand by an exchange of information between the two nations; in each nation there will be individuals who are authorised

Fig. 13.8 The sequence of processes leading from the individual to the international order

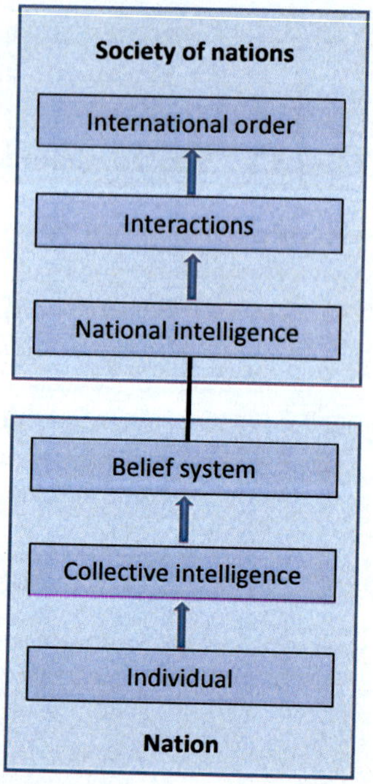

to exchange information *on behalf of the nation*. The information will be in the form of a *message* and relate to a specific *subject matter*. The content of such messages may subsequently be published and become common knowledge, and intergovernmental organisations also contribute to this common knowledge, which we might look upon as a global version of circulating information. But it is significantly different to the circulating information in a society of individuals in that the information of greatest importance—the messages—are not included. What was previously Class 3 information is now represented by messages, identity elements are replaced by messages, and they form the primary focus of our treatment of the interaction between nations. Messages are more complex than information items; they are collections of information items combined and structured to convey messages intended to have a certain effect, as discussed in the following section.

In Sect. 5.3 we introduced and discussed the individual's interaction with society by disregarding the specific physical realisation of the interaction, such as face-to-face, via mass media, via social media, etc., and representing it instead with the abstract concept of the circulating information, as illustrated in Fig. 5.7. In the society of nations, it is, in some way, exactly the opposite: the interaction is strictly on a nation-to-nation basis, and the circulating information and the access to it play only a supporting role, similar to the "common information" in Fig. 5.7. This is illustrated in Fig. 13.9. And instead of the interaction resulting in an evolution of society's belief system, it results in an evolution of society's power structure.

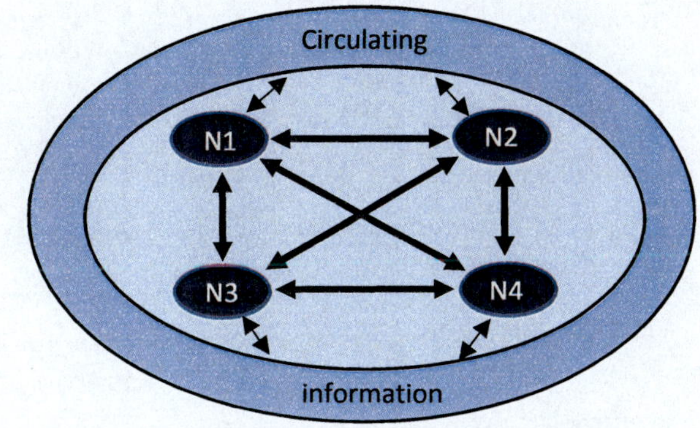

Fig. 13.9 Information exchange in the society of nations (four shown for illustration only). The primary exchange is the nation-to-nation exchange of messages; the information exchange with the circulating information (the information environment) is secondary

In the high-level picture of the society of nations presented in Fig. 13.9 the nation is a "black box", disregarding any internal structure of the nations. Accordingly, the interaction is limited to that which can be attributed to a nation as a whole, effectively, to government-to-government interaction; all other international interactions, such as between organisations, institutions, and corporations, and even between individuals, are disregarded. When we, in the following sections, start to develop this picture in more detail, the approach will be to first develop a model of how nations process the nation-to-nation interactions and then, subsequently, consider the effects of internal structural elements on the operation of the model, both through their direct interaction with government and through their international interactions.

13.2.3.2 The Nation

When we now turn to developing a model of the nation as a member of the world society, we are confronted by a couple of questions or issues, and the first is: Why do we want to develop a model; what is the purpose? With any model, the purpose is simplification; to catch essential features without becoming bogged down in detail, and when we consider this, we can see that the situation is very different to the national case and the model of the individual we developed earlier. Individuals are quite similar in their physical features, such as sensory and information-processing organs. Nations vary greatly, as we saw above, so that building a model based on averages of their features and capabilities makes little sense. But the *process* involved in exchanging messages is very similar irrespective of the features of the nations involved, so our model should be a model of the process, representing the essential steps involved.

Messages are exchanged between nations utilising a variety of *channels* differing in formality and confidentiality, ranging from the normal diplomatic channel involving ministers of foreign affairs and ambassadors through official spokespersons to various classified channels. For outgoing messages, this front-end process needs to consider the audience of the message and what impact it wants to achieve

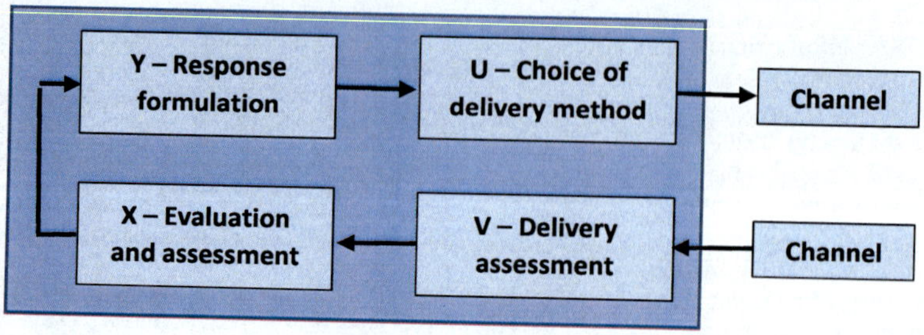

Fig. 13.10 The four basic processes involved in a nation's processing of messages

in different parts of the audience. Also, choosing the most effective of the means available for delivering the message. For incoming messages, the process evaluates all aspects of the delivery method, from body language to choice of location and background images.

Once received, a message must be analysed and evaluated, drawing on existing information and resources within the foreign affairs department. The outcome of this process is an agreed interpretation of the message, which is normally used to decide on a response, but may also simply be taken on notice. Generating a response will generally involve several government departments and may require approvals, for example, from Parliament.

These four main processes involved in the processing of messages within a nation are shown in Fig. 13.10.

Process X is located in the government's Foreign Affairs or State department, headed by a Minister or a Secretary. In a country such as Australia a Minister has a dual role: On the one hand elected by the people and member of Parliament—the legislative arm of government. On the other hand, a member of Cabinet—the executive arm of government, thus ensuring that the executive is directly responsible to the people. In the US, the Secretary is not elected and not a member of Congress, so that the executive arm of government is not directly responsible to the people. In either case, the Foreign Affairs department or State department is supposed to be apolitical and its officers, including Ambassadors, career personnel judged solely by their professional performance and regard for the good of the nation. However, it is obvious that either of these arrangements are susceptible to corruption, in particular if one party stays in power over a long period, during which there is significant natural turnover of personnel.

Process Y is where the executive arm of government, headed by Cabinet, considers the information and recommendations provided by Process X and decides how to respond to the incoming message.

Processes U and V would generally be handled by diplomatic personnel, either stationed in the nation with which the message exchange takes place or, in Process U, sent as an emissary. As there are many possible channels, there are many different persons that may be involved in the two processes and, depending on the skill, knowledge, and political and personal beliefs of these persons, there is always the possibility that the message becomes distorted in these processes.

In contradistinction to the development of the model of the individual, the next level of the model of the nation does not develop the main functions into subfunctions, but details some of the information involved in their operation. The reason for this is that the subfunctions involved in processing and generating this information are highly specialised (and some of them classified) and well beyond my general knowledge. Under this perspective, the central process is Process X, and we can identify four groups of knowledge involved in the evaluation of the incoming message:

- *Historical information.* This is a database of relevant general information about the previous state of the world of nations, such as economic, military, and political data, agreements, alliances, treaties, and the like.
- *Current information.* Detailed information about the current state of nations, about important persons and recent shifts in nation-internal power. Also, information about the state of the relationships between nations—alliances, trade agreements, meetings, missions, etc.
- *Strategies.* Based on all available relevant information, including successes and failures, strategies for responding to messages should be continually questioned as to their current relevance and be updated as required. Without strategies, creating successful responses takes too long. We might first think of strategies in the form of military strategies and associated with names like Clausewitz and Sun Tzu, but they are mostly subordinate to political strategies, not least in the sense of "war is simply the continuation of political intercourse with the addition of other means" (von Clausewitz, 1832). An example of such a geopolitical strategy is the US strategy of turning a competitor into an enemy, a strategy that involves military power, but not necessarily war. Another example is the long-term US strategy of thwarting the rise of China, initiated over a hundred years ago.
- *Belief system and ideology.* This is perhaps the most difficult group of information to handle, as it requires a great deal of introspection and honesty, in order for it not to simply turn into propaganda. This is the core of soft power in international relations; persuasion from a position of asserted moral values embedded in an ideology, such as the supremacy of Western culture.

13.2.3.3 The Interaction—The Structure and Intent of Messages

The intent of a message is to improve the position of the issuing nation ('the issuer') in the society of nations by making the recipient nation ('the recipient') take a particular action, such as changing its behaviour, joining an alliance, ceding a piece of land, refraining from developing nuclear weapons, etc., and the types of messages are accordingly manyfold. A question is then how effective a message is, to what extent the intent is realised, something that can only be assessed after the message-response process has played out. In performing an analysis, we might expect that the effectiveness will vary by the type of message, so that it would be useful to have a *taxonomy* of messages, and a possible structure of such a taxonomy has a top-level subdivision into three groups of types:

Purpose. The purpose is for the recipient to (a) stop doing something, such as aggression, developing ABC WMDs, manipulating exchange rates, etc.; (b) start doing something, such as accepting refugees, participating in an initiative by the issuer, buying goods from the issuer, etc.

Format. The format can be (a) in the form of information, transmitted by any means, or (b) in the form of an action, such as an air strike, an assassination, a tariff, a personal sanction, etc., in the sense of "delivering a message". We will consider that all messages are of the type (a), as actions would almost always be preceded by such messages. The format is then a question of the formality and means of delivery of the message.

Incentive. The incentive can be (a) positive, in the sense of a benefit to the recipient of complying, directly in monetary terms or in international standing and influence; (b) negative, as a punishment for non-compliance, such as a loss of support, and sanctions of various sorts. To the extent that the incentive relates to the future, it involves a degree of uncertainty, and in the case of a negative incentive it may take the form of an unspecified threat.

13.2.4 The Action View and the Physical Domain

13.2.4.1 Adaptive Action

As was the case within a nation, where a change in society's belief system had to be reflected in adaptive action to have any effect, messages exchanged between nations will only have an effect if they lead to a change in the international order or, what is equally important, maintain the existing order in the face of opposition. The difference is that within a nation there is an established process for effecting this adaptive action—the common infrastructure consisting of the political system and the government functions, whereas in the society of nations there is no such common infrastructure. It is up to each nation to take whatever action it considers most advantageous, if any, as a consequence of an exchange of messages and the receipt of new information.

The action can be in the form of changed trade relations, either positive (increased trade) or negative (tariffs, sanctions), movement of people (immigration, refugees, border controls), aid, and, of course, direct military action. But the action does not have to be directly between the issuer and recipient nations; it can often be via an intergovernmental organisation, such as the UN or NATO, and the ability to influence such organisations and cause them to act in the interests of a nation is an important aspect of international power, as was mentioned in Sect. 11.5.1.

We can distinguish two groups of intergovernmental organisations:

- those with broad participation and support of most governments; essentially the organisations in the United Nations system;
- those supported by groups of nations with the purpose of strengthening their position vis-à-vis other nations—economically or militarily; these are the various alliances and alignments;

and two groups of international nongovernmental organisations:

- formal organisations defined by a particular purpose, often humanitarian, as the Red Cross, or professional, such as the International Commission of Jurists and Engineers Without Borders, or belief, such as the various churches;
- informal groupings and networks based on belief, ideology, or a common interest, such as the Transnational Capitalist (TNC) group (Robinson, 2014) and the network centred on the Koch brothers in the US (MacLean, 2017).

The following sections describe some of the more significant of these organisations.

13.2.4.2 The United Nations System

The Main Bodies of the UN. The main bodies of the United Nations are the General Assembly, the Security Council, the Economic and Social Council, the Trusteeship Council, the International Court of Justice, and the UN Secretariat. All were established under the UN Charter when the Organization was founded in 1945. The United Nations is part of the UN system, which, in addition to the UN itself, comprises many funds, programmes and specialized agencies, each of which have their own area of work, leadership and budget. The programmes and funds are financed through voluntary rather than assessed contributions. The Specialized Agencies are independent international organizations funded by both voluntary and assessed contributions. The UN coordinates its work with these separate UN system entities, which cooperate with the Organization to help it achieve its goals.

For the UN itself, the budget for 2021, in million US$, is shown in Table 13.2. (For all UN expenditure in 2019, as shown in Table 1.3, see *Budgetary and financial situation of the organizations of the United Nations system*, in https://unsceb.org/reports/, accessed 20.06.2022).

Programmes and funds. There are four programmes and two funds:

The *United Nations Development Programme* (UNDP) works in nearly 170 countries and territories, helping to eradicate poverty, reduce inequalities and build resilience so countries can sustain progress. As the UN's development agency, UNDP plays a critical role in helping countries achieve the Sustainable Development Goals.

The *United Nations Environment Programme* (UNEP) established in 1972, is the voice for the environment within the United Nations system. UNEP acts as a catalyst, advocate, educator, and facilitator to promote the wise use and sustainable development of the global environment.

Table 13.2 The UN programme budget for 2021, in US$ million (UN doc. A/75/6)

Part I Overall policymaking, direction and coordination	400
Part II Political affairs	837
Part III International justice and law	84
Part IV International cooperation for development	252
Part V Regional cooperation for development	298
Part VI Human rights and humanitarian affairs	189
Part VII Public information	97
Part VIII Common support services	299
Part IX Internal oversight	21
Part X Jointly financed administrative activities and special expenses	82
Part XI Capital expenditures	19
Part XII Safety and Security	124
Part XIII Development Account	14
Part XIV Staff assessment	279
Total	2995

The *United Nations Population Fund* (UNFPA) is the lead UN agency for delivering a world where every pregnancy is wanted, every birth is safe, and every young person's potential is fulfilled.

The mission of the *United Nations Human Settlements Programme* (UN-HABITAT) is to promote socially and environmentally sustainable human settlements development and the achievement of adequate shelter for all.

The *United Nations Children's Fund* (UNICEF) works in 190 countries and territories to save children's lives, to defend their rights, and to help them fulfil their potential, from early childhood through adolescence.

The *World Food Programme* (WFP) aims to eradicate hunger and malnutrition. The world's largest humanitarian agency, WFP helps almost 100 million people in approximately 88 countries with assistance every year through food or cash distributions and more. The World Food Programme was awarded the Nobel Peace Prize in 2020.

Specialised agencies. The UN specialized agencies are autonomous international organizations working with the United Nations. All were brought into relationship with the UN through negotiated agreements. Some existed before the First World War. Some were associated with the League of Nations. Others were created almost simultaneously with the UN. Others were created by the UN to meet emerging needs.

The *Food and Agriculture Organization* leads international efforts to fight hunger. It is both a forum for negotiating agreements between developing and developed countries and a source of technical knowledge and information to aid development.

The *International Civil Aviation Organization* develops standards for global air transport and assists its 192 Member States in sharing the world's skies to their socio-economic benefit.

The *International Fund for Agricultural Development*, since it was created in 1977, has focused exclusively on rural poverty reduction, working with poor rural populations in developing countries to eliminate poverty, hunger, and malnutrition; raise their productivity and incomes; and improve the quality of their lives.

The *International Labour Organization* promotes international labour rights by formulating international standards on the freedom to associate, collective bargaining, the abolition of forced labour, and equality of opportunity and treatment.

The *International Monetary Fund* fosters economic growth and employment by providing temporary financial assistance to countries to help ease balance of payments adjustment and technical assistance. The IMF currently has $28 billion in outstanding loans to 74 nations.

The *International Maritime Organization* has created a comprehensive shipping regulatory framework, addressing safety and environmental concerns, legal matters, technical cooperation, security, and efficiency.

The *International Telecommunication Union* is the United Nations specialized agency for information and communication technologies. It is committed to connecting all the world's people—wherever they live and whatever their means. Through its work, it protect and support everyone's fundamental right to communicate.

The *United Nations Educational, Scientific and Cultural Organization* focuses on everything from teacher training to helping improve education worldwide to protecting important historical and cultural sites around the world.

The *United Nations Industrial Development Organization* is the specialized agency of the United Nations that promotes industrial development for poverty reduction, inclusive globalisation, and environmental sustainability.

The *World Tourism Organization* is the United Nations agency responsible for the promotion of responsible, sustainable and universally accessible tourism.

The *Universal Postal Union* is the primary forum for cooperation between postal sector players. It helps to ensure a truly universal network of up-to-date products and services.

The *World Health Organization* is the directing and coordinating authority on international health within the United Nations system. The objective of WHO is the attainment by all peoples of the highest possible level of health. Health, as defined in the WHO Constitution, is a state of complete physical, mental, and social well-being and not merely the absence of disease or infirmity.

The *World Intellectual Property Organization* protects intellectual property throughout the world through 23 international treaties.

The *World Meteorological Organization* facilitates the free international exchange of meteorological data and information and the furtherance of its use in aviation, shipping, security, and agriculture, among other things.

The *World Bank* focuses on poverty reduction and the improvement of living standards worldwide by providing low-interest loans, interest-free credit, and grants to developing countries for education, health, infrastructure, and communications, among other things. The World Bank works in over 100 countries, and the World Bank Group comprises the following organisations:

Table 13.3 Budget for 12 peacekeeping missions from July 2021 to June 2022

Mission/Operation	Total appropriation
MINURSO (Mission for the Referendum in Western Sahara)	$60.91 million
MINUSCA (Multidimensional Integrated Stabilization Mission in the Central African Republic)	$1.12 billion
MONUSCO (Stabilization Mission in the Democratic Republic of the Congo)	$1.12 billion
MINUSMA (Multidimensional Integrated Stabilization Mission in Mali)	$1.26 billion
UNAMID (African Union-United Nations Hybrid Operation in Darfur)	$45.72 million
UNDOF (Disengagement Observer Force)	$65.51 million
UNFICYP (Peacekeeping Force in Cyprus)	$57.67 million
UNIFIL (Interim Force in Lebanon)	$510.25 million
UNISFA (Interim Security Force for Abyei)	$280.58 million
UNMIK (Interim Administration Mission in Kosovo)	$44.19 million
UNMISS (Mission in South Sudan)	$1.20 billion
UNSOS (Support Office in Somalia)	$560.07 million
UNLB (United Nations Logistics Base at Brindisi)	$65.69 million
RSCE (Regional Service Centre in Entebbe)	$40.27 million
Peacekeeping support account	$356.41 million
Total	$6.37 billion

- International Bank for Reconstruction and Development (IBRD)
- International Centre for Settlement of Investment Disputes (ICSID)*
- International Development Association (IDA)
- International Finance Corporation (IFC)
- Multilateral Investment Guarantee Agency (MIGA)*.

*International Centre for Settlement of Investment Disputes (ICSID) and Multilateral Investment Guarantee Agency (MIGA) are not specialized agencies in accordance with Articles 57 and 63 of the Charter but are part of the World Bank Group.

Peacekeeping. The largest single of the UN's activities is its peacekeeping activity. This activity operates under the authority of the Security Council and is managed by the Department of Peace Operations, led by the Under-Secretary-General for Peacekeeping Operations. The current peacekeeping missions are listed in Table 13.3, together with their funding.

13.2.4.3 International Organisations

There are a large number of non-governmental organisations that operate internationally and that in this informal way contribute to both the cohesion and the structure of the world society.

Table 13.4 Summary of all contributions to ICRC in 2021

1. Governments	1,634,561,460
2. European commission	122,718,897
3. Supranational and international organizations	11,935,757
4. National societies	29,110,129
5. Public sources	6,890,632
6. Private sources	62,090,445
Grand total	1,867,307,320

The International Committee for the Red Cross (ICRC) is a unique institution (neither an IGO nor an NGO) based in Switzerland, which has a mandate to protect the victims of international and internal armed conflicts under the Geneva Conventions and their Additional Protocols.

ICRC, national societies from 188 countries and their International Federation form the International Red Cross and Red Crescent Movement. All within the Movement share common fundamental principles but are not linked hierarchically. In conflict situations ICRC takes the lead role and directs the work of its partners. The ICRC works closely with the National Red Cross and Red Crescent Societies and the International Federation to ensure a concerted, efficient and rapid response to conflict or violence.

The International Red Cross and Red Crescent Movement is an international humanitarian movement with approximately 97 million volunteers, members and staff worldwide, which was founded to protect human life and health, to ensure respect for all human beings, and to prevent and alleviate human suffering. It is the largest humanitarian network in the world.

The International Federation of Red Cross and Red Crescent Societies has yearly budget of about 500 million CHF; a summary of the contributions to the ICRC is shown in Table 13.4.

Médecins Sans Frontières (MSF) translates to Doctors without Borders. This organisation provides medical assistance to people affected by conflict, epidemics, disasters, or exclusion from healthcare. Its teams are made up of tens of thousands of health professionals, logistic and administrative staff—bound together by our charter. Its actions are guided by medical ethics and the principles of impartiality, independence and neutrality. It is a non-profit, self-governed, member-based organisation.

MSF was founded in 1971 in Paris by a group of journalists and doctors. Today, it is a worldwide movement of nearly 65,000 people.

In 2020, more than 7 million individual donors and private institutions (private companies and foundations) provided 97.2% of the €1.9 billion raised. Funding relies largely on individuals donating small amounts. Table 13.5 breaks down these categories further still:

The purpose of including these monetary values in this and the preceding section is to show that they are a very small fraction of the global GDP and therefore, as

Table 13.5 MSF expenditure for 2020, in millions of euros

Programmes	1081
Programme support	202
Awareness-raising and access campaign	43
Other humanitarian activities	26
Total social mission	1352
Fundraising	250
Management and general administration	77
Other expenses	327
Total	1679

already indicated in the Introduction, are at best very embryonic components of a common world infrastructure.

13.2.4.4 Treaties and Alliances

The organisations in the previous section are essentially independent businesses, each with its own staff, management, and continuous operation. The organisations listed below are quite different in that they are mainly agreements between nations to participate in common activities when the need arises or to individually observe certain rules. Some organisations, such as NATO and OECD, have substantial permanent staff and facilities, others, such as Five Eyes, are simply information-sharing agreements without any permanent organisational body.

NATO The North Atlantic Treaty Organization, also called the North Atlantic Alliance, is an intergovernmental military alliance between 30 North American and European countries. The organization implements the North Atlantic Treaty that was signed on 4 April 1949.

CSTO The Collective Security Treaty Organization is a Russia-led military alliance of seven former Soviet states that was created in 2002. The CSTO's purpose is to ensure the collective defence of any member that faces external aggression.

SCO The Shanghai Cooperation Organisation, or Shanghai Pact, is a Eurasian political, economic, and security alliance, the creation of which was announced on 15 June 2001 in Shanghai, China by the leaders of China, Kazakhstan, Kyrgyzstan, Russia, Tajikistan, and Uzbekistan. Since then, the organisation has expanded its membership to eight countries when India and Pakistan joined SCO as full members on 9 June 2017.

OECD The Organisation for Economic Co-operation and Development is an intergovernmental economic organisation with 38 member countries, founded in 1961 to stimulate economic progress and world trade. It does not have any political or security purpose; its nature is mostly that of a think tank and research organisation.

ASEAN The Association of Southeast Asian Nations is a regional intergovernmental organization comprising ten countries in Southeast Asia, which promotes intergovernmental cooperation and facilitates economic, political, security, military, educational, and sociocultural integration among its members and other countries in Asia.

ANZUS The Australia, New Zealand, United States Security Treaty is the 1951 collective security non-binding agreement between Australia and New Zealand and, separately, Australia and the United States, to co-operate on military matters in the Pacific Ocean region, although today the treaty is taken to relate to conflicts worldwide.

AUKUS A trilateral security pact between Australia, the United Kingdom and the United States, aimed primarily at confronting China in the Indo-Pacific.

Five Eyes An intelligence alliance comprising Australia, Canada, New Zealand, the United Kingdom and the United States.

QSD The Quadrilateral Security Dialogue, colloquially the Quad, is a strategic security dialogue between the United States, India, Japan and Australia.

OSCE The Organization for Security and Co-operation in Europe is the world's largest security-oriented intergovernmental organization. Its mandate includes issues such as arms control, promotion of human rights, freedom of the press, and fair elections. All 57 participating States enjoy equal status, and decisions are taken by consensus on a politically, but not legally binding basis.

EU The European Union is a political and economic union of 27 member states that are located primarily in Europe. It was founded in 1993, but it is still very much a work in progress, moving perhaps towards a true federal structure.

OAS The Organization of American States is a continental organization that was founded on 30 April 1948, for the purposes of solidarity and cooperation among its member states within the Western Hemisphere.

AU The African Union is a continental body consisting of the 55 member states that make up the countries of the African Continent, focused on increased cooperation and integration of African states to drive Africa's growth and economic development.

G20 The Group of Twenty is an international forum for the governments and central bank governors from 19 countries and the European Union (EU). Founded in 1999 with the aim to discuss policy pertaining to the promotion of international financial stability.

BRICS An association of five major emerging national economies: Brazil, Russia, India, China, and South Africa.

APEC The Asia-Pacific Economic Cooperation is an inter-governmental forum for 21 member economies in the Pacific Rim that promotes free trade throughout the Asia-Pacific region.

CPTPP The Comprehensive and Progressive Agreement for Trans-Pacific Partnership is a free trade agreement (FTA) between Australia, Brunei Darussalam, Canada, Chile, Japan, Malaysia, Mexico, Peru, New Zealand, Singapore, and Vietnam. The CPTPP was signed by the 11 countries on 8 March 2018 in Santiago, Chile.

RCEP The Regional Comprehensive Economic Partnership is a proposed free trade agreement in the Indo-Pacific region between the ten member states of the Association of Southeast Asian Nations (ASEAN), namely Brunei, Cambodia, Indonesia, Laos, Malaysia, Myanmar, the Philippines, Singapore, Thailand, and Vietnam, and five of ASEAN's FTA partners—Australia, China, Japan, New Zealand, and South Korea.

MIKTA An informal partnership between Mexico, Indonesia, South Korea, Turkey and Australia. It is led by the Foreign Ministers. It was created in 2013 on the sidelines of the United Nations General Assembly in New York City and aims to support effective global governance.

ALBA An intergovernmental organization based on the idea of the social, political and economic integration of the countries of Latin America and the Caribbean.

13.2.4.5 EU as a Hybrid Structure

The European Union (EU) is the only attempt, so far, to create a society greater than the nation. It can be viewed as a "small" world society in that it encompasses only 27 out of the world's 200 or so nations, and less than 6% of the world's population'. In terms of our two basic structures—either a society of individuals or a society of nations—it is a hybrid, in that it has two governing bodies: one representing the nations and consisting of the European Council and the Council of the European Union, and one representing individuals, the European Parliament. In addition, there is an executive body, the European Commission.

The main substance of this society is its common market, with a system of laws that apply to all member states in the areas of trade, agriculture, fisheries, and regional development. Passport controls have been abolished for travel between the 22 member states that belong to the Schengen area, and 19 member states have a common currency, the euro. Through its Common Foreign and Security Policy, the union is attempting to establish a significant role in external relations and defence, but there is no European defence force, and national security remains the sole responsibility of each member state.

However, members of the EU are not prevented from joining other groupings and alliances and, in particular, 22 of its members are also members of NATO, which in practice means that issues of foreign policy and defence are dominated by the US. Through the European External Action Service, the EU maintains missions (delegations) in many states, but so do its member states. And while the two are required to cooperate, there is nothing preventing individual member states from pursuing international objectives of their own. This provides a means for countries outside the EU to weaken the cohesion of the EU in external matters.

For the period 2019–2022, the main objectives of the EU are:

- Protecting citizens and freedoms
- Developing a strong and vibrant economic base
- Building a climate-neutral, green, fair and social Europe
- Promoting European interests and values on the global stage.

This shows that the focus of the EU is its internal market, in promoting an effective and efficient cooperation between its member states, rather than promoting any further integration. In the view of a society presented in this book, the EU appears as an incomplete, or weak, society.

13.3 Central Issues in the World Society

13.3.1 Power in International Relations

There is a very considerable body of work devoted to the subject of International Relations, which can be grouped according to the approach taken to characterising and analysing the relations, as describe in, for example, in https://en.wikipedia.org/wiki/International_relations_theory, accessed 20.06.2022, and also in *International Relations Theory*, (S. McGlinchey, R. Walters & C. Scheinpflug, editors), E-International Relations Publishing, 2017. A prominent such group goes under the name of the English School (ES), and as there is a recent book (Buzan & Schouenborg, 2018) providing a detailed and readable account of the current state of ES theory as well as its relationship to other approaches, and with a correspondingly extensive bibliography; that account will, without prejudice to other approaches, be used to compare the IR view of society to our high-level view and model.

In ES theory, the approach is based on the concept of a Global International Society (GIS), a society arising out of the Treaty of Westphalia and forming an international society by virtue of all the societies involved in that treaty subscribing to a common set of *primary institutions*: sovereignty, territoriality, war, international law, diplomacy, balance of power, and great power management. As this society expanded through conquest and colonisation, it became the GIS. The expansion is described in terms of two abstract models, the monocentric model, in which one subglobal international society overawes and absorbs the rest, and the polycentric model, in which various subglobal international societies expand and interact until they form a global one. The components of this GIS are units, structure, and binding forces (i.e., a classical system description), with the latter described in terms of belief, coercion, and calculation, and the structure described in terms of four non-exclusive arch-types of society—like-units model (LUM), regional/subglobal model (RSM), hierarchy/privilege model (HPM), and functional differentiation model (FDM).

The units were originally nations, but in recent times non-government organisations (NGOs) and transnational organisations (TNCs) are increasingly becoming significant actors in the GIS, and even individuals can be identified as actors. Thus, in addition to the interstate domain there is a transnational and an interhuman domain, and the relations between them, as well as how to relate the ES's core concept of primary institutions to the IR 'institutionalism' that focuses on secondary institutions (intergovernmental and nongovernmental organizations and regimes), are discussed.

The first point of comparison is the definition of "society". In ES it is a collection of nations bound together by the acceptance of a set of primary institutions; in our view it is a system of interacting individuals bound together by a common identity that is reflected in in a common infrastructure. That is, while the GIS is reflected in behaviour, our concept of society requires a material manifestation.

The second point of comparison is in what drives the evolution of society. The ES is not concerned with this; it presents a tool for analysing the GIS and provides a description of the changes taking place. In our view, the evolution of society is driven by the collective intelligence—the operation of interacting human brains. The primary structure of society is the structure of this operation.

The third point of comparison is the level of detail. Through its multi-dimensional analysis framework ES provides a rich picture of the GIS; our view is at a high level of abstraction and focused on the single issue of society's progress in one dimension—the operation of the collective intelligence. And, as a fourth point of comparison, we can note that although the GIS is not the same as the world society of nations [as appreciated in (Buzan & Schouenborg, 2018)], that world society is related to the LUM, and the embryonic world society formed by competing groups of nations is related to the RSM.

Irrespective of the IR theory adopted, and with it the choice of significant actors (states, NGOs, TNCs, etc.), our view is that the actions taken by these actors are determined by the information available to them at the time of action. Therefore, when we now turn to the issue of power in IR, we need to develop the relationship between action and information and, as always, we start with a very high-level view of what is meant by "a projection of power". Referring to the definition of power cited in Sect. 11.5, "A has power over B to the extent that A can get B to do something that B would not otherwise do", there are two actions directly involved: the action involved in A's projection of power (a1) and the action B must undertake to conform to A's desire (a2). However, as A and B are elements of the same society, there is a third action involved: the action taken by the rest of society in response to the changed situation (a3).

With these actions are associated costs and benefits (as negative costs), and for simplicity we assume that these are all accounted for in monetary terms over a time period for which the estimates make sense, and at an agreed discount rate. These costs and benefits, as estimated by A prior to undertaking the projection of power, will determine whether A goes ahead with the action a1 or not. There is the benefit A expects will accrue from the change in B's behaviour (c1), the cost of undertaking the action a1 (c2), and the cost or benefit to A of society's reaction (c3).

However, this is assuming that A is successful in making B change its behaviour, this is implied in the statement "A has power over B", but to fully understand such a confrontation, A must allow for the possibility that it is not successful, in which case c1 will be zero. To understand that possibility, A must estimate the cost to B of agreeing to the action a2 (c4), of mounting a successful resistance (c5), and of the benefit to B of society's reaction to the successful resistance (c6).

The information associated with such a projection of power consists of several largely independent bodies of information. One group contains the information available to A that causes A to want to change B's behaviour, consisting of information about the system of which A and B are elements that, when evaluated by A, identifies an opportunity for A to improve its position within the system (as measure by whatever is important to A, such as power or security) by changing B's behaviour. This group has two bodies of information: information about B (i1), and information about the environment in which the interaction with B would take place (i2), that is, information about that part of the system, S, that would be affected by the change in the status of A and of B due to the interaction.

Another group contains the information required to define the action a1, which includes A's intended interaction with B (i3), which can range from diplomatic persuasion through economic pressure to the threat of military action, with the information ranging from propaganda to detailed plans for military operations, and information about A's capabilities as they relate to the projection of power being considered (i4).

The three players in this scenario are the two actors, A and B, and the rest of society, S. All three are both sources and users of information, and the relationships between bodies of information and the three players are indicated in Fig. 13.11.

The relationships between the cost estimates and the required information are as follows:

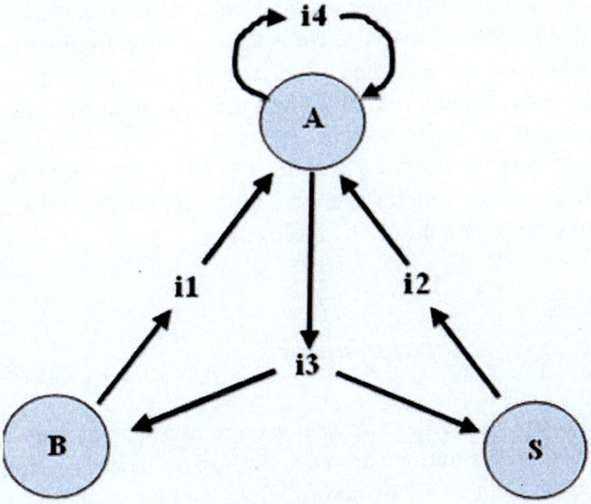

Fig. 13.11 The three players in A's projection of power are the two actors, A and B, and the affected part of society, S. The relationships between them are shown in the form of bodies of information, ix, x = 1, 2, 3, whereas i4 is the knowledge A has about itself in relation to this power play

c1: i1, i2
c2: i1, i2, i3, i4
c3: i2, i3
c4: i1
c5: i1, 13
c6: 12

As initial and simplified decision-making rules, A will proceed with the action if

$$c1 + c2 - c3 < 0; \tag{13.1}$$

and B will resist it if

$$-c4 + c5 + c6 < 0; \tag{13.2}$$

The action a1 will consist of a combination of providing information extolling the reasonableness and desirability of the change—usually a mixture of selected facts and suggestive misinformation, of applying economic pressure, of providing support to subversive activities within B, and, as a (hopefully) last resort, of military action. In designing the information component, A will use its knowledge of B, i1, to optimise the information conveyed to B as part of the action, in accordance with the process illustrated in Fig. 7.2. The actor A will also design a1 so that the cost to B of resisting, c5, is as high as possible, while at the same time minimising the value of the expression in Eq. 13.1.

The cost components consist of not only actual costs, i.e., items inherently expressed in monetary units, but of numerous items to which a monetary value must be assigned, including prestige, national sovereignty and security, access to raw materials, and human lives. The actors A and B may put different values of such items, and A needs to take this into account when estimating c4. Of course, all the cost components are estimates, subject to uncertainty regarding their accuracy and completeness, so that there is a probability distribution attached to each component, which might be expressed as a confidence interval.

And, finally, the simplified picture presented here assumes that the players all act rationally, in the sense of using all available information and obeying the rules expressed in Eqs. 13.1 and 13.2. In reality, there is almost always an amount of irrationality, ranging from wishful thinking to deliberate suppression or falsification of data in pursuit of an ideology.

13.3.2 Global Problems

The third argument for why we are at a special stage of evolution, as introduced in Sect. 1.2, is that we are now starting to experience the effects of human activity on Nature, not as small perturbations on Nature, but as a strong nonlinear interaction between global society and Nature, to the extent that they can no longer be considered distinct entities but form a system of two closely coupled elements. And as such, it this system is displaying emergent properties that we perceive as more or less independent global problems, and three of these are climate change, pandemics, and immigration. They are global because they affect all parts of the globe and because their mitigation and possible eventual elimination require global effort and cooperation. And they are problems for two (not completely independent) reasons: One is that the professional (scientific, engineering) definition and understanding of the issues involved are complex and only partially complete. The other is that the economic and social impacts, both of the issues themselves and of their possible mitigation, vary greatly among nations. The first component of the problem requires international cooperation and funding of research and engineering and is the smaller component. It is the second component that poses the greatest challenge, as it impacts not only the world economy, but also the relative positions of nations within the international power structure.

Climate change. Climate change manifests itself as a number of measurable effects: Rising average global temperature, rising ocean temperature, melting of polar ice caps and glaciers, more frequent and extreme weather events (storms, floods), and associated effects, such as an increase in forest fires. The causes, or drivers, of climate change are multiple and complex, but the major one is the effect of emissions, or Green House Gasses (GHG), on the planetary energy balance—essentially, the balance between the energy received from the sun and the energy lost to the universe (which is very cold, around −270 °C). The most significant source of emissions is the production of carbon dioxide (CO_2) by combustion of fossil fuels (coal, oil, gas), but another source is the methane (CH_4) produced by cattle in their digestive process (rumination), as well as by the thawing of the Siberian tundra. But there are other processes involved, many of them naturally occurring, such as the absorption of CO_2 by vegetation and the reflection of sunlight by snow and ice, and therefore our interference with these processes, as in land clearing, is also involved. The science of global warming is complex, but well developed (see e.g., https://en.wikipedia.org/wiki/Greenhouse_gas or https://ipcc.ch/report/ar6/wg1/, both accessed 20.06.2022); somewhat less well developed is the science of the effects on climate, as in cycles of extreme heat and draught (https://en.wikipedia.org/wiki/Climate_change, accessed 20.06.2022). What must be done to limit the damaging effects is also well known—primarily stopping the combustion of fossil fuel and convert to renewable energy—but the whole issue has significant economic implications, both locally and on a global scale, as well as many social issues (employment, retraining, relocation, etc.).

Pandemics. Pandemics are not something new; they have been with us at least from the Black Death, a bubonic plague epidemic around 1350 that wiped out about half of the population in Europe. In the last hundred years we have had a number of influenza pandemics, notably the "Spanish Flu" following the First World War, and currently we are experiencing the COVID-19 pandemic. Advances in medical science and technology is now enabling the development and production of effective vaccines within a timeframe of a year, which, together with government control of people's movements and mandating protective equipment and procedures, have greatly reduced the impact of this pandemic. As of March 2022, the number of deaths exceeded six million out of a total of around 440 million cases, but of these deaths, a majority were unvaccinated persons. Again, the economic and social impacts have been very significant, not least, the psychological impact of prolonged isolation and the loss and separation of family members, and few regions of the world have escaped some damage.

Immigration. Migration, the movement of large bodies of people into new territories has been a feature of *homo sapiens* ever since they moved out of Africa and from there covered every part of the globe. First as nomads, then as kingdoms spreading out through conquest. Only with the appearance of nation-states with defined borders did the concept of *immigration* arise—the movement of citizens of one nation across borders into another nation to settle there. The reasons for such movements include civil unrest, religious and racial intolerance, and famine; generally the hope of a higher standard of living. The greatest such movement in the last two centuries was the Great Atlantic Migration from Europe to North America. Immigrants decided to move to another country based on their own assessment of the options open to them, including the conditions offered by the countries willing to accept them. Another type of movement is that of the *refugee*, a forced movement with no other options, caused by violent persecution and war, or by natural disaster. With the increased mobility and the increased awareness of both inequality and people's plight, immigration and refugees have become pressing global issues and important political issues in all Western nations.

There are several reasons for raising these three issues at this point in our narrative. One is that they are all symptomatic of the increase in human activity due to increasing population and increasing economic activity taking place in a fixed environment—Planet Earth—and thus, practically by definition, global issues. Two, because individual nations are variously able to handle them, there is the temptation for a nation—or a group of nations—to use these and related issues as means of improving its position relative to other nations and thereby decrease the stability of the global society, something we consider in the next section. And three, because of their global nature, the solutions to these problematic issues need to be on a global scale, and according to our model and the role of the common infrastructure as the reflection of society's will, the most effective and efficient approach is through establishing corresponding elements of a global infrastructure. This is the subject of Sect. 13.4.

13.3.3 Stability and Tension

13.3.3.1 The Relationship Between Stability and Tension

In Sect. 11.5.4 we illustrated the concept of stability by a ball constrained by a curved surface, a well, Fig. 11.15. This picture, with the ball representing society peacefully at rest in its equilibrium position and with any deviation from this position being

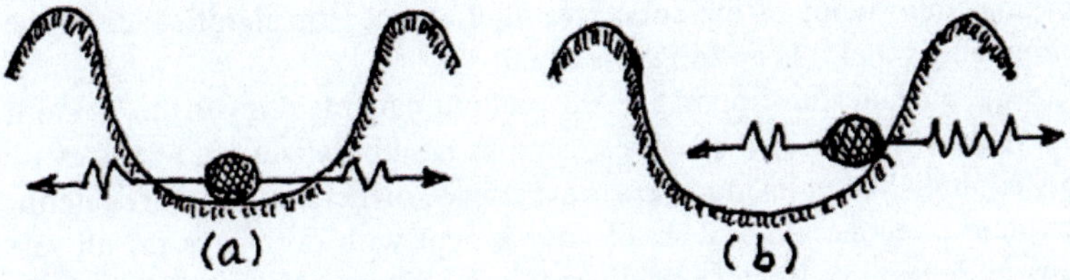

Fig. 13.12 The imbalance in the forces causing tension in society reduces society's stability

caused by external forces only, corresponds to an ideal society, one in which the transformation cycle operates unimpeded, and the common infrastructure is correspondingly developed. In a real society there is always some tension, as noted in Sect. 11.5.3, and this normal level of tension (or, normal society temperature) could be illustrated by adding two springs, as shown in Fig. 12.13a. But if the forces causing the tension get out of balance, as shown in (b), then the stability is reduced, (as can be easily shown by considering the balance of the forces acting on the ball). The effect of the imbalance depends on the strength of the common infrastructure (represented by the steepness of the well) relative to the imbalance.

In the case of the world society as it is today, there is, effectively, no common infrastructure, so that any imbalance would, in this simplified picture, lead to a runaway situation. There are various mechanisms, both diplomatic and economic, that tend to counteract any imbalance, but as the tension rises, as is currently the case, these mechanisms become less effective and reliable (Fig. 13.12).

13.3.3.2 Intranational Stability

Nation-states vary greatly in size, from China, with a population of almost one-and-a-half billion, to island states in the Pacific with only a couple of thousand inhabitants, and, correspondingly, they vary greatly in structure and complexity. So, when we now look at *the* nation-state, as a typical representative of the society of interest, it will be a better representative of some than of others; it will generally correspond to a developed nation, and in the first instance, to what goes under the label of a representative democracy.

The single most significant characteristic of the current evolution of society is—despite a certain government redistribution effort and (theoretically) progressive taxation—the increasing importance of capital and the associated inequality in both wealth and income, and thereby in power. Now, the situation where a small section of the population dominates the activities of society and reaps the benefits of them is nothing new; that has been the case since the beginning of historical times, with ruling families and nobility. But there has also been an unmistakeable trend towards greater equality and participation in the affairs of society; first in the form of a merchant class, then the trades through their guilds, then the manufacturers and the supporting professions, and finally the workers themselves. It can be seen as an increasing importance of the collective intelligence; the significance of the current development is that it is defying this trend.

The driver of this development is the predominant ideology in the Western democracies; it goes under various names, such as neoliberalism, neo-conservativism, or simply capitalism, but its main tenets are private ownership, a self-regulating, profit-driven market economy, and small government with user-pays for all services, as we saw in Sect. 11.4. Besides being an inherently unstable system, due to exponential growth, there are two immediate effects of this ideology: One is the pressure on finding new investment opportunities for the accumulating capital, reducing the individual to a form of consumption slavery. The other is suppressing any opposition to what is a highly non-equilibrium situation by subverting the political process; what appears as representative democracy is increasingly simply a reflection of economic reality. We are thus faced with an amazing development: a system that advantages a shrinking proportion of the population, and that is being criticised and mistrusted by an increasing part of the population, is still preferred by the majority.

There are a couple of reasons for this development, and one is the fact that, despite many grumblings, life is pretty good for the majority of the population. Taking Australia as an example, most of us have a place to live, enough to eat, adequate clothing, and the basic modern conveniences, such as a washing machine, refrigerator, TV, and mobile phone, and while many might wish they could afford more (due in large part to advertising), it is not enough to send us to the barricades. Another reason is that not only does our consumer society provide numerous distractions, but the barriers to contributing effectively to change are surprisingly high. One might think that with the means of communication at our disposal, it would be relatively easy to be heard, and indeed, there is nothing stopping us from posting our opinions and insights on social media and on websites, where they are freely available to everyone, but it is exactly this ease of posting that is reducing the Internet's effectiveness as a forum for discussion and development of ideas as part of the political process. And anyone that has tried to contribute through membership in the local branch of a political party will know that this is a very time-consuming and inefficient approach, and not a realistic option for most people with a busy professional life. The result is that an effective involvement in politics is really only possible for those who want to make it their career, which brings us back to the political party as a business.

We might think that running political parties as businesses should not be such a bad thing; after all, many companies are run very successfully by providing products to society's members in a competitive market. The difference is that in the political arena the competition is missing—there is no other process competing with the party system. The "competition" between Liberal and Labor, or between Republican and Democrat, has little effect on the party as a business, and so there is little incentive for change. What can change, and is slowly changing, are the policies of the parties; these changes are partly in response to the demands of the electorate, but also in response to special interest groups and, significantly, in response to the wishes of such groups as the TNC—all evaluated relative to the party as a business. Given this, it is unlikely that Western societies will experience catastrophic internal fluctuations, such as happened in the French and Russian revolutions, in the foreseeable future.

If we look beyond the liberal democracies of Western nations, we see a very different picture. With the exception of China, which has the longest continuous existence of any nation-state, most of the other nations were created either by the colonial powers, as in the case of Middle Eastern and African nations, or by the beak-up of forced configurations, such as in the case of the Ottoman empire, the Soviet Union, and Yugoslavia. These states often have significant internal ethnic and religious strain, and little experience with democratic government, with the result that there has been a rising number of intrastate armed conflicts since the end of WW2. The previously referenced book, *On Building Peace: Rescuing the Nation-State and Saving the United Nations*, by von der Schulenburg (2017), describes and analyses these intrastate armed conflicts in great detail, based on both on-site experience and a long career in peace-building efforts by the UN. His analysis shows how the current legal and institutional framework for maintaining peace is inadequate, as it is focused on avoiding and containing interstate conflicts, and he proposes how the UN could be restructured and revitalised to be effective on the intrastate level. But equally important is the strengthening of the nation-state as the building block of the global society, something that has come under attack as a result of a misconstrued understanding of globalisation.

These intrastate conflicts have been—and still are—exacerbated by two factors: the interests of the arms industry in promoting them, and the competing interests and power struggles of the major powers, with the result that many of these intrastate conflicts have taken on the form of proxy wars between the major powers. They are symptomatic of the West endeavouring to hold on to its dominant position, both economically and ideologically, and therefore, devastating as some of these conflicts are, such as the current ones in Syria and Yemen, they are in themselves not a threat to the stability of evolution. That threat arises on the international level, which is the subject of the next section.

13.3.3.3 International Stability

An unstructured collection of nations interacting on an equal basis and without any common infrastructure is not in a stable state. Due to their competing nature, differences in history, resources, climate, and technology, no matter how small, will eventually lead to some nations dominating others within their environment, resulting in a structuring of the collection into groups of nations, each dominated by one or a few nations and defined by one or more organisations of which the nations in the group are members.

However, this structure is not quite as clear-cut as that of a few separate groups. The relations between nations are defined by numerous organisations of various types, some of which were listed in Sect. 13.2.4. Some of these, in particular the ones belonging to the UN, have practically universal membership, whereas most vary greatly in the size of their membership. Some have broad military, political, and economic agendas; others are concerned with highly specific issues. In addition to intergovernmental relations anchored in organisations, there are also a very large number of relations in the form of agreements (often bilateral) that are handled by the respective departments within the nations (e.g., bilateral free trade agreements). And the relationships between nations are not defined only by formal instruments; they are formed by common beliefs and cultural traits, such as language, as well as by historical connections, ranging from antiquity to recent wars. So, in short, the structure of the system of nations is highly complex.

As noted earlier, if we consider the view of evolution presented in Chap. 2, with the evolution of the *genus* society as the current stage in this general process, we would expect the next *species* to emerge to be a world society with all the nations in the world as its elements. As in all other societies that evolved, this extended reach of the interaction would result in a further structuring, and the world society would be defined by a new level of governance, perhaps mirroring the national executive, legislative, and judiciary triple function structure. Within this structure, each nation-state would be able to develop for the benefit of its population and to be a valuable contributor to the world society, just as individuals should be able to do within a nation.

The first attempt at such a structure was the League of Nations, founded in 1920, with the main objective of preventing war between nations. It was never very effective, not least because the US never joined, and was replaced by the United Nations (UN) after the end of the Second World War. With its numerous agencies, programs, and affiliated organisations, the UN has been, and is, quite successful in promoting and sustaining international cooperation in various fields, and as such must be seen an initial, but substantive part of such a world-wide structure. However, with regard to maintaining stability between its member states, the UN was completely sidelined by two situations, as has been very well discussed and documented in the book by Michael von der Schulenburg referenced earlier—*On Building Peace* (von der Schulenburg, 2017). The first was the Cold War arising as a result of WW2, with two superpowers—the Soviet Union and the US—confronting each other in a precarious balance of nuclear weapons power known as Mutual Assured Destruction (MAD). Correspondingly, the world became divided into two spheres of influence, with numerous skirmishes, both covert and overt, on the fault line between them, with the two most serious ones being the Korean and Vietnam wars, and within this scenario there was clearly no scope for effective UN action. The second arose after the fall and disintegration of the Soviet Union, with the US emerging as the sole, and greatly strengthened, superpower. Emboldened by this success, the US set about providing "peace and order" on its own terms, while aggressively expanding its sphere of influence through such means as NATO, an organisation that should have been disbanded following the end of the Cold War. Again, this state of affairs made it practically impossible for the UN to have any decisive influence on the relationship between nations.

Already in the Introduction it was suggested that the wealth and dominant position of the West was, to a significant extent, due to the subjugation and exploitation of a largely defenceless third of the world over the last 500 years or so. This phase of the evolution of the world society came to an end around the middle of the last century, in part as a result of two world wars that devastated much of the West, but from which the United States eventually arose as the sole superpower. The nations that had, prior to 1900, made up what would be considered the West became politically and militarily subservient to the US, and much of the rest of the world was brought under US domination by economic means (sanctions or grants) or by direct military intervention (as in Iraq), as well as by an extensive repertoire of proxy wars and covert operations; a number of the latter, such as those in Guatemala, El Salvador, and Nicaragua, are documented in the book *How the World Works*, by Chomsky (2011). Thus, the world society is currently in a highly non-equilibrium state, where the supremacy of the US-led West is maintained by military and economic force against what in this book is presented as the natural evolution of the world society, and such a non-equilibrium situation always carries with it the possibility of a collapse, if the forces supporting this balancing act should no longer be able to maintain it.

However, it is not the existence of the current non-equilibrium situation that is of greatest concern; if there was a will to let it settle gradually back into an equilibrium state that would be possible; albeit not without some significant adjustments for many nations. The greatest concern is the US effort to maintain and strengthen its hegemonial position by polarising the world into two camps: "Either you're with us, or you're against us." This was already a clear strategy following WW2, with the Soviet Union and communism defining the polarisation; any nation that only wanted to be itself, such as Vietnam, soon got to feel the consequences. With the collapse of the Soviet Union and its influence over Eastern Europe and Central Asia, one could have expected the polarising momentum to have faltered. But Russia could still be made into a threat justifying the strengthening of the Western alliance and the use of NATO to push eastwards and enforce US policy. Following the end of the Cold War and the dissolution of the Warsaw Pact, one could have hoped for a return to a world of mutually interacting nation-states, based on non-exclusive bilateral relationships under the umbrella of a world-wide framework, anchored in the UN. But this was not a view compatible with the US imperial vision of itself.

To maintain the commitment and active participation of the members of this group, it is important to make the threat tangible and represented by an identifiable enemy. A purely ideological argument can easily reveal shades of grey; what is needed is a clear, black-and-white distinction between the good and the bad, between us and them. The Soviet Union was an obvious target in this regard, but with its collapse in 1991, the identification became less convincing. However, by aggressively pushing the borders of NATO steadily closer to Russia, despite assurances provided in connection with the reunification of Germany (new members: Czech Republic, Hungary, and Poland in 1999, Bulgaria, Estonia, Latvia, Lithuania, Romania, Slovakia, and Slovenia in 2004, Albania and Croatia in 2009, Montenegro in 2017, and North Macedonia in 2020), the Russian response in the form of a build-up of forces on its borders could be interpreted as threatening. When Russia quite predictably annexed Crimea in 2017 in response to the looming inclusion of Ukraine in the Western alliance, and to avoid the Sevastopol Naval Base, the main base of the Black Sea Fleet, falling under "enemy" control, there was much feigned surprise, indignation, and horror in the West, and imposition of sanctions. And when Russia intervened in Syria to protect its only Mediterranean military base after the US promoted regime change by supplying the rebels with more than a billion dollars in arms, here was another proof of the Russian danger.

According to the view of evolution presented in this essay, with society as an information-processing system, the risk of instability is measured by the deviation of the operation of the collective intelligence from its optimal operating point. That operating point is characterised by several parameters, some of which were discussed in Chap. 6, but the by far most volatile measure, and therefore the one determining the evolution in the short term, is the extent to which information is unrestrictedly available. The struggle to maintain the current non-equilibrium situation and the hegemonial position of the West is therefore, as we would expect, mirrored in the effort to ensure that the information available to not only the people on the Western world, but to as large a part of the world population as possible, presents a narrative that justifies that position. And not only justifies, but makes the western ideology, with its mixture of economics, ethics, and religion, as presented in Sect. 7.2.2, appear as the only viable one, as the final socioeconomic framework on which the further existence of humanity is to be based. We can discern two distinct components of this effort. The first one is a "filtering" or "editing" of information, in the sense of emphasizing, deleting, and associating items of information to create an impression in the reader's or viewer's mind that is favourable to the West and/or detrimental to what it considers its adversaries. For example, including the lifting around one billion people out of poverty (< US$1.9 per day) in a list of achievements of capitalism, even though the majority of these were in China. Or suppressing the results of US-led airstrikes in Syria. Pictures of the devastation in Syria only started appearing very frequently in Australian newspapers once they could be assigned to Russian airstrikes. In the recent Houthi drone strikes on Saudi oil facilities the fact that the drones and rockets were manufactured in Iran was taken as proof that the attack was launched from Iran; whereas when a school or hospital in Yemen is hit by Saudi air strikes, it is not pointed out that the rockets are of US manufacture and that the strikes therefore are a US responsibility.

The second component is the extensive reporting of incidents in countries not aligned to the West that are contrary to western ideals of justice and freedom, even though they may be justified under the rules and customs of these countries. Conversely, when such actions occur in countries aligned with the West, such as Saudi Arabia or Egypt, there is little, if any, notice taken.

And now a new polarising factor has arisen—China—which has become the subject of the same kind of aggressive containment policies previously reserved for Russia. But while Russia is still a major power militarily and through the size of the area within its borders, it is no comparison with China when it comes to economic power, with a GDP of only US$1.6 trillion as compared with China's US$13.6 trillion (World Bank, 2018). From both its size—population and economy—and its history, it is evident that China is currently the main challenger to the hegemonial status of the West, and the following section is dedicated to examining the current relationship between China and the West, where "the West" consists of a "hard core" made up of the leader—the US—and Israel, the UK, Australia and Japan, and a "ring" of nations bound to that core and the US through various security and economic treaties, such as NATO, as was illustrated in Fig. 13.1. The role of other nations will feature only to the extent that they are drawn into the contest between these two parties.

13.3.4 Tension Between China and the West

The rise of China was the subject of Sect. 12.1.4, and to understand why this is such a threat to Western hegemony, it is necessary to understand that China is the leading nation in an economic (and power) development that will take place over the next few decades. Any prediction of this growth will depend on numerous assumptions and, in the short term, on the impact of the COVID-19 pandemic (see World Bank report *Global Economic Prospects*, June 2020, and IMF report *Economic Outlook Update*, June 2020). A long-term prediction was provided in the 2017 PwC report *The Long View: How will the global economic order change by 2050?*, and although somewhat speculative and presenting a best-case scenario for growth, the general trend is likely to be correct. Table 13.6 gives the predicted PPP GDP values for the top 32 nations in the years 2030 and 2050, together with the actual values for 2016. It is interesting to note how, in 2050, the three largest economies—China, India, and the US—form a group at the top, with a big step down to the fourth largest. To be in this group, a nation either must have a large population, or have a high productivity—*per capita* GDP—which can, to some extent, be seen as a measure of the standard of living. According to the PwC projection, the *per capita* GDP of the US in 2050 will still be just over twice that of China, and more than three times that of India. This is also reflected in the relationship between the G7 (representing the West) and the E7 (representing the emerging countries and comprising China, India, Brazil, Turkey, Russia, Mexico and Indonesia); the *per capita* GDP of the G7 will be about twice that of the E7 in 2050.

The picture that emerges is that the West, as the group of nations related to, and led by, the US, still has a dominating position, but that this position, as measured by both GDP and *per capita* GDP, will be significantly reduced over the next thirty years. However, an issue that is not considered in the PwC methodology is the influence of military power. The dominant position of the West was built on military power, and the same is true of the leadership position of the US. In the past, military power was also used in the competition between the nations of the West, but only in the twentieth century was that competition allowed to cause a serious reduction of the dominant position of the West. A question—and perhaps the central question as far as the stability of social evolution is concerned—is then: To what extent is the relationship between nations currently and in the foreseeable future determined by military power? And in the present context—the formation and competition between groups of nations—this question can be partitioned according to the entities involved in the relationships: between nations within a group, between a group and uncommitted nations, and between groups.

Between nations within a group, the relationship is maintained by identifying a threat—a common enemy. It is also promoted by creating a common infrastructure,

Table 13.6 Projected rankings of economies based on GDP at PPP (in constant 2016 $bn)

GDP PPP rankings	2016 rankings		2030 rankings		2050 rankings	
	Country	GDP	Country	Projected GDP	Country	Projected GDP
1	China	21,269	China	38,008	China	58,499
2	US	18,562	US	23,475	India	44,128
3	India	8721	India	19,511	US	34,102
4	Japan	4932	Japan	5606	Indonesia	10,502
5	Germany	3979	Indonesia	5424	Brazil	7540
6	Russia	3745	Russia	4736	Russia	7131
7	Brazil	3135	Germany	4707	Mexico	6863
8	Indonesia	3028	Brazil	4439	Japan	6779
9	UK	2788	Mexico	3661	Germany	6138
10	France	2737	UK	3638	UK	5369
11	Mexico	2307	France	3377	Turkey	5184
12	Italy	2221	Turkey	2996	France	4705
13	South Korea	1929	Saudi Arabia	2755	Saudi Arabia	4694
14	Turkey	1906	South Korea	2651	Nigeria	4348
15	Saudi Arabia	1731	Italy	2541	Egypt	4333
16	Spain	1690	Iran	2354	Pakistan	4236
17	Canada	1674	Spain	2159	Iran	3900
18	Iran	1459	Canada	2141	South Korea	3539
19	Australia	1189	Egypt	2049	Philippines	3334
20	Thailand	1161	Pakistan	1868	Vietnam	3176
21	Egypt	1105	Nigeria	1794	Italy	3115
22	Nigeria	1089	Thailand	1732	Canada	3100
23	Poland	1052	Australia	1663	Bangladesh	3064
24	Pakistan	988	Philippines	1615	Malaysia	2815
25	Argentina	879	Malaysia	1506	Thailand	2782
26	Netherlands	866	Poland	1505	Spain	2732
27	Malaysia	864	Argentina	1342	South Africa	2570
28	Philippines	802	Bangladesh	1324	Australia	2564
29	South Africa	736	Vietnam	1303	Argentina	2365
30	Colombia	690	South Africa	1148	Poland	2103
31	Bangladesh	628	Colombia	1111	Colombia	2074
32	Vietnam	595	Netherlands	1080	Netherlands	1496

From the 2017 PwC report The Long View: How will the global economic order change by 2050?

in the form of communications networks, interoperable hardware and software, and by staging periodic exercises, and it is protected by sanctions against any member that is not carrying its weight or that tries to break away. But above all it is sustained by a carefully edited version of information—not untruthful, just a very limited slice of the truth that reinforces the group ideology.

Between the group and uncommitted nations, the relationship is one of enticement—mainly economic—and of a constant bombardment with information favourable to the image of the group and unfavourable to other groups. Between groups, the relationship is one of exerting pressure, and each group will consider the approach and solution to any common issue or problem in the light of what advantage it could bring to its position relative to other groups.

Therefore, for the US to maintain its position as the dominant player on the world stage, it must stay in command of an intact and, if possible, increasing Western alliance, and prevent China from forming an alliance of its own. Conversely, China is trying to gain wider international influence and form its own group of allied nations, and so the next level of description of the society of nations is as two groups of nations, one supporting the US and one supporting China. To this end Table 13.7 presents some data for the organisations listed in Sect. 12.2.4.3 and additionally indicates whether the organisation should be considered to be in the US or the China sphere of influence.

In viewing the above information on intergovernmental organisations as an indicator of the relative international strength and influence of the US and China, there are three points to consider:

(a) In all the organisations where the US is a member, but China is not (the eight indicated in Table 13.7), calculate the sum of {GDP; Mil. exp.} over all members excluding the US GDP = 2.1E+13 and Mil = 7.34E+11), and sum over the organisations. Then do the same for China (the three indicated in Table 12.7, with China GDP = 1.45E+13 and Mil = 2.4E+11). The result is that the US sphere of influence is roughly three times that of China, both in GDP and in military expenditure.
(b) The monetary values in Table 13.7 are in US$; this is appropriate for comparison on the international scene. For a comparison of local, national values, PPP dollars are more appropriate (as we shall see below).
(c) Table 13.7 does not include the major Chinese initiative—the Belt and Road Initiative—as this is not an intergovernmental organisation. It is a Chinese managed initiative, but with participation of many nations, and is a major factor in Chinese international influence.

In the previous section we stated that an important component in maintaining the current hegemonial position of the West is the extensive reporting of incidents in countries not aligned with the West that are contrary to western ideals of justice and freedom. In the case of China, two issues stand out in this regard, and the first is the role of religion and religious freedom. These are areas where the West sees an opportunity, cloaked as a moral obligation, to denounce any action on the part of the Chinese state that can be construed as violating the normative Western view of

Table 13.7 Main characteristics of the most relevant intergovernmental organisations

Acronym	Members	Mil. exp	Population	GDP	Sphere
NATO	29	1.00E+12	9.81E+08	4.25E+13	US
CSTO	6	6.44E+10	1.91E+08	1.97E+12	
SCO	8	3.91E+11	3.19E+09	1.95E+13	China
ASEAN	10	4.07E+10	6.60E+08	3.17E+12	
ANZUS	3	6.77E+11	3.58E+08	2.30E+13	US
AUKUS	3	7,24E+11	4.21E+08	2.57E+13	US
Five Eyes	5	7.49E+11	4.62E+08	2.76E+13	US
QSD	4	7.78E+11	1.84E+09	3.08E+13	US
OSCE	56	1.05E+12	1.28E+09	4.59E+13	US
EU	27	2.31E+11	4.47E+08	1.56E+13	
OAS	35	7.35E+11	1.01E+09	2.88E+13	US
AU	54	3.56E+10	1.30E+09	2.42E+12	
G20	20	1.15E+12	2.89E+09	5.27E+13	
G7	7	9.08E+11	7.69E+08	3.69E+13	US
BRICS	5	4.09E+11	3.18E+09	2.11E+13	China
APEC	20	1.15E+12	2.89E+09	5.27E+13	
CPTPP	11	1.32E+11	5.08E+08	1.12E+13	
RCEP	15	4.09E+11	2.27E+09	2.58E+13	China
MIKTA	5	1.03E+11	5.59E+08	6.17E+12	
ALBA	9	6.84E+08	5.77E+07	6.91E+11	

The population and GDP values are from The World Bank data base, military expenditure data is from the SIPRI database, the monetary values are in US dollars, mostly for the year 2019, and the "Sphere" column indicates whether the organisation should be considered to be in the US or the China sphere of influence

these areas. While useful in stoking anti-China feelings, this activity is unfounded and ignores the salient facts, which include the following:

1. Throughout its long history, organised religion in the form of a Church, has seldom been a significant part of Chinese society. Apart from early periods, when Buddhism was influential in some regions and Buddhist monasteries became very wealthy and powerful, the Chinese adopted a Confucian, pragmatic philosophy as a "state religion", and would, if anything, be justified as regarding Christianity and Islam to still be mired in superstition.
2. Through its role in the West's arrogant treatment of China in the nineteenth century, the Chinese could rightfully view Christianity as a Trojan horse for Western interference in China's internal affairs.
3. The history of the Christian Church is not one to inspire confidence in its benefits and behaviour. When it had the power to do so, it ran what we would today

consider a terrorist operation; eagerly participating in the enslavement and extermination of indigenous people and amassing great riches in the process, banning the publication of any text not approved by it, and suppressing scientific knowledge and any opposition or contrary opinion by torture and burning at the stake, with the Spanish Inquisition alone burning more than 10,000 persons. And even today, the revelations of sexual misconduct by the clergy is shameful. So it is understandable and reasonable that China is sceptical as to the advisability of allowing the Christian Church to gain influence.

The second issue is that of freedom—both the freedom *to* and the freedom *from*. Regarding the first, any society places numerous restrictions on the freedom to act, and if we would measure these restrictions in terms of the number of laws, rules, regulations, taxation office determinations, and so on, it is not clear that the Chinese are any less free than, for example Australians. It is just that the restrictions are different, and it is in the framework of our restrictions that we judge some of those in China to be objectionable. The treatment of persons who advocate Western ideals in China is decried as violations of human rights, whereas people who promote the PRC in a country like Australia are seen as security threats. The difference between the two frameworks can be seen as a reflection of what we introduced, in Sec, 2.3, as a primary characterisation of society: how the members of society understand the relative importance of their dual roles as individuals and as elements of society as a system. This determines the extent to which they are willing to forego some aspects of personal freedom in exchange for the benefits arising out of increased integration.

Regarding the freedom *from*, we would probably first think of freedom from starvation, poverty, and ill health; then freedom from slavery, persecution, and oppression; and then, increasingly, from mental limitations, such as censorship. One concept—the standard of living—tries to combine many of these, but that inevitably involves both a selection of characteristics and their weighting; a simple selection is income per person and poverty rate. The Index of Economic Freedom is an annual index and ranking created by The Heritage Foundation and The Wall Street Journal, but it is firmly anchored in the neoconservative ideology, in that it assumes that the strength of basic institutions that protect the liberty of individuals to pursue their own economic interests result in greater prosperity for the larger society. The simplest measure is the per capita GDP, which is about US$60,000 for the US and about US$17,000 for China, with most Western European nations in the range US$50,000–US$60,000. A slightly more sophisticated measure is that proposed in (Aslaksen 2018) as a financial restriction, with 0 being best and 1 being worst; on this measure the US is rated 0.512, China 0.68, and Western European nations in the range from 0.418 (Switzerland) to 0.634 (Portugal). In any case, as far as the freedom from want is concerned, it is clear that the average person in the West is considerable better off than the average person in China; to what extent this can be attributed to their respective ideologies is a different matter, given our previous consideration of their histories. Of greater significance is that the rate of improvement of any of these measures over the last twenty years or so has been significantly greater in China than in the West.

An illustration of the selective handling of information about China and the Chinese is provided by a comparison with the handling of information about Israel and the Jews. Both the Jews and the Chinese were demonised as second-rate humans or “Untermenschen” and caricaturised as such in Western media. They were persecuted in the century leading up to WW2 and then subjected to unspeakable atrocities in the war—the Jews to the Holocaust, with its 8 million victims, and the Chinese to the Japanese terror with its 20 million victims. Both would have wowed that this would not be allowed to happen again and took corresponding steps—Israel by acquiring nuclear weapons, encroaching on Palestinian land, and annexing the Golan Heights; China by modernising and expanding its defence capability, strengthening its grip on such border provinces as Tibet, and by annexing islands in the South China Sea. In the Australian media there is a report about some information item—interview, book, film, presentation, exhibition—relating to the persecution and discrimination of the Jews at least once a week, and there are also numerous items about Nazism and Hitler, and about antisemitism. But any similar items about the persecution and discrimination of the Chinese, or about Japanese imperialism and Hirohito, are practically non-existent, and the terms anti-sinoism or sinophobia are hardly ever used. The implication is that whatever happened to the Chinese was well deserved, and any criticism of the Chinese is not based on prejudice but should be accepted as factual.

To get back to the question of international stability, the point is that the West and China are quite different—in their histories, in their current stages of development, in their ideologies—and they are competitors for power and for the greatest slice of the cake. But, being together on the same planet, they are also highly dependent on each other and, in the long run, must fit into a common framework together with all the other nations in the world; this is the next species of the genus society. The current polarisation, in the form of an intense propaganda effort to paint China as a dangerous enemy instead of simply another competitor, driven mainly by the West for the purpose of clinging to its hegemonial position, greatly increases the possibility that the path to this next species will involve a violent fluctuation. A most appropriate quotation in this regard is Singapore’s PM Lee Hsien Loong’s advice to Australia’s PM Scott Morrison:

> You don’t have to become like them, neither can you hope to make them become like you. You have to be able to work on that basis, that this is a big world in which there are different countries and work with others who are not completely like-minded but with whom you have many issues where your interests do align and where your mutual cooperation is necessary.
>
> (Press conference 11.06.2021, https://www.pm.gov.au/media/press-conference-istana-singapore, accessed 20.06.2022.)

13.4 Looking Ahead

13.4.1 The Dynamics of Society

Society has evolved from its simplest forms as groups of nomads and cave-dwellers to the present-day nation-states and the world society, and along the way a constant feature has been *conflict*. Almost always accompanied by violence and destruction, but also new concepts of society and of the relationship between the individual and society. The most well-known conflicts are the power struggles between similar societies, and the great wars and battles resulting from them. These conflicts did not usually result in any significant advance on the nature of the society; one king and his nobles replaced another king and his nobles, one religion and church hierarchy replaced the previous ones, and so on. One type of conflict had a greater impact on the dynamics of society—the *class conflict*, which became the cornerstone of the political economic theory of Karl Marx:

> The history of all hitherto existing society is the history of class struggles. Freeman and slave, patrician and plebeian, lord and serf, guild-master and journeyman, in a word, oppressor and oppressed, stood in constant opposition to one another, carried on an uninterrupted, now hidden, now open fight, a fight that each time ended, either in a revolutionary re-constitution of society at large, or in the common ruin of the contending classes. (Marx & Engels, 1848)

Marx was influenced by the particular time in which he observed and wrote about society, the full force of the industrial revolution and the importance of the means of production. The examples given in the quote above are all based on the control of manual labour, and while the early stages of the industrial revolution might have seemed to perpetuate this, he seems to have misunderstood the role of the rapidly developing technology. He does recognise its existence,

> The bourgeoisie cannot exist without constantly revolutionising the instruments of production, and thereby the relations of production, and with them the whole relations of society. (*op. cit.*)

but not the resulting demand for better educated industrial personnel in design, production, operation, and maintenance of the means of production, focusing instead on the simplest form of mass production:

> Owing to the extensive use of machinery, and to the division of labour, the work of the proletarians has lost all individual character, and, consequently, all charm for the workman. He becomes an appendage of the machine, and it is only the most simple, most monotonous, and most easily acquired knack, that is required of him. (*op. cit.*)

The demands of a society based on rapidly developing technology has, in the West, created a much more finely structured society than the two-component class structure considered by Marx, and has reduced the purely manual, uneducated labourer to a minority without a distinct class consciousness and that would not be the source of a revolution. There is no reason not to expect the same development to take place in the rest of the world.

The basic driver of change is the development of new technology, but the manner in which technology is applied and the purpose of application has undergone a major change in the days since Karl Marx. From initially being about improving the throughput and efficiency of the means of production, it became increasingly also about creating new products, and the post-war years in the US were a testimony to this change, with many new products, from nylon stockings to washing machines and all other sorts of household gadgets creating a new lifestyle. But besides driving change within the nations where the technology was developed, from research to manufacturing facilities, it resulted in a separation of the world's nation into those nations, essentially all in the West, and the Rest, so that technology is today a main driver of the dynamics of the world society, and why intellectual property rights is such a concern of the West.

13.4.2 Two Approaches

Predicting the evolution of the world society is difficult, and any prediction is likely to be wrong. In the previous chapters we developed a limited picture of where we currently stand and of the forces that drive the evolution, so that, if we, as before, denote the state of society by **S**, we have some knowledge of its values over the past and of the current value of **dS**/dt. But, as we noted in Sect. 10.1.1, we cannot, with any confidence, integrate **dS**/dt to determine the future values of **S**, due to the many random influences. Instead, there are two approaches we can take. The first utilises our idea of the evolution of society as the current stage of a general process of evolution and that this process is characterised by increasing strength of the interaction between the individuals but is otherwise a "classical" view of international relations, as outlined in Sect. 13.3.1, with power as the driving factor. The increasing interaction strength in the world society is readily observable, no matter what level we consider. At the basic level of the individual, it is expressed in the increase in international travel, as shown in Fig. 10.16, and in the increase in international information exchange via the Internet, as indicated in Sect. 8.4.4. It is also expressed in the proportion of international news in the mass media. At the level of business, it can be observed in the increase in global trade, as shown in Fig. 10.15, and in the number and size of multinational corporations and the rise of a Transnational Capitalist class, as we saw in Sect. 6.5. At the level of nations, the increased interaction is reflected in the number of intergovernmental organisations, of which the United Nations and the European Union are two of major ones. As indicated earlier, a uniform society, as is assumed by our high-level model, is not stable under increasing interaction, due mainly to the competitive nature of the interaction and the human need for affirmation. The resulting structure creates a certain tension between the elements (rich vs poor, powerful vs weak, etc.), and if this tension is allowed to increase beyond a certain point, the structure can become unstable and result in a violent collapse. It is a race between the rise in tension due to the structuring and the effort put into measures that reduce the tension, as illustrated by Fig. 2.2, and this is the approach we take in the next section to assess the immediate future of the world society.

The other approach is based on the idea that the evolution of society is driven by the transformation cycle and that the direction of evolution with the greatest probability of survival results from having ideal operating conditions for the collective intelligence. These conditions were the subject of Chap. 6, and by projecting current trends forward we can obtain an indication of whether we will be progressing towards or away from the ideal conditions, and perhaps an estimate of the speed of change. This is the subject matter of Sect. 13.4.4.

In both approaches the global problems mentioned in Sect. 13.3.2 will play a role, and in both the role has two aspects. One is unifying, in that the global nature of the problems invite cooperation; the other is segregating, in that the solutions to the problems can be designed to have winners and losers, at least on a relative scale.

13.4.3 Structure and Tension

13.4.3.1 The Concept of the West

As stated briefly already in Sect. 1.2 and described in more detail in Sect. 3.1, we are currently at a special point in the evolution of the world society (the society) as a result of an explosive development of that part of the society we identify as the West. Accordingly, we can, as a first approximation and with a high level of abstraction consider society to consist of two parts: the West, and the Rest. The nations included in the West is a matter of definition and depends on exactly what one wants to emphasize—cultural, economic, political, or military commonality. For our purpose it will be most convenient to define the West as consisting of the nations shown in Table 13.8.

Within this structure, the two parts are different in two respects. One is the difference in what we in Sect. 6.3.2 identified as the *Quality of Life* (QoL), a

Table 13.8 The nations included in the concept of the West

Australia	Germany	Poland
Austria	Greece	Portugal
Belgium	Hungary	Saudi Arabia
Canada	Iceland	Slovakia
Croatia	Ireland	Slovenia
Czech Republic	Israel	Spain
Denmark	Italy	Sweden
Finland	Netherlands	Switzerland
France	New Zealand	United Kingdom
	Norway	United States

composite measure that characterises The Good Society. The difference in this measure expresses itself as one component of the tension between the two parts. The other measure is the difference in *power*; this is the difference that is required to counteract the tension and to maintain what is a highly non-equilibrium situation. But this exercise of power by the West implies a loss of sovereignty by the nations of the Rest, which adds a further component to the tension. Both measures are open to somewhat different definitions and also become increasingly complex as one tries to make the definitions more precise. In the following section we develop some indicators of tension and power for this first approximation, much as we did in Chaps. 6 and 10, and now look specifically at their projections into the near future.

13.4.3.2 Global Indicators

As a start, we determine the GDP and the population of the two parts over the recent past in the form of their values in the years 1990, 2000, 2010, and 2020 (1990 was chosen as the first date, as it was the culmination of US hegemony after the fall of the Berlin Wall and the ensuing collapse of the Soviet Union). All data input in this section are, unless otherwise noted, from the World Bank Open Data (Table 13.9).

These values are displayed as curves in Fig. 13.13, and it is clear that these values do not allow a projection beyond 2020 with any confidence. However, what we can do is to use the projections included in the 2017 PwC Report; the projection for GDP on a PPP basis was shown in Table 13.6. The projection for GDP at Market Exchange Rates (MER) in 2016 $US is shown in Table 13.10, and from this we can extract the values for the nations belonging to the West and those belonging to the Rest and form the ratio of West/Rest, as shown in Table 13.11. These are, of course, not all the nations in either group, but the most significant ones. So, we can make the assumption that the relative values of the Ratio would be approximately correct even if we had been able to include all the nations, and fit the value for the year 2016 to the dotted curve in Fig. 13.13, which is about 1.40. The result is shown in Fig. 13.14.

While Fig. 13.14 shows the Rest overtaking the West around 2026 as far as GDP is concerned, the projection looks quite different when we consider the populations of the two groups of nations, as shown in Table 13.12.

We can then determine the *per capita* GDP at market values in current $US, as shown in Table 13.13.

The ratio of the populations in the West and in the Rest shown in Table 13.12 can be approximated with a high degree of accuracy by the expression.

Table 13.9 GDP of the West and of the Rest, in billions of current US dollars

	1990	2000	2010	2020
West	18,246	26,104	41,122	48,936
Rest	4,454	7,712	25,366	35,811
West/Rest	4.10	3.38	1.62	1.37

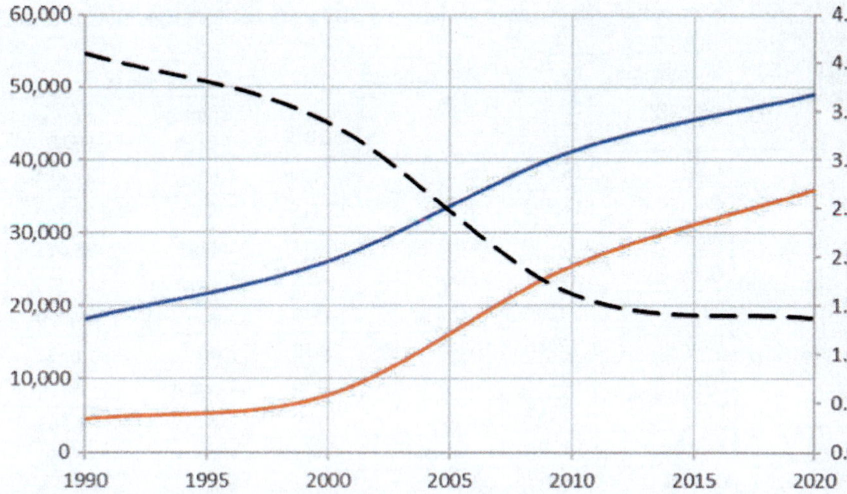

Fig. 13.13 GDP of the West (upper curve) and of the Rest (lower curve), in billions of current US dollars at market prices (left-hand scale). The dashed curve is the ratio of the two curves, West/Rest (right-hand scale)

$$\text{Population ratio} = 0.2012 - 0.001484 \cdot \text{x},$$

with x being the number of years from 1990 (although, of course, this linear approximation can only be valid for a limited extended time period). With this, and the GDP ratio shown in Fig. 13.14, we arrive at the values for the per capita GDP ratio shown in Table 13.14 and displayed in Fig. 13.15.

Gross Domestic Product is an indicator of both standard of living and of international power. Another indicator of power is military expenditure, and the values presented in Table 13.15 are based on what is probably the best openly available data.

The last global indicator refers to agricultural land, which the World Bank defines as follows:

> Agricultural land refers to the share of land area that is arable, under permanent crops, and under permanent pastures. Arable land includes land defined by the FAO as land under temporary crops (double-cropped areas are counted once), temporary meadows for mowing or for pasture, land under market or kitchen gardens, and land temporarily fallow. Land abandoned as a result of shifting cultivation is excluded. Land under permanent crops is land cultivated with crops that occupy the land for long periods and need not be replanted after each harvest, such as cocoa, coffee, and rubber. This category includes land under flowering shrubs, fruit trees, nut trees, and vines, but excludes land under trees grown for wood or timber. Permanent pasture is land used for five or more years for forage, including natural and cultivated crops.

The data provided by the World Bank is in % of land area, which is defined as a country's total area, excluding area under inland water bodies, national claims to continental shelf, and exclusive economic zones. In most cases the definition of

Table 13.10 Projected rankings of economies based on GDP at Market Exchange Rates (in constant 2016 $bn)

GDP MER rankings	2016 rankings		2030 rankings		2050 rankings	
	Country	GDP	Country	Projected GDP	Country	Projected GDP
1	US	18,562	China	26,499	China	48,853
2	China	11,392	US	23,475	US	34,102
3	Japan	4730	India	7841	India	28,021
4	Germany	3495	Japan	5468	Indonesia	7275
5	UK	2650	Germany	4347	Japan	6779
6	France	2488	UK	3530	Brazil	6532
7	India	2251	France	3186	Germany	6138
8	Italy	1852	Brazil	2969	Mexico	5563
9	Brazil	1770	Indonesia	2449	UK	5369
10	Canada	1532	Italy	2278	Russia	5127
11	South Korea	1404	South Korea	2278	France	4705
12	Russia	1268	Mexico	2143	Turkey	4087
13	Australia	1257	Russia	2111	South Korea	3539
14	Spain	1252	Canada	2030	Saudi Arabia	3495
15	Mexico	1064	Spain	1863	Nigeria	3282
16	Indonesia	941	Australia	1716	Italy	3115
17	Turkey	830	Turkey	1705	Canada	3100
18	Netherlands	770	Saudi Arabia	1407	Egypt	2990
19	Saudi Arabia	638	Poland	1015	Pakistan	2831
20	Argentina	542	Netherlands	1007	Spain	2732
21	Poland	467	Iran	1005	Iran	2586
22	Nigeria	415	Argentina	967	Australia	2564
23	Iran	412	Egypt	908	Philippines	2536
24	Thailand	391	Nigeria	875	Vietnam	2280
25	Egypt	340	Philippines	871	Bangladesh	2263
26	Phillipines	312	Thailand	823	Poland	2103
27	Malaysia	303	Pakistan	776	Argentina	2103
28	Pakistan	284	Malaysia	744	Malaysia	2054
29	South Africa	280	Bangladesh	668	Thailand	1995
30	Colombia	274	Vietnam	624	South Africa	1939
31	Bangladesh	227	Colombia	586	Colombia	1591
32	Vietnam	200	South Africa	557	Netherlands	1496

From the 2017 PwC report The Long View: How will the global economic order change by 2050?

Table 13.11 GDP values extracted from Table 13.10 by summing for the West and for the Rest nations included in Table 13.10

	2016	2030	2050
West	39,693	51,322	75,698
Rest	24,900	57,399	137,447
Ratio	1.59	0.89	0.55

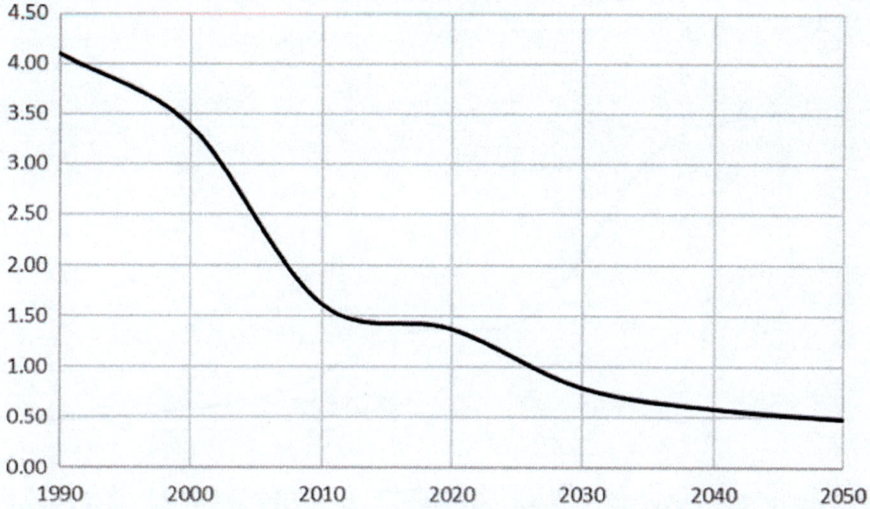

Fig. 13.14 The ratio of GDP at market values at current $US of the West and the Rest, extrapolated by using the data in Table 13.10

Table 13.12 Population of the West and of the Rest, in millions

	1990	2000	2010	2020
West	892.2	950.3	1,003.9	1,062.8
Rest	4,387.8	5,164.0	5,918.0	6,698.8
West/Rest	0.203	0.184	0.170	0.159

Table 13.13 *Per capita* GDP in current US dollars, and the resulting ratio

	1990	2000	2010	2020
West	20,451	27,469	40,961	46,044
Rest	1015	1493	4286	5346
West/Rest	20.15	18.39	9.56	8.61

inland water bodies includes major rivers and lakes. Combining this data with data on land area and population, we obtain the *per capita* values shown in Table 13.16.

Table 13.14 The conversion of the GDP ratio into a *per capita* GDP ratio by dividing by the population ratio

	1990	2000	2010	2020	2030	2040	2050
GDP ratio	4.10	3.38	1.62	1.37	0.79	0.58	0.48
Population ratio	0.203	0.184	0.170	0.159	0.142	0.127	0.112
Per capita GDP ratio	20.20	18.37	9.53	8.62	5.56	4.57	4.29

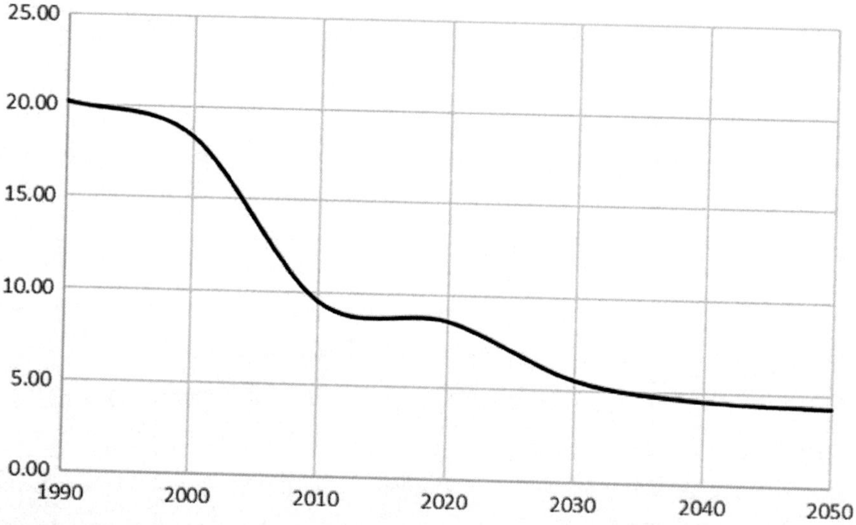

Fig. 13.15 The ratio of the *per capita* GDP in the West and in the Rest, extrapolated from previous values using the data in Table 13.10

Table 13.15 Military expenditure in current US dollars and exchange rates, based on data from SIPRI (Stockholm International Peace Research Institute, available at milex.sipri.org/sipri, accessed 20.06.2022)

	1990	2000	2010	2020
West	20,451	27,469	40,961	46,044
Rest	1015	1493	4286	5346
West/Rest	20.15	18.39	9.56	8.61

Table 13.16 Agricultural land, in hectares (10,000 m^2) *per capita*, for the West and the Rest, and the ratio of West/Rest

	1990	2000	2010	2020
West	1.44	1.37	1.19	1.10
Rest	0.86	0.68	0.60	0.54
West/Rest	1.68	2.00	1.97	2.04

13.4.3.3 Discussion of the Global Indicators

There are a couple of things we can immediately note about the economic indicators provided in the previous section. The first is the simultaneous strong increase, in Fig. 13.13, in the GDP of the West and the Rest in the period 2000–2010. It was the result of an aggressive outsourcing and globalisation policy in the West, providing a great amount of lower-skilled work to the Rest while shifting the West's production into higher-value products. It is also an example of the saying "A rising tide lifts all boats", but while the lift is necessarily the same for each boat, this is by no means the case when the saying is meant to refer to other entities, as we shall see in the next two sections.

The second thing to note is the gradual reduction in the relative dominance of the West, and with it in the tension arising from it. Figure 13.14 indicates that the GDP of the Rest will exceed that of the West by about 2026, but Fig. 13.15 Shows that while there has been a significant reduction in the West/Rest ratio of *per capita* GDP, which we may consider to be a measure of the Quality of Life (QoL), over the sixty years since 1990, the rate of reduction is now slowing down, and this raises the question of whether this rate of reduction is enough to avoid significant confrontations between the two parts. It becomes a race between the increasing recognition of the differences and the resulting demands for equality and the finding of means to accommodate the tension in a peaceful manner, in the sense illustrated in Fig. 2.2.

And this brings us to an important difference between the West and the Rest that is not directly addressed by the economic indicators, but which certainly plays a role in the rate of change of the relationship between the two parts, and that is what is most accurately identified as *ideology*. Section 11.4 described how the West evolved over the last five hundred years to become a distinct part of the world society in culture, religion, and political and economic thought; and developing a view of itself as the forerunner in the ultimate transformation of society, of which modernity is the most recent phase. Growing up in the West, we become so inured in this ideology that anything else is considered backwardness, and our economic superiority is the main proof of this. This nexus between ideology and economic hegemony has two effects: slowing down the rate of the equalisation of the QoL, and introducing an ideological component, in addition to the economic one, into the tension between the two parts.

The military expenditure values in Table 13.15 show the dominant position of the West. This is even more so than these values show, in that the majority of the nations in the West are integrated in a single military alliance—NATO—as listed in Sect. 13.2.4.4, whereas there is no comparable alliance within the Rest; not even between its main members. We might also note that, with the monetary values in $US, the lower cost-basis in the Rest allows a relatively more numerous military establishment, whereas the greater expenditure in the West is an indicator of a higher level of weapons technology.

The dominant military power of the West can be used to enforce the unequal status quo and avoid instability, as has been repeatedly the case, but it also acts to maintain a simmering resentment and tension. The existence of tension in the world society, both local and global, and the potentially disastrous effects should it rise above a certain level, have been the concern of many people and organisations, but the only tangible result so far has been the UN, as described in Sect. 13.2.4.2. Any realisation of a global police force is relegated to a very distant future; the importance of national military forces and the principle of "might is right" will be with us for a long time yet.

Agricultural land area is included as an indicator because food production is a critical issue in many nations, with some being totally dependent on imports. The West seems to be clearly better off than the Rest, but the values must be treated with caution, as many other variables than land area enter into the productivity of the land; above all water. Water is a scarce resource in many areas, and its variability, in the form of floods and prolonged draughts, presents a further problem. The access to water is already a contentious issue in some places and is likely to become an increasing source of localised tensions.

The two-part view of society presented above is highly simplified in that the parts are presented as homogeneous. Most of the issues and processes that drive the evolution of society can only be identified and understood once we recognise that the nations making up these two parts are very different, as are the relationships and interactions between them. Or, in other words, that the two parts are highly structured, and it is this structure we address in the next two sections.

13.4.3.4 Detailed Indicators

At this level of detail, the two parts are populated by nations, and we look for characteristics of these populations that can be indicators of more detailed properties of the two parts. The nations are treated as homogeneous entities, i.e., without any internal structure, and defined by their external parameters only.

The first indicator of the structured nature of the two parts is a purely statistical one, in that we look at the distribution of one of the main characteristics, the *per capita* GDP as an indicator of the QoL. The mean, standard deviation, and relative variation (standard deviation divided by mean) of the two distributions are presented in Table 13.17.

Table 13.17 The distribution of the populations of the two parts in terms of the mean, standard deviation, and relative variation (standard deviation divided by mean) of their nations' *per capita* GDP at market prices in current $US

		1990	2000	2010	2020
West	Mean	3207	3914	6127	6639
	Std deviation	10,051	10,882	15,446	16,447
	Relative variation	3.13	2.78	2.52	2.48
Rest	Mean	2602	3781	7509	7895
	Std deviation	6904	7850	11,959	14,848
	Relative variation	2.65	2.08	1.59	1.88

Table 13.18 The dominance of the US and of China within their part (West and Rest) of the world society, as measured by the ratio of GDP at market values in current $US

	1990	2000	2010	2020
US	0.485	0.647	0.574	0.749
China	0.355	0.811	1.417	2.754

Table 13.19 The dominance of the US and of China within their part (West and Rest) of the world society, as measured by the ratio of military expenditure in current $US

	1990	2000	2010	2020
US	1.279	1.261	1.669	1.584
China	0.029	0.152	0.291	0.615

Table 13.20 The dominance of the US and of China within their part (West and Rest) of the world society, as measured by the ratio of soft power, represented by the *per capita* GDP at market prices in $US

	1990	2000	2010	2020
US	1.166	1.323	1.196	1.381
China	0.313	0.642	1.060	1.952

A more significant structural feature is that both distributions are dominated by a single nation—the US in that of the West and China in that of the Rest. Dominance is measured primarily in terms of the economy (GDP) and military strength (measured, as above, by expenditure), but also by soft power. The latter is more difficult to define in a measurable way, but a central component of any composite measure would be the per capita GDP measured in $US at market prices, as this projects a desirable lifestyle. In the following Tables 13.18, 13.19 and 13.20 these measures are presented, based on the same World Bank and SIPRI data as before.

13.4.3.5 Discussion of the Detailed Indicators

From the data in Table 13.17 we see that neither of the two parts (West and Rest) constitutes a homogeneous collection of nations: the QoL shows a very considerable relative variation in both cases. While this relative variation is no greater than what one can find within a nation, there is a significant difference. In a nation there is a common infrastructure that represents a shared investment, that provides benefits to everyone (not necessarily equally), and that promotes a sense of belonging; all of which results in a certain tolerance of inequality. In the world society there is no such common infrastructure, and each nation conducts its relations to other nations based almost exclusively based on the benefit to itself; benevolence and altruism are not significant factors in international relations.

It is useful to consider dominance within the two parts to be a combination of two types. One is when the dominant nation is much more powerful than any of the other nations within its part, in which case the dominance is characterised by the tension between the dominant nation and the other nations. A benevolent and sophisticated hegemon will use its power sparingly and, above all, to adjudicate between the other members of the part, so that they perceive the hegemony as providing not only subjugation but a significant benefit through this harmonisation, and the tension is low. A hegemon focussed only on its own immediate benefit and applying its power indiscriminately will result in great tension. The other type is when the hard power of the hegemon is not much greater than that of the other members of the part, and the hegemony is based on soft power arising from the cultural superiority of the hegemon; its society is an ideal for others to aspire to. What is important here is the *perception* of superiority, which arises as a combination of actual superiority—the hegemon's society does contain many desirable features—and of the *projection* of an image of a superior society, a process that is best characterised as marketing on a global scale of items that include both products (high-tech, luxury, abundance) and beliefs (political system, importance of personal freedom, solidarity, role of government),

As we have noted earlier, the evolution of society over the next thirty years will be shaped by the competition between the US and China for global dominance, albeit that dominance will take a different form in the two cases. First of all, we need to recognise that the rise to hegemony has been and, in the case of China is, very different. In the case of the US, it arose first out of the colonial empire of the West and then out of the destructive infighting among the other members of the West. As a result, the hegemony was, until recently, one of great power, both economic and military, and also of a desirable culture and lifestyle. This was described briefly in Sect. 13.1.3, and Fig. 13.1 illustrated how the hegemony extended beyond the West to the whole world.

The economic dominance (see Table 13.18) has stagnated, and the cultural component is in decline (political and racial upheaval, gun violence, declining life expectancy). Also, what might be termed the ideology of the West—a narrative starting with the ancient Greeks, progressing via the Romans, the Renaissance, the Enlightenment, and finding its culmination in the US and the triumph of capitalism—has been a very strong unifying force, but even that is starting to crumble under more detached scrutiny and greater knowledge of other ancient cultures. So, going forward the emphasis will increasingly be on maintaining and, if possible, extending their military dominance. A recent action in this regard was the AUKUS alliance and the related plans for turning northern Australia into a major US military base. And currently the dominance is being given a boost through the incredibly stupid and deplorable behaviour of Russia in its attack on Ukraine. The Russian bogeyman became reality, NATO members closed ranks under US leadership, Europe demonstrated its helplessness, other nations (Finland, Sweden) are considering joining NATO, and the US defence industry got a shot in the arm. (And, of course, the defence industry is an excellent means of transferring wealth from the less wealthy (taxpayers) to the wealthier (shareholders).) All without the loss of a single US soldier or any damage to US property—Sun Tzu would have applauded.

A side benefit of the Russian action is its demonstration of a successful strategy and the opportunity it provides for casting aspersions on China; we can now expect to see the same prodding of China, for example, in the form of increasing recognition of a sovereign Taiwan, to see if President Xi will "spit the dummy". This was discussed in an article in *Pearls & Irritations* on 01.12.2021, with the title *Strategists admit West is goading China into war*, by Nury Vittachi and Phill Hynes, (https://johnmenadue.com/strategists-admit-west-is-goading-china-into-war/, accessed 20.06.2022), where they make reference to the just published book *The Strategy of Denial*, by Colby (2021). It contains the complete playbook for a repeat of Ukraine on Taiwan.

In the case of China, the rise to dominance is very recent and mainly economic in nature, as shown in Tables 13.18 and 13.19. The rise in military power has been directed almost exclusively at strengthening its borders and access lanes, in particular maritime access towards the East, in consideration of the West's previous incursions as well as of China's dependence on external inputs of raw materials, from agricultural products to iron ore. Going forward, China would no doubt like to increase its soft power, but in making Chinese society a desirable model for other countries it faces the problems associated with being off-set in history by a couple of centuries as compared with the West. Forging a coherent and unified society under the scrutiny of today's humanitarian rules is very different to, for example, forging the US under the view on the treatment of the native American Indians prevalent at that time. China's economic and military strength will continue to increase, but if China will become a hegemon, in the all-encompassing meaning of the word, within the Rest, or ultimately in the world, is impossible to predict. Firstly, there is no narrative, such as the ideology of the West, *within* which China can take a leading role. Of course, China will try to develop such an ideology, a "global great power symbiosis theory", as its as is described, e.g., in Yang (2021), but this is a daunting task. It will take great restraint and persistence by China, it will require a considerable time to diffuse into the Rest, supported by successful applications, and it will be opposed at every turn by the West.

Secondly, the unifying effect of complicity in despoiling a great part of the world will not be available, although it may be possible to create a narrative of solidarity in the rise of the Rest in relation to the West as a component of China's "great power diplomacy" based on the principle of "striving to shine", as detailed in Yan (2021). Thirdly, other nations within the Rest will also want to rise and shine as individual nations, providing a challenge to Chinese hegemony. It may well be that, in the long term, the world society will be multipolar, with a range of "hegemonies".

And finally, the US has a track record of providing opportunity and a better life to many generations of people from all over the world. It has not been just theory and promises, as e.g., in the Soviet Union; it has been a very visible performance. China will not be able to provide this; not only does it already have too many people, but it is endowed with an ancient history as an inward-looking civilisation and an associated uncritical evaluation of its status in relation to the rest of the world, resulting in a degree of arrogance as well as a lack of experience in foreign affairs. It will be admired from afar for its achievements, but it will not be saying "Give me your tired, your poor, Your huddled masses yearning to breathe free, The wretched refuse of your teeming shore".

Thus, within this view of the world as composed of two parts, the West and the Rest, and with the two dominant players, the US and China, the medium-term development might be as follows: The West will maintain its dominant position and use it to slow down China's attempt to develop a similarly coherent core of nations under its influence. The US will also use every opportunity to criticise China and to, where possible, impose sanctions, thus heightening tensions. China itself will continue to evolve in economic and military strength, and in its internal structure so as to meet the demands of an increasingly educated and assertive population. It will also continue to strengthen commercial ties within the Rest, and its existing ties with the West, excepting the US, will probably not diminish significantly. The attempts by the US to goad China into a devastating war *a la* Ukraine are unlikely to succeed; despite a clumsy performance in international relations, the Chinese leadership is much smarter than Putin.

A propos the war in Ukraine, Russia could become an unwelcome political headache for China. On the one hand, their shared revolutionary heritage makes them appear to the Rest as the valiant defenders of the downtrodden masses against the capitalist West, and a very real connection is their extensive border in the East and associated trade and development activities in this region. On the other hand, there is the erratic behaviour of Putin, which it is then easy for the West to reflect onto China. And the poor performance of Russia in looking after its own people is not something China wants reflected onto socialism in general.

13.4.4 Intelligence and Transformation

13.4.4.1 Information and the Two-Part Society

This second approach to exploring the likely evolution of society in the short term is based on the view of society as an information-processing system developed throughout this book. We start by reflecting this view onto the high-level structure of society as consisting of two parts—the West and the Rest—and consider these to form a society of two individuals, each characterised by its identity. In this two-member society, the interaction between the parts, in the form of an information exchange, is also the circulating information. As before, we are focussing on the class 3 information, and the processing of this information—the collective intelligence—is performed by the two parts, and the result of this processing is changes to the identities of the two parts. However, the original assumption about the individuals of the society being identical is, of course, not applicable; it is the difference between them that gives rise to a structure of the society and an associated tension, as discussed in Sect. 11.2.1. And we can apply the model and results of Sect. 11.2.2.1 and, in particular, Eq. 11.7, which shows that the tension between the two parts increases with the strength of the interaction between them and decreases with how much they have in common, as measured by the set of identity elements common to both parts.

The identity of the West can be simplified to be represented by the single ideology built up over the last 500 years, as described in Sect. 11.4. The West believes that its rise to a position of superiority and hegemony is due to its culture and ideology; conversely, its current dominant position is seen as a vindication of its ideology. This mutual reinforcement means that the West is not interested in a proper interchange of information and processing of the received information with an open mind; the West's efforts will endeavour, on the one hand, to prevent the "infection" of its ideology by dissenting principles and concepts entering from the Rest and, on the other hand, to present the ideology in a form that will entice members of the Rest to convert and join the West. And until recently this effort was quite successful, both because the West was able to dominate the interaction and because the Rest does not have a single ideology accepted by all, or at least the majority, of its member nations. However, this is clearly a highly non-equilibrium situation and one where the operating conditions of the collective intelligence are far from ideal. It is an approach that is bent on frustrating a "public discourse" on the international level, and the result is, predictably, a rising tension between the two parts.

With the rise of China as the major power within the Rest, the longstanding tension between the West and the Rest, most visible in the envy and resentment of the Rest as they become aware of the difference in living standards and of how they are being exploited by the West, is being crystallised into a rivalry between the US and China. This rivalry will take on quite different forms depending on China's intentions and on the US response, as indicated in Table 13.21.

China's intention may be to simply progress its economic rise both internally and internationally and obtain a position in the world commensurate with its size, and with the US as one of its major competitors. This is not a multipolar society; there is no polarisation through alliance building, it is a society of large and small nations with a web of bilateral trade and other agreements (e.g., the extradition of criminals). As within a nation, larger economies will have more weight and influence than small nations, but there is no coercion. The US could accept this without any loss of sovereignty, ability to compete, or Quality of Life, just as e.g., Switzerland and Sweden are presently doing very well. But the US might also find this loss of its hegemonial status unacceptable and resist it through imposition of sanctions and the

Table 13.21 A simplified picture of the future relationship between China and the US

China's intention	US response
(1) Economic rise to first among equals	(a) Accept as economic competitor
	(b) Deny by military power
(2) Hegemonial rise to local domination	(a) Accept, but maintain hegemony in the West
	(b) Deny by military power

like, backed by military power, much as was the case with the rise of Japan prior to Pearl Harbour.

However, China's intention could be to be to rise as a hegemon in its region, with the same ability to deny others access as the US has in the West. This would entail projecting its military power into the region and forcing the nations there into an alliance under Chinese leadership. The US might accept this, resulting in a bipolar, and eventually perhaps multipolar world order, but it might, as above, find this unacceptable and oppose it.

13.4.4.2 The Case of Australia

Turning to Australia as a prominent member of the emerging anti-China alliance led by the US, and the county with which I have first-hand experience, the assessment of China's intentions is of the greatest importance. If it is assessed to be according to option 1 in Table 13.21, Australia's policy should be to take an independent stand and act to dilute the tension between the US and a rising China, encouraging the US to adopt option a. Australia's defence policy should focus on the defence of Australia rather than on any capability to threaten other nations, and Australia's diplomatic efforts should be directed towards the nations in our region, where Australia would have the opportunity to have a significant impact based on its successful multicultural society.

If, however, Australia's assessment is that China's intention is to become the regional hegemon and force nations in the region into a submissive relationship, Australia must assess the likely longer-term response of the US. Will it persevere in denying China regional hegemony, which could come at a considerable cost, or will it in the end settle for option a? In the former case Australia would have to pursue its current policy of strengthening the Western anti-hegemonial alliance; in the latter case Australia should start to extricate itself from that alliance in order to minimise the probability of being used as a sacrifice in a proxy war between China and the US and end up in a position vis-à-vis China considerably worse than just hegemonial submission.

This brings us back to the subject of this section—the role of information. Assessment is based on information and, given the importance of this assessment for the future of Australia, the assessment process should be an application of the collective intelligence and its public discourse. Of course, different members of society will have a different level of knowledge and expertise, but it is essential that this information is shared and subjected to public scrutiny and evaluation. In the assessment considered here there are many different viewpoints, interests, and beliefs involved and, in particular, given the dominant role of the US, it involves the special interests of the US military-industrial complex. The military plays a vital role in US society, both in its history and in its present-day self-image, and the industry associated with defence and armaments is both a significant component of US industry and a main drive of technological research and development. The *raison d'être* of the military-industrial complex is armed conflict, or at least the constant threat of it, and so the establishment of favourable conditions for this complex is significant input to the assessments. As a little example, the share price of Halliburton, the company of which Dick Cheney was CEO before becoming Vice-President under George W. Bush in 2000, rose from $7 at the start of the Iraq war in 2003 to $36 at the onset of the GFC in 2008.

The situation in Australia, and possibly in many of the nations within the West, is that there is practically no public discourse regarding foreign policy. Both major parties—Liberal-National and Labor—are committed to supporting and being subservient to US foreign policy and to offer support for its international actions, even to the extent of participating in a war crime like the Iraq war. And as these two parties and their supporters effectively control the mass media—newspapers, radio, and TV—there is little opportunity for any divergent public opinion to develop. Of course, there are numerous sources of independent, informed, and thoughtful opinion, both within Australia and internationally, including in the US, but the readership in Australia is too small and scattered to have any significant political weight. How many people have read Noam Chomsky's political work or heard his lectures on the subject of US international relations? How many read John Menadue's public policy journal, *Pearls & Irritations*? Most people get their information from the mass media, and whenever they meet and exchange information, it simply becomes a reinforcement of this common belief rather than a dialectical development of further understanding.

Although the dissenting voices in Australia are a small minority, they could still be enough to act as a seed for a re-evaluation of our current national identity were it not for the fact that the mass media are heavily influenced by the US and that Australian foreign policy is in effect an extension of US foreign policy. The view of the world presented in the Australian mass media is a view that supports US interests, and the effort that goes into maintaining this view completely swamps any dissenting effort. An example of this that has had some limited exposure is the influence of US defence companies in determining the nature of the redeveloped Australian War Memorial and using it as a platform for lobbying for their commercial interests (Time to reclaim remembrance, by Sue Wareham, in *Pearls & Irritations*, https://johnmenadue.com/time-to-reclaim-remembrance/, accessed 20.06.2022). In discussing this situation, it is important to distinguish two aspects of the relationship between the US and Australia (and other states within the West). One is where the people of Australia have a genuine admiration for the US—for its culture, lifestyle, and achievements in overcoming the hated regimes of Nazi Germany under Hitler and of Imperial Japan under Emperor Hirohito, and as a result wish to be close to the US. The other aspect is that the US coerces Australia into a submissive relationship by manufacturing an elaborate alternative reality, one that poses dire threats to Australia by entities including communism, Russia, and now China, and to which Australia's only salvation is integration into the US military structure. This is becoming increasingly evident in the tenor of the description of our relationship with China, with wordings such as "threat", "war", and "aggression", and ascribing malicious intent to most of China's external actions, even though there is no evidence to support this view. On the contrary, it is the West that has engaged in the overwhelming majority of military actions against other nations since WW2. That we accept this view is a tragedy for Australia, as it prevents us from playing a vital non-aligned and stabilising role in our part of the world. Instead, we are contributing to a dangerous polarisation, and by painting China as an aggressive adversary we are increasing the probability of this coming true. And the adversarial stance, driven by the military-industrial complex, is a tragedy for the US also; this vibrant, innovative, and dynamic society would have nothing to fear from China, and its leadership role would be enhanced beyond the negative attitude of defence.

13.4.4.3 The Importance of Common Infrastructure

The concept of a common infrastructure, first introduced in Sect. 1.2, can be seen as the reflection, in the physical domain, of society's belief system in the information domain. It consists of all those institutions and organisations that transforms a collection of individuals into a society, and it provides services that the individuals have agreed should be a public good. That is, the individuals have given up their right and ability to provide these services, which can be seen as a loss of individual freedom. And the extent and autonomy of the common infrastructure is a central measure of the strength and cohesion of the society. If the individuals are hesitant in giving up their rights to provide such services, as is, for example, the case with education in Australia, this diminishes the cohesion of the society.

The common infrastructure we are most familiar with is the infrastructure of the nation in which we reside. We experience its workings every day through the various services it provides, from education to public transport. It is normally structured into two or more levels, such as federal, state, and local community, but at any level the common infrastructure consists of certain components: A political system that provides a concrete representation of the society's belief system in the form of legislation—expressing what society ought to be. A government that converts the legislation into a structure of organisations that provide the services envisaged by the legislation, and the means of managing these. And, finally, the organisations themselves, concerned with providing the required services as cost-effectively as possible.

Turning now to the world society with nations as the individuals as the next stage in the evolution of society, there is no reason to expect that it will not require a common infrastructure with the same or similar components, and it is clear that the road to reaching this goal is both long and full of obstacles and dangers. Reaching a consensus on the principles and beliefs that would define such a society—its belief system—will require a differentiated approach, one that takes into account the different economic, social, and cultural makeup of each nation. And agreeing to a realisation of this belief in the form of a common infrastructure might take a different path and require a different timeframe in each nation. But before any of this can take place, there needs to be a consensus that we want such a world society of nations and a consensus about the extent of the common infrastructure. What defines a society can vary greatly, from a single function, such as defence, to the multitude of functions in a developed nation, but the defining criterion is that the individuals surrender the corresponding function from their sovereignty.

As far as the world society is concerned, there is currently really no infrastructure that deserves the name of *common* in the sense that it is backed by an irreversible commitment of all nations, resulting in a corresponding mandate to act. As we indicated in Sect. 12.2.4.2, the United Nations performs much valuable work, but it is at the pleasure of its member nations, as we have seen so often and most recently in the international response to the COVID-19 pandemic. The principle of national sovereignty is made explicit in clause 7 of Article 2 of the UN Charter: "Nothing contained in the present Charter shall authorize the United Nations to intervene in matters which are essentially within the domestic jurisdiction of any state or shall require the Members to submit such matters to settlement under the present Charter". However, the member states might agree to make an exception in certain cases where a common infrastructure can be demonstrated to be essential in meeting an existential threat, and such a stepwise approach to creating a world society could be initiated by the following two cases.

Pandemics. There is at present no organisation that can be identified as a component of a common infrastructure for the world society. The closest might have been the World Health Organisation (WHO), but the current COVID-19 pandemic has demonstrated the weakness of WHO's advisory role, its susceptibility to political interference and manipulation, as well as its dependence on contributions that are all, in the end, voluntary. However, with the motivation for establishing a common infrastructure being that the benefits must outweigh the perceived disadvantage of the loss of sovereignty, and with not only the benefits, but the pressing necessity for world-wide action demonstrated by the COVID-19 pandemic, separating out the part of WHO concerned with communicable diseases with the potential to turn into pandemics might be the best candidate for expansion into a common infrastructure component. The greater part of WHO would be left unchanged, as would the health systems of the individual nations, but those functions and facilities of the national health systems specifically targeting communicable diseases would be transferred to the new supranational organisation. The UN could develop a constitution for common infrastructure, and once it was ratified by the General Assembly (UNGA), its first component—the Communicable Diseases Organisation (CDO)—could be defined and created, consisting initially almost completely of existing entities. The UNGA would act as the representative of the population of the world society, much as a Parliament, Assembly, or Congress does for a nation, and it should be managed along the lines outlined in Sect. 12.3. The additional cost would be minimal; the two major changes would be that the contributions of each nation would not be voluntary, but in the form of a supranational tax, and that, in addition to providing advice and information to the national bodies, the CDO would be able to issue instructions that nations would be required to follow and for the execution of which they would answer only to the UNGA. This would include inoculation regimes and the production and distribution of vaccines and protective material.

There would be many obstacles to overcome in realising this proposal. The first would be the West's opposition to relinquishing its privileged position and not be able to prioritise itself in the distribution of vaccines. The only way to surmount this issue would be to ensure that there was a standing capability to produce vaccine and other medical supplies required to handle an outbreak of a pandemic distributed throughout various parts of the world, so that there would be no need to shift supplies from the West in order to meet global demand. This distribution of manufacturing capability is already partly in place; the main "black spot" is Africa. This change would no doubt be opposed by the major pharmaceutical companies who would lose a substantial part of what is now a profitable manufacturing business, but then, pandemics were not supposed to be a source of profit?

A much more significant problem would arise with regard to patents and intellectual property rights generally, as is demonstrated by the opposition of "big pharma" to any such changes. The most reasonable approach would be to split out research and development into communicable diseases from the pharmaceutical companies and structure them into a centrally administered distributed research enterprise as part of the CDO. A considerable part of this research is currently publicly funded anyway, and one would expect an increase in efficiency due to the elimination of duplication and secrecy, so that the financing of this step should not be an insurmountable problem.

A further issue is that many nations see fast spreading and lethal diseases as potential weapons, either to be used as such or, mostly, to be defended against. The defence aspect would be covered by the CDO if its research component is adequately funded; it would constantly search for new pathogens and mutations of known ones and develop means of combatting them. This activity would at the same time make the offensive aspect more difficult and costly, as it would mean having to "outsmart" the CDO, and as such would be a reason to object to establishing the CDO for any nation contemplating the use of biological weapons (although this reason would, of course, not be stated publicly). One would have to hope that this issue would not cause enough resistance to derail the establishment of the CDO.

Global warming. A second area for cooperation on a supranational scale and the formation of an element of a common infrastructure is that of the environment—in particular, global warming, as it is central to all the components of the environment—climate change and associated natural disasters, extinction of species, loss of habitable areas and associated flow of immigrants, and more. Forming such an element would be a more challenging step, as it would be more intrusive into the economies of individual nations, and the actions required and agreed upon, such as the phasing out of coal-fired power stations, would have greatly differing impacts on different nations.

The influence of human activity on our environment is a very complex subject matter, involving almost all aspects of our existence, and while there is general agreement that our current exploitation and modification of the environment cannot continue, the complexity makes it difficult to obtain clear, simple, and convincing prescriptions for change. A way forward is to try to divide the subject matter into segments whose mutual interaction is weak enough to allow them to, as a first approximation, to be handled as independent issues. One such issue is the global warming due to the increase of CO_2 in the atmosphere. The theory is well developed, future

Table 13.22 Electricity production in TWh by type of fuel (https://ourworldindata.org/grapher/electricity-production-by-source?country=~OWID_WRL, accessed 20.06.2022)

	2010	2015	2020
Coal	8347	9135	9215
Gas	4688	5380	6017
Hydro	3409	3870	4312
Nuclear	2687	2502	2636
Wind	346	829	1587
Solar	31	254	835
Oil	971	1027	793
Other renewables	409	568	713

projections can be made with reasonable accuracy, and the experimental verification, in the form of the correlation between temperature rise and CO_2 level is also convincing. But even this limited issue becomes complex when we ask what determines the CO_2 level, what processes generate and absorb CO_2. Again, we can separate out one issue that is well defined and understood and that can be considered as an independent subject matter, and that is the generation of CO_2 through the generation of electricity by the combustion of fossil fuel—coal, gas, and oil—and among these, coal-fired power stations are the main culprits. Coal is also the most significant fuel, as shown in Table 13.22, which also shows that the use of coal for power generation is still increasing.

To shut down a coal-fired power station one must first provide a substitute source of power, e.g., photo-voltaic panels or wind turbines, as well as a means of transmitting and storing energy from this intermittent source, then shut down and decommission the power plant, and then do likewise for the coal mine. In many cases this can be considered as a stand-alone project, and the proposal is that the management and financing of such project be assigned to another small part of the common infrastructure, say, a Power Renewal Authority (PRA) within the UN system. At least initially, priority would be given to smaller power stations in poorer countries, and the Authority would have the capacity of starting four projects a year, or, with a typical project duration of five years, the capacity to handle twenty projects at one time. However, the projects would have much in common from the administrative side, so that it should be possible to achieve a high degree of rationalisation and automation, resulting in high efficiency.

The cost of these renewal projects would be borne by all UN members, with due consideration of the degree of culpability for the current increase in global temperature and of the current ability to pay, perhaps in a scheme similar to the one used in determining the contributions to the UN budget. And, as with the CDO and as part of the common infrastructure, the PRA would have supranational authority in all matters relating to these projects, including sanctions as a means of enforcement.

Again, there would be many objections raised to this proposal, primarily related to the reluctance to relinquish control over the amount of money and over how it is spent. As most of the money would come from the West, this proposal would entail a diminishing of the West's dominant position. The major international engineering and construction companies, as well as property developers, financing companies, and the like, would also object to this limitation on their ability to use their market power; it would be a small step in the direction outlined in Sect. 12.3. But, given the recognition of the urgency of reducing CO_2 emissions, it might be given some consideration?

13.4.4.4 A Return to Education

Throughout this book the two central themes have been the role of *information*, as the determinant of actions, and that the *essence of society*, as the most basic characteristic of the difference between two societies, is the relationship between individual and society. The overwhelming majority of the information exchange is concerned with our daily lives, and its processing results in those changes to society we might think of as surface changes. But some of the information, the class 3 information, is concerned with the relationship between individual and society, and its processing by the *collective intelligence* results in changes to the essence of society; that is, in the evolution of society.

The single most significant process involved in the exchange of information, in terms of extent and effectiveness, is the formal and structured process of transmitting information from one generation to the next we identify as *education* and, in particular, primary and secondary education, due to its general and society-building nature. We dedicated a whole chapter to it—Chap. 9—and it is fitting that it forms the final section of the book.

As argued in Sect. 13.2.1, a society is, in the physical domain, characterised by its common infrastructure. A step on the way to building a world society would be to include material in the school curriculum that explains the concept of a common infrastructure. It is something all children experience in the family, at school, in sports organisations: that for the interaction between the members of the group to function efficiently, the members have to give up some of their independence and autonomy. The relationships are governed by rules, and there must be an authority that can enforce the rules. It is not enough to know what is right, to agree on rules that express this knowledge, and to commit to observing those rules; that commitment needs to be realised in the form of a common infrastructure that supports and enforces the rules. It means that the members give up some of their individual rights, some of their freedom.

In the Western society, the concept of freedom tends to be promoted in a simplistic way, as the freedom of a stand-alone individual, as epitomised by the character of Howard Roark in "The Fountainhead" by Ayn Rand, and this is, to a certain extent, reflected in education's focus on the individual and on individual attainment. But while an isolated individual might be completely free to choose, it does not have much to choose from. It is being a member of society that offers up a greatly expanded set of opportunities to choose from, albeit within a framework of rules. The evolution of society can be seen as the transformation of freedom as an individual into freedom as a member of society. And while it might appear to be too abstract a concept to include in a school curriculum, its practical realisation is all around us, reflecting humanity's experience with society-building over the last 10,000 years. It should not be too difficult to find numerous examples, from family-sized societies to the world society, that could be woven into the curriculum, from primary school to high school.

There are several organisations that are concerned with global aspects of education, prominent among them OECD's Programme for International Student Assessment (PISA) and UNESCO's Development Program. The data provided by these two organisations are basically different. The PISA assessment measures 15-year-old students' performance in mathematics, science, and reading, and the individual results are processed into a national score for each of the three areas, as well as an overall score for each nation. Nothing is said in this assessment about the existence of other subjects, nor about the degree of focus on these three subjects in the curriculum. The UNESCO assessment measures the proportion of the population over 25 years of age that has attained an ascending sequence of educational levels according to the International Standard Classification of Education (ISCED), where level 1 is primary education, level 2 is lower secondary education, and level 3 is upper secondary education. The levels are generally correlated with the number of years of schooling, with the three levels together typically amount to 12 years of schooling, with less scrutiny of the curriculum and of student performance. However, UNESCO's involvement in education goes beyond this testing, and it is noteworthy that they acknowledge the importance of information, stating "Since wars begin in the minds of men and women, it is in the minds of men and women that peace must be built. Education transforms lives and is at the heart of UNESCO's mission to build peace, eradicate poverty and drive sustainable development."

The circular relationship between education and changes to society is well known, and under the economic and political pressures of the current state of the world the focus in the West has been, and is, overwhelmingly on whatever it takes to maintain Western dominance under US leadership based on a neoliberal ideology. This is reflected in the current emphasis on STEM subjects and on student-oriented learning, promoting individual success in a competitive environment, neglecting the development of an understanding of society and the forces that drive its evolution under the increasing level of interaction between its members, leading potentially to the situation depicted in Fig. 2.2.

The relationship between education and changes to society was the subject of a Forum of the Great Transition Initiative in May 2021 with the title *The Pedagogy of Transition: Educating for the Future*, and in the opening essay Stephen Sterling states the problem as follows:

> Our educational systems are implicated in the multiple crises before us, and without meaningful rethinking, they will remain maladaptive agents of business as usual, leading us into a dystopian future nobody wants. (https://greattransition.org/gti-forum/pedagogy-transition-sterling, accessed 20.06.2022)

And further on in the same essay, he echoes the sentiments expressed in our Chap. 9 as follows:

> In recent decades, the dominance of neoliberal thinking in economics, politics, and wider society has usurped previous conceptions and traditions of education as a public service for the public good. A narrowly instrumental view of education, modeled to serve the perceived demands of a globalizing economy and culture, now defines and shapes learning. This turn is reflected in an increasingly market-driven educational system maintained by a proliferating "global testing culture." The system fosters competition, homogenization, and standardization in both national and international spheres. These developments rest on a conviction that education should serve a growth-oriented economy, fallaciously equated with the social good. Over time, this neoliberal wave has subtly but powerfully displaced more educationally defensible practices informed by liberal, holistic, and humanistic philosophies regarding the nature and purpose of education.

The focus of this Forum is on sustainability, but throughout the contributions there are numerous mentions of the importance of social knowledge and understanding, and of education's role in providing this. For example, the contribution *How Does Culture Shape Education*, by I. Silova, H. Komatsu, and J. Rappleye, considers the issue of "self-construal", which is very close to our concept of social integration, Sect. 10.1.3, and advocates "a cultural paradigm shift that entails a much-needed reconfiguration of the cultural foundations of Western modern(ist) schooling".

Among thoughtful and informed people there is a feeling that the world is not heading in the right direction, and that the global problems we considered earlier are the symptoms of a more fundamental malaise. In this book I have tried to show that at the most basic level the problem is an inadequate, or in part completely absent, understanding of society; of what it is, of the mutual relationship between it and its members, and of the processes that drive its dynamics. Society is a framework for the support of the collective intelligence, in the same sense as the body supports the brain. The body and the brain are in a symbiotic relationship, as are society and individuals; the difference is that the body, with its capabilities, needs, and limitations has remained unchanged, whereas society has undergone great change, a change that is accelerating. As society changes, individuals have to adapt, but in doing so they change society further, which requires further adaptation, and so on; a dynamic process without end. Will we manage to stay on this merry-go-round, or will we let it destroy us? Are we progressing along the path illustrated in Fig. 2.2, with problems and dysfunction spiralling out of control? Hopefully not; hopefully our sense of what it takes for society to survive will be allowed to steer us away from disaster. This sense of what the individual's behaviour needs to be is based on inherited knowledge—the "sprouts" or "potentialities" we noted earlier—which is then further developed during the individual's formative years into the individual's identity, as described in Storey (2006), characterised by such concepts as compassion, equality, and solidarity. If we want to characterise the identity by a single variable, it would be *social integration*, as introduced in Sect. 10.1.3, and as with all human characteristics, its value is given by a distribution over the population, with the value decreasing from left to right. The situation is then that in the West, due to its development over the last 500 years, the dominant ideology, as described in Sect. 11.4 and reflected in both political and economic policies, is one that favours the right-hand side of the distribution. And because of the formative process there is a feed-back or reinforcing aspect to this influence of the ideology, resulting in a distribution similar to curve c in Fig. 5.15.

In the view of society and its evolution presented in this book the Western ideology is an anachronism due to very special circumstances and clinging to it is just going to make the adjustment that much more painful. The Wild West mentality and the vision of the conquistador or the rugged frontiersman relying on his own strength and his own gun is unsuited to a modern nation, as is so clearly demonstrated in the US. And the Westphalian vision of sovereign states doing their own thing, trading and competing mainly peacefully, but settling any serious differences with small wars conducted as a sport between gentlemen, is also well past its use-by date, as all the wars since WW1, including the current one in Ukraine, have demonstrated. It is not difficult to see in which direction society *should* develop, but the obstacles in the way of achieving it are numerous and well known. A central obstacle is understanding that any society involves exchanging some individual freedom for the benefits provided by the society, no matter what the size and form of the society are; it is valid from family to world society and from the US to Iran. It is another expression of the essence of society. What varies is whether the loss of freedom is by consensus or by force, how the benefits are distributed, and so on.

To promote this understanding education is the obvious tool, both because of the captive audience and because the setting is already a mini-society with certain restrictions on personal freedom. The curriculum modules must avoid any reference to or criticism of particular national realisations of society and focus on the relationships that exist in any society and on the consequences of such relationships for the individual, with the type of society ranging from family to the world society. The class setting also presents numerous opportunities to demonstrate interaction and cooperative behaviour; for example, in grading some exercises on a class-wide basis, providing an incentive for the better students to help the rest. Given UNESCO's already strong involvement in education worldwide, it could be given the mandate, and the funds, to develop and introduce such modules and monitor their execution and provide mentoring and other assistance where required.

Appendix—List of Nations

Country name	Population (1000)	Size (sq km)	Arable land (ha *per capita*)	GDP (*per capita* PPP int'l $)
Aruba	107	180	0.019	38,897
Afghanistan	38,928	652,860	0.207	2088
Angola	32,866	1,246,700	0.159	6538
Albania	2838	27,400	0.213	13,818
Andorra	77	470	0.011	
United Arab Emirates	9890	71,020	0.004	
Argentina	45,377	2,736,690	0.881	20,768
Armenia	2963	28,470	0.151	13,284
American Samoa	55	200	0.054	
Antigua and Barbuda	98	440	0.042	18,942
Australia	25,687	7,692,020	1.240	52,518
Austria	8917	82,520	0.150	55,097
Azerbaijan	10,110	82,654	0.211	14,452
Burundi	11,891	25,680	0.107	771
Belgium	11,556	30,280	0.074	51,968
Benin	12,123	112,760	0.244	3506
Burkina Faso	20,903	273,600	0.304	2279
Bangladesh	164,689	130,170	0.048	5083
Bulgaria	6927	108,560	0.495	24,367
Bahrain	1702	780	0.001	43,181
Bahamas, The	393	10,010	0.021	32,454

(continued)

E. W. Aslaksen, *The Evolution of Society*,
https://doi.org/10.1007/978-3-031-23054-7

(continued)

Country name	Population (1000)	Size (sq km)	Arable land (ha *per capita*)	GDP (*per capita* PPP int'l $)
Bosnia and Herzegovina	3281	51,200	0.310	15,612
Belarus	9399	202,980	0.602	20,200
Belize	398	22,810	0.235	6456
Bermuda	64	54	0.005	85,264
Bolivia	11,673	1,083,300	0.395	8367
Brazil	212,559	8,358,140	0.266	14,836
Barbados	287	430	0.024	13,577
Brunei Darussalam	437	5270	0.012	65,662
Bhutan	772	38,140	0.125	11,508
Botswana	2352	566,730	0.115	16,921
Central African Republic	4830	622,980	0.386	980
Canada	38,005	8,965,590	1.044	48,073
Switzerland	8637	39,516	0.047	71,352
Channel Islands	174	198	0.022	
Chile	19,116	743,532	0.065	25,068
China	1,402,112	9,424,703	0.086	17,312
Cote d'Ivoire	26,378	318,000	0.140	5458
Cameroon	26,546	472,710	0.246	3773
Congo, Dem. Rep	89,561	2,267,050	0.140	1131
Congo, Rep	5518	341,500	0.105	3639
Colombia	50,883	1,109,500	0.121	14,565
Comoros	870	1861	0.079	3313
Cabo Verde	556	4030	0.092	6377
Costa Rica	5094	51,060	0.051	21,032
Cuba	11,327	103,800	0.257	
Curacao	155	444		25,530
Cayman Islands	66	240	0.003	76,748
Cyprus	1207	9240	0.087	38,458
Czech Republic	10,699	77,200	0.234	41,737
Germany	83,241	349,380	0.141	53,694
Djibouti	988	23,180	0.002	5782
Dominica	72	750	0.084	10,434
Denmark	5831	40,000	0.413	60,398
Dominican Republic	10,848	48,310	0.083	17,937

(continued)

(continued)

Country name	Population (1000)	Size (sq km)	Arable land (ha *per capita*)	GDP (*per capita* PPP int'l $)
Algeria	43,851	2,381,741	0.178	11,268
Ecuador	17,643	248,360	0.057	10,896
Egypt, Arab Rep	102,334	995,450	0.030	12,608
Euro area	342,950	2,682,452	0.185	46,759
Eritrea	0	101,000	0.200	
Spain	47,352	499,603	0.254	38,335
Estonia	1331	43,470	0.520	38,395
Ethiopia	114,964	1,129,300	0.148	2423
Finland	5531	303,920	0.407	51,090
Fiji	896	18,270	0.187	11,601
France	67,392	547,557	0.270	46,227
Faroe Islands	49	1396	0.062	
Micronesia, Fed. Sts	115	700	0.018	
Gabon	2226	257,670	0.153	15,191
United Kingdom	67,215	241,930	0.091	44,916
Georgia	3714	69,490	0.083	14,863
Ghana	31,073	227,540	0.158	5596
Gibraltar	34	10		
Guinea	13,133	245,720	0.250	2817
Gambia, The	2417	10,120	0.193	2278
Guinea-Bissau	1968	28,120	0.160	1949
Equatorial Guinea	1403	28,050	0.092	17,942
Greece	10,716	128,900	0.199	28,464
Grenada	113	340	0.027	15,893
Greenland	56	410,450		
Guatemala	16,858	107,160	0.053	8854
Guam	169	540	0.006	
Guyana	787	196,850	0.539	19,706
High income	1,214,930	34,946,338	0.278	50,572
Hong Kong SAR, China	7482	1050	0.000	59,238
Honduras	9905	111,890	0.106	5421
Croatia	4047	56,590	0.197	28,504
Haiti	11,403	27,560	0.096	2925
Hungary	9750	91,260	0.442	33,084
Indonesia	273,524	1,877,519	0.098	12,073

(continued)

(continued)

Country name	Population (1000)	Size (sq km)	Arable land (ha *per capita*)	GDP (*per capita* PPP int'l $)
Isle of Man	85	570	0.275	
India	1,380,004	2,973,190	0.116	6454
Ireland	4995	68,890	0.093	93,612
Iran, Islamic Rep	83,993	1,628,760	0.180	13,116
Iraq	40,223	434,128	0.130	9764
Iceland	366	100,830	0.343	55,216
Israel	9217	21,640	0.043	41,855
Italy	59,554	297,730	0.111	41,840
Jamaica	2961	10,830	0.041	9222
Jordan	10,203	88,780	0.020	10,356
Japan	125,836	364,500	0.033	42,197
Kazakhstan	18,754	2,699,700	1.628	26,729
Kenya	53,771	569,140	0.113	4452
Kyrgyz Republic	6592	191,800	0.204	4965
Cambodia	16,719	176,520	0.241	4422
Kiribati	119	810	0.017	2418
St. Kitts and Nevis	53	260	0.095	24,537
Korea, Rep	51,781	97,520	0.027	43,124
Kuwait	4271	17,820	0.002	
Lao PDR	7276	230,800	0.220	8234
Lebanon	6825	10,230	0.019	12,289
Liberia	5058	96,320	0.104	1428
Libya	6871	1,759,540	0.258	10,847
St. Lucia	184	610	0.016	12,944
Liechtenstein	38	160	0.057	
Sri Lanka	21,919	61,864	0.063	13,225
Lesotho	2142	30,360	0.204	2405
Lithuania	2795	62,630	0.755	38,735
Luxembourg	632	2430	0.102	118,360
Latvia	1902	62,090	0.672	32,019
Macao SAR, China	649	33		57,807
St. Martin (French part)	39	50		
Morocco	36,911	446,300	0.208	7296
Monaco	39	2		
Moldova	2618	32,885	0.621	13,002
Madagascar	27,691	581,800	0.114	1593

(continued)

(continued)

Country name	Population (1000)	Size (sq km)	Arable land (ha *per capita*)	GDP (*per capita* PPP int'l $)
Maldives	541	300	0.008	13,766
Mexico	128,933	1,943,950	0.189	18,833
Marshall Islands	59	180	0.034	4200
North Macedonia	2083	25,220	0.201	16,927
Mali	20,251	1,220,190	0.336	2339
Malta	525	320	0.019	42,640
Myanmar	54,410	652,790	0.206	4794
Montenegro	622	13,450	0.015	20,567
Mongolia	3278	1,557,255	0.419	12,101
Northern Mariana Islands	58	460	0.018	
Mozambique	31,255	786,380	0.192	1297
Mauritania	4650	1,030,700	0.091	5257
Mauritius	1266	2030	0.059	20,539
Malawi	19,130	94,280	0.198	1568
Malaysia	32,366	328,550	0.026	27,887
Namibia	2541	823,290	0.327	9382
New Caledonia	272	18,280	0.022	
Niger	24,207	1,266,700	0.789	1263
Nigeria	206,140	910,770	0.174	5187
Nicaragua	6625	120,340	0.233	5570
Netherlands	17,441	33,670	0.059	59,229
Norway	5379	365,108	0.151	63,198
Nepal	29,137	143,350	0.075	4009
Nauru	11	20		14,100
New Zealand	5084	263,310	0.100	44,252
Oman	5107	309,500	0.016	28,448
Pakistan	220,892	770,880	0.144	4877
Panama	4315	74,177	0.135	26,776
Peru	32,972	1,280,000	0.107	11,879
Philippines	109,581	298,170	0.052	8390
Palau	18	460	0.017	18,315
Papua New Guinea	8947	452,860	0.035	4326
Poland	37,951	306,170	0.290	34,265
Puerto Rico	3194	8870	0.016	35,279
Korea, Dem. People's Rep	25,779	120,410	0.092	

(continued)

(continued)

Country name	Population (1000)	Size (sq km)	Arable land (ha *per capita*)	GDP (*per capita* PPP int'l $)
Portugal	10,306	91,606	0.089	34,496
Paraguay	7133	397,300	0.681	13,013
West Bank and Gaza	4803	6020		5690
French Polynesia	281	3520	0.009	
Qatar	2881	11,490	0.005	89,949
Romania	19,286	230,080	0.446	31,946
Russian Federation	144,104	16,376,870	0.842	28,213
Rwanda	12,952	24,670	0.094	2214
South Asia	1,856,882	4,770,754	0.114	6099
Saudi Arabia	34,814	2,149,690	0.102	46,762
Sudan	43,849	1,849,234	0.474	4244
Senegal	16,744	192,530	0.202	3481
Singapore	5686	709	0.000	98,526
Solomon Islands	687	27,990	0.031	2619
Sierra Leone	7977	72,180	0.207	1739
El Salvador	6486	20,720	0.098	8499
San Marino	34	60	0.059	63,420
Somalia	15,893	627,340	0.073	875
Serbia	6908	87,460	0.370	19,231
South Sudan	11,194	631,928		
Sao Tome and Principe	219	960	0.019	4274
Suriname	587	156,000	0.108	17,016
Slovak Republic	5459	48,080	0.247	31,832
Slovenia	2100	20,136	0.088	39,593
Sweden	10,353	407,310	0.251	54,563
Eswatini	1160	17,200	0.154	8854
Sint Maarten (Dutch part)	41	34		36,190
Seychelles	98	460	0.002	25,700
Syrian Arab Republic	17,501	183,630	0.275	
Turks and Caicos Islands	39	950	0.027	22,282
Chad	16,426	1,259,200	0.336	1603
Togo	8279	54,390	0.336	2224
Thailand	69,800	510,890	0.242	18,236

(continued)

(continued)

Country name	Population (1000)	Size (sq km)	Arable land (ha *per capita*)	GDP (*per capita* PPP int'l $)
Tajikistan	9538	138,790	0.077	3858
Turkmenistan	6031	469,930	0.332	16,195
Timor-Leste	1318	14,870	0.122	3356
Tonga	106	720	0.194	6647
Trinidad and Tobago	1399	5130	0.018	25,031
Tunisia	11,819	155,360	0.225	10,262
Turkey	84,339	769,630	0.240	28,119
Tuvalu	12	30		4653
Tanzania	59,734	885,800	0.240	2780
Uganda	45,741	200,520	0.161	2297
Ukraine	44,135	579,400	0.737	13,057
Uruguay	3474	175,020	0.573	22,795
United States	329,484	9,147,420	0.483	63,544
Uzbekistan	34,232	440,555	0.123	7378
St. Vincent and the Grenadines	111	390	0.018	12,770
Venezuela, RB	28,436	882,050	0.090	
British Virgin Islands	30	150	0.034	
Virgin Islands (U.S.)	106	350	0.009	
Vietnam	97,339	310,070	0.073	8651
Vanuatu	307	12,190	0.068	2915
Samoa	198	2830	0.166	6777
Kosovo	1775			11,368
Yemen, Rep	29,826	527,970	0.039	
South Africa	59,309	1,213,090	0.208	12,096
Zambia	18,384	743,390	0.219	3450
Zimbabwe	14,863	386,850	0.277	2895
World	7,752,841	1.30E + 08	0.184	17,109

References

Acheson, D. (1969). *Present at the creation—My years in the state department*. W.W. Norton & Company.

Adams, G. R., & Marshall, S. K. (1996). A developmental social psychology of identity. *Journal of Adolescence, 19*(4), 429–442.

Archick, K. (2017). *The European Union: Current challenges and future prospects*. Congressional Research Service, https://crsreports.congress.gov/product/pdf/R/R44249/15. Accessed on June 20, 2022.

Arendt, H. (1978). *The life of the mind*. Harcourt Brace Jovanovich.

Arendt, H. (1958). *The human condition*. Chicago University Press.

Aristotle, *Politics*, Jowett, B. (transl.). Available from https://ebooks.adelaide.edu.au/a/aristotle/a8po/book3.html.

Ashby, W. R. (1956). *An introduction to cybernetics*. Chapman & Hall.

Aslaksen, E. W. (1996). *The changing nature of engineering*. McGraw-Hill Australia.

Aslaksen, E. W. (2013a). *The system concept and its application to engineering*. Springer Verlag.

Aslaksen, E. W. (2013b). The engineering paradigm. *International Journal of Engineering Studies, 5*(2), 129–154.

Aslaksen, E. W. (2015). The future of engineering. *Journal and Proceedings of the Royal Society of New South Wales, 148*(457), 159–165.

Aslaksen, E. W. (2017a). Technology, society, and survival, Chapter 12. In A. Fritsche, & S. J. Oks, (Eds.), *The future of engineering: Philosophical foundations, ethical problems and application cases*. Springer. This material was presented at fPET-2016 in Nürnberg

Aslaksen, E. W. (2017b). Engineers and the evolution of society, Chapter 9. In D. P. Michelfelder, B. Newberry, & Q. Zhu, (Eds.), *Philosophy and engineering*. Springer Nature.

Aslaksen, E. W. (2017c). Technology and the practice of engineering, Chapter 13. In S. H. Christensen, et al., (Eds.), *The engineering-business Nexus*. Springer.

Aslaksen, E. W. (2018a). *The social bond*. Springer Nature.

Aslaksen, E. W. (2018b). An engineer's approach to the philosophy of engineering, Chapter 8. In C. Mitcham, et al. (Eds.), *Philosophy of engineering, east and west*. Springer.

Aslaksen, E. W. (2020). *The stability of society*. Lecture Notes in Networks and Systems 113, Springer Nature.

Aslaksen, E. W. (2021). *Measures of social evolution*. SpringerBriefs in Sociology, Springer Nature.

Aslaksen, E. W., & Belcher, W. R. (1992). *Systems engineering*. Prentice Hall of Australia Pty Ltd.

Atherton, N. E. (1984). *From compass to computer: A history of electrical and electronics engineering*. San Francisco Press.

Axelrod, R. (1997). The dissemination of culture: A model with local and global polarization. *Journal of Conflict Resolution, 41*(2), 203–226.

E. W. Aslaksen, *The Evolution of Society*,
https://doi.org/10.1007/978-3-031-23054-7

Baldwin, N. (1995). *Edison, inventing the century*. Hyperion. A discussion of Edison's West Orange laboratory can be found in Hollister-Short, G. & James, F. A. J. L. (Eds.), (1991). *History of technology*, Mansell, London.

Bannon, L., Barry, U., & Holst, O. (Eds.). (1982). *Information technology: Impact on the way of life*. Tycooly Int'l Publishing Ltd.

Bates, T. R. (1975). Gramsci and the Theory of Hegemony. *Journal History of Ideas, 36*(2), 351–366.

Bellah, R. N., Madsen, R., Sullivan, W. M., Swindler, A., & Tipton, S. M. (1991). *The good society*. Knopf Inc.

Bergman, T. J., & Beehner, J. C. (2015). Measuring social complexity. *Journal of Animal Behaviour, 103*, 203–309.

Bieri, J. (1955). Cognitive complexity-simplicity and predictive behaviour. *Journal of Abnormal and Social Psychology, 61*, 263–268.

Biesta, G. (2003). How general can Bildung Be? Reflections on a modern educational ideal, Chapter 4 in Lövlie (*op.cit.*).

Bishop, C. A. (2009). *Internationalizing the right to know: Conceptualizations of access to information in human rights law*. PhD dissertation, University of North Carolina at Chapel Hill. https://core.ac.uk/download/pdf/210596739.pdf. Accessed on June 20, 2022.

Bishop, C. A. (2012). *Access to information as a human right*. LFB Scholarly Publishing LLC.

Bode, H. (1967). The systems approach. In *Applied science—technological progress*. Report to Committee on Science and Astronautics, US House of Representatives.

Bohman, J. (1996). *Public deliberation*. MIT Press.

Bohman, J., & Rehg, W. (Eds.). (1997). *Deliberative democracy*. MIT Press.

Boushay, H. (2019). A new economic paradigm. In *Democracy: A journal of ideas*, no, 53.

Britz, J. J. (2008). Making the global information society good: A Social justice perspective on the ethical dimensions of the global information society. *Journal America Society for Information Science and Technology, 59*(7), 1171–1183.

Broadberry, S., Campbell, B. M. S., Klein, A., Overton, M., & van Leeuwen, B. (2015). *British economic growth, 1270–1870*. Cambridge University Press. http://www.cambridge.org/gb/academic/subjects/history/economic-history/british-economic-growth-12701870?format=PB. Accessed on June 20, 2022.

Bruns, A. (2008). Life beyond the public sphere: Towards a networked model for political deliberation. *Information Polity, 13*(1–2), 71–85, http://dl.acm.org/citation.cfm?id=1412688.1412691. Accessed on June 20, 2022.

Bruns, A. (2019). Digital public spheres in Australia. In A. Bruns, B. McNair, & A. K. Schapals (Eds.), *Digitizing democracy* (pp. 133–146). Routledge.

Bruns, A., & Highfield, T. (2015). *Is Habermas on Twitter?* Social Media and the Public Sphere.

Buchanan, R. A. (1992). *The power of the machine*. Penguin Books. In particular Chapter 11, Technology and People.

Buckley, W. (1968). Society as a complex adaptive system, Chapter 59. In W. Buckley, (Ed.), *Modern systems research for the behavioural scientist*. Aldine Publishing Company.

Buzan, B., & Schouenborg, L. (2018). *Global international society—A new framework for analysis*. Cambridge University Press.

Byrne, K. B. (Ed.). (1985). *Responsible science—The impact of technology on society*. Nobel conference XXI, Harper & Row.

Card, C., & Krueger, A. B. (1995). *Myth and measurement: The new economics of the minimum wage*. Princeton University Press.

Champlin, D. P., & Knoedler, J. T. (2005). Whither, or wither, the good society? *Journal of Economic Issues, 39*(2), 455–463.

Channell, D. F. (1989). *A history of engineering science*. Garland.

Chomsky, N. (1980). *Rules and representations*. Columbia University Press.

Chomsky, N. (2011). *How the world works*. Hamish Hamilton.

Cicero, M. T., De re publica, book 2. In F. Barham, (transl.), *The political works of Marcus Tullius Cicero* (vol. 1) (Treatise on the Commonwealth), Available from http://oll.libertyfund.org/titles/cicero-treatise-on-the-commonwealth--5#toc_list. Accessed on June 20, 2022.

Clark, A., & Chalmers, D. (1998). The extended mind. *Analysis, 58*(1), 7–19.

Cohen, A. M. (2009). *Technology and the early modern self*. Palgrave Macmillan.

Colby, E. A. (2021). *The strategy of Denial—American defense in an age of great power conflict*. Yale University Press.

Crawford, M. B. (2015). *The world beyond your head: On becoming an individual in an age of distraction*. Farrar.

Credit Suisse, *Global Wealth Report 2015*. Available at https://www.credit-suisse.com/about-us/en/reports-research/global-wealth-report.html. Accessed on June 20, 2022.

Cruddas, J., & Nahles, A. (2009). *Building the good society*. Compass, Available at www.compassonline.org.uk. Accessed on June 20, 2022.

da Vinci, L., There is a vast literature concerned with the life and work of Leonardo da Vinci, and quite a bit of it is available on the Internet. Perhaps the best starting point is simply Wikipedia, i.e., http://en.wikipedia.org/wiki/Leonardo_da_Vinci. Accessed on June 20, 2022.

Dahl, R. A. (1957). The concept of power. *Behavioral Science, 2*(3), 201–215.

Darwin, C. (1874). *The descent of man* (2nd ed.). John Murray. Available from https://www.gutenberg.org/files/2300/2300-h/2300-h.htm. Accessed on June 20, 2022.

Darwin, C. (1887). *The life and letters of Charles Darwin*. In F. Darwin, (ed.), John Murray, vol. 1, p. 266.

Dawkins, R. (2006). *The selfish gene* (30th anniversary edition, pp. 189–201). Oxford University Press.

Derrida, J. (1994). *Spectres of Marx: The state of the debt, the work of mourning and the new international*. transl. Kanuf, P., Routledge.

Dessauer, F. (1956). *Streit um die Technik*. Verlag Josef Knecht.

Detter, D., & Fölster, S. (2015). *The public wealth of nations*. Palgrave Macmillan.

Djankov, S., McLiesh, C., Nenova, T., & Shleifer, A. (2003). Who owns the media. *Journal of Law and Economics, 46*(2).

Dorrestijn, S. (2012). Technical mediation and subjectivation: Tracing and extending Foucault's philosophy of technology. *Philosophy & Technology, 25*, 221–241.

Downs, A. (1957a). An economic theory of political action in a democracy. *Journal of Political Economy, 65*(2), 135–150.

Downs, A. (1957b). *An economic theory of democracy*. Harper.

Durkheim, E. (1970). *Education and sociology*, Fox, S. D. (transl.), The Free Press.

Durkheim, E. (1997). *The division of labour in society*, trans. Halls, W. D., intro. Coser, L. A., Free Press.

Duverger, M. (1964). *Political parties* (3rd ed.). Methuen & Co Ltd.

Eberhard, W. (1960). *A history of China* (2nd ed.). Rutledge & Keegan Paul Ltd.

Elias, N. (1978). *What is sociology?* Columbia University Press.

Ellul, J. (1992). Technology and democracy. in L. Winner (Ed.), *Democracy in a technological society*. Kluwer Academic Publishers.

Epstein, D., & O'Halloran, S. (1999). *Delegating powers*. Cambridge University Press.

Erikson, E. H. (1968). *Identity: Youth and crisis*. W.W. Norton Company.

Etzioni, A. (2002). The good society. *Seattle Journal for Social Justice, 1*(1), Article 7, Available at https://digitalcommons.law.seattleu.edu/cgi/viewcontent.cgi?article=1264&context=sjsj. Accessed on June 20, 2022.

Ewuoso, C., & Hall, S. (2019). Core aspects of ubuntu: A systematic review. *South African Journal of Bioethics and Law, 12*(2).

Farazmand, A. (2010). Bureaucracy and democracy: A theoretical analysis. *Public Organization Review, 10*, 245–258.

Farrelly, H., & Shalizi, C. (2012). *An outline of cognitive democracy*. Available at https://oromianeconomist.files.wordpress.com/2012/05/democracy.pdf. Accessed on June 20, 2022.

Feenberg, A. (2013). The mediation is the message: Rationality and agency in the critical theory of technology. *Techné, 17*(1), 7–24.
Fitzgerald, C. P. (1954). *China—A short cultural history* (4 ed., first published in 1935), Frederick A. Praeger, Inc.
Fitzgerald, C. P. (1964). *The birth of communist China.* Penguin Books.
Fletcher, R. (1995). *The limits of settlement growth.* Cambridge University Press.
Foucault, M. (1977). *Discipline and punish: The birth of the prison.* Vintage Books.
fPET, forum on Philosophy, Engineering, and Technology, https://philosophyengineering.com/
Freedman, M. (2006). Ideology and political theory. *Journal of Political Ideologies, 11*(1), 3–22.
French, J. R., & Raven, B. (1959). The bases of social power. In Cartwright, D., & Zander, A. (Eds.), *Group dynamics.* Harper & Row.
Fritsche, A., & Oks, S. J. (Eds.). (2018). *The future of engineering: Philosophical foundations.* Springer.
Fukuyama, F. (1992). *The end of history and the last man.* The Free Press.
Furter, W. F. (Ed.). (1980). *History of chemical engineering.* American Chemical Society.
Galbraith, J. K. (2001). What happened to the Good Society? *Challenge, 44*(4).
Garnham, N. (1982). The information society is also a class society, Chspter 24. In L. Bannon, U. Barry, & O. Holst (Eds.), *Information technology: Impact on the way of life.* Tycooly Publishing Ltd.
Garnham, N. (2001). The information society: Myth or Reality?, *Bugs, Globalism and Pluralism Conference*, Montreal September 19–22.
Gilbert, M. (1989). *On social facts.* Routledge.
Gilbert, M. (2006). *A theory of political obligation.* Oxford University Press.
Gill, T. O. (2008). Structural complexity and effective informing. *International Journal of Emerging Transdisciplines, 11.*
Gramsci, A. (1971). In Q. Hoare, & G. N. Smith, (Eds.), *Selections from the Prison Notebooks.* International Publishers.
Gregory, C. E. (1980). *A concise history of mining.* Pergamon Press.
Grodi, R. H. (1995). *The experience of personal freedom.* Doctoral Dissertation, The Union Institute. ProQuest dissertation 9519593.
Gunther, R., & Diamond, L. (2001). Types and functions of parties. In R. Gunther & L. Diamond (Eds.), *Political parties and democracy* (pp. 3–39). Johns Hopkins University.
Gur-Ze'er, I. (2003a). Bildung and critical theory in the face of postmodern education, Chapter 5 in Lövlie (op.cit).
Gur-Ze'er, I. (2003b). Bildung and critical theory in the face of postmodern education, Chapter 5 in (Lövlie *et al.*, 2003b).
Guterres, A. (2020, Tackling the inequality pandemic: A new social contract for a New Era. *18th Nelson Mandela Annual Lecture*, https://www.nelsonmandela.org/content/page/annual-lecture-2020. Accessed on June 20, 2022.
Habermas, J. (1970). Technology and science as ideology. In *Toward a rational society.* Shapiro, J. (transl.), Beacon Press.
Habermas, J. (1984). *The theory of communicative action*, T. McCarthy (ed.) (transl.), Heineman.
Habermas, J. (2006). Political communication in media society: Does democracy still enjoy an epistemic dimension? The impact of normative theory on empirical research. *Communication Theory, 16*, 411–426.
Habermeier, H.-U. (2007). Education and economy—An analysis of statistical data. *Journal of Materials Education, 29*(1–2), 55–70.
Hahn, R. A. (1973). Understanding beliefs. *Current Anthropology, 14*(3), 207–229.
Hailey, J. (2008). *Ubuntu: A literature review.* Tutu Foundation.
Hanna, T. M. (2013). Challenging corporate domination: The public ownership approach. *The Good Society, 22*(1), 26–43 (Penn State University Press).
Hanushek, E. A., & Wössmann, L. (2010). Education and economic growth. In P. Peterson, E. Backer, & B. McGaw, (Eds.), *International encyclopedia of education* (vol. 2, pp. 245–252).

Harrison, M. (1998). The economics of World War II: An overview. In *The economics of World War II: Six great powers in international comparison* (pp. 1–42). Cambridge University Press.

Harvie, D. I. (2004). Eiffel—The Genius Who Reinvented Himself. Sutton Publishing.

Hassan, A., & Barber, S. J. (2021). The effects of repetition frequency on the illusory truth effect. *Cognitive Research, 6*, 38, Springer Open, Available at https://cognitiveresearchjournal.springeropen.com/track/pdf/https://doi.org/10.1186/s41235-021-00301-5.pdf. Accessed on June 20, 2022.

Hegel, G. W. F. (1966). Who thinks abstractly. In Kaufmann, W. (Ed. and Trans.), *Hegel: Texts and commentary* (pp. 113–118). Anchor Books.

Heidegger, M, (1977). The question concerning technology. In Krell, D. F. (Ed.), *Basic writings: Second edition, revised and expanded*. Harper Collins. Also available online at http://simondon.ocular-witness.com/wp-content/uploads/2008/05/question_concerning_technology.pdf. Accessed on June 20, 2022.

Hepp, A., Breiter, A., & Hasebrink, U. (Eds.). (2018). *Communicative configurations—Transforming communications in times of deep mediatization*. Palgrave Macmillan.

Herman, E. S., & Chomsky, N. (1988). *Manufacturing consent: The political economy of the mass media*. Pantheon Press.

Herold, B. (2016). Issues A-Z: Technology in education: An Overview. *Education Week*. http://www.edweek.org/ew/issues/technology-in-education/. Accessed on June 20, 2022.

Hilbert, M., & López, P. (2011). The world's technological capacity to store, communicate, and compute information. *Science, 332*(6025), 60–65.

Hill, D. R. (1984) *A history of engineering in classical and medieval times*. Open Court Publishing Company, La Salle, Ill. This well-known and excellent account covers the period from 600 BC to 1450 AD.

Hill, S. (1988). *The tragedy of technology*. Pluto Press.

Hills, R. L. (2002). *James Watt*. Landmark Publications.

Hirschi, T. (1969). *Causes of delinquency*. University of California Press.

Hofstadter, D. R. (2007). *I am a strange loop*. Basic Books.

Huber, J. D., & Shipan, C. R. (2002). *Deliberate discretion?* Cambridge University Press.

ISO/IEC 15288, *Systems engineering—System life cycle processes* (1st edn 2002, 2nd edn 2008).

Ivanhoe, P. J., & Van Norden, B. W. (Eds.). (2001). *Readings in classical Chinese philosophy*. Seven Bridges Press.

Jabr, F. (2012), Does thinking really hard burn more calories?. *Scientific American*, 18 July, Available at https://www.scientificamerican.com/article/thinking-hard-calories/. Accessed on June 20, 2022.

Kahneman, D. (2002). Maps of bounded rationality. *Nobel Prize Lecture*, Available at http://nobelprize.org/nobel_prizes/economics/laureates/2002/kahneman-lecture.html. Accessed on June 20, 2022.

Kamke, E. (1944). *Differentialglechungen, Losungsmethoden und Losungen*, 3. Auflage. Reproduced by Chelsea Publishing Company, New York, 1959. (Solution given in item 8.5.).

Kant, I, *Kritik der reinen Vernunft*, originally published in 1781, with a second edition in 1787. A later complete reprint is that published by Th. Knaur Nachf, Berlin, and the standard English translation is that by N. Kemp-Smith, Macmillan Press Ltd. A very good discussion of Kant's ideas is given by Wolff, R. P. (1963) *Kant's theory of mental activity*, Cambridge MA.

Kappeler, P. M. (2019). A framework for studying social complexity. *Behavioural Ecology and Sociobiology, 73*(13).

Karácsony, G. (2021). How democracy can win again, available from *Social Europe* at https://socialeurope.eu/how-democracy-can-win-again. Accessed on June 20, 2022.

Kasarda, J. D. (1974). The structural implications of social system size. *American Sociological Review, 39*, 19–28.

Kelly, M. (2013). The economy: Under new ownership, *yes! Magazine*, February 20. https://www.yesmagazine.org/issue/issues-how-cooperatives-are-driving-the-new-economy/2013/02/20/the-economy-under-new-ownership. Accessed on June 20, 2022.

Kim, Y. J., & Terada-Hagiwara, A. (2010). A survey on the relationship between education and growth with implications for developing Asia. *ADB Economics Working Paper Series* No. 236.

Kitcher, P. (2011). *The ethical project*. Harvard University Press.

Kivisto, P. (2013). *Social theory: Roots and branches* (5th ed.). Oxford University Press.

Kölln, A.-K. (2014). The value of political parties to representative democracy. *European Political Science Review, 7*(4), 593–613.

Kovářová, M., & Filip, M. (2015). Integrating the differentiated: A review of the personal construct approach to cognitive complexity. *Journal of Constructivist Psychology, 28*(4), 342–366.

Kuhn, T. S. (1962). *The structure of scientific revolutions.* The University of Chicago Press.

Kuznets, S. S. (1955). Economic growth and income inequality. *American Economic Review, 45*, 1–28.

Lakoff, G. (2002). *Moral politics: How liberals and conservatives think* (2nd ed.). Chicago University Press. See also https://press.uchicago.edu/books/excerpt/2016/lakoff_trump.html. Accessed on June 20, 20220.

Landemore, H. (2017). *Democratic reason: Politics.* Princeton University Press.

LaPalombara, J., & Weiner, M. (1966). The origin and development of political parties, Chapter 1. In J. LaPalombara, & M. Weiner, (eds.), *Political parties and political development.* Princeton University Press.

Lenski, G. (1970). *Human societies: A macrolevel introduction to sociology.* McGraw Hill Book Company.

Levy, G. (2004). A model of political parties. *Journal of Economic Theory, 115*, 250–277.

Li, Y. (2017). *The analysis on educational expenditure of China.* In 3rd International Conference on Management Science and Innovative Education, ISBN: 978-1-60595-488-2.

Lötter, H. P. P. (2000). *Christians and poverty.* Unpublished doctoral dissertation, University of Pretoria, Pretoria, South Africa.

Lövlie, L, Mortensen, K. P., & Nordenbo, S. E. (Eds.). (2003). *Educating humanity: Bildung in postmodernity.* Blackwell Publishing. The essays were first published as Volume 36, Issue 3 of the *Journal of the Philosophy of Education*, 2002.

Lundberg, H. (2013). Karl Mannheim's sociology of political knowledge. *E-International Relations*, https://www.e-ir.info/2013/10/26/karl-mannheims-sociology-of-political-knowledge/. Accessed on June 20, 2022.

Luria, S. E. (1985). The single artificer, Chapter 3. In, K. B. Byrne (Ed.), *Responsible science—The impact of technology on science, Nobel conference XXI*. Harper & Row.

Lynch, A. (1996). *Thought contagion.* Basic Books.

MacKenzie, D., & Wajcman, J. (Eds.). (1999). *The social shaping of technology* (2nd edn.). Open University Press.

MacLean, N. (2017). *Democracy in Chains—The deep history of the radical right's stealth plan for America.* Scribe Publications.

Malone, T. W., & Bernstein, M. S. (Eds.), *Handbook of collective intelligence.* The MIT Press.

Mannheim, K. (1927). The problem of generations, Chapter 7. In P. Kecskemeti, (Ed.), *Karl Mannheim: Essays*, (1952) (pp. 276–322), Routledge.

Mannheim, K. (1936). *Ideology and Utopia.* Routledge and Kegan Paul. (This is a translation of the German 1929 edition, plus two new chapters.)

Mannheim, K. (1986). *Conservatism.* Routledge & Kegan Paul.

Marcia, J. E. (1966). Development and validation of ego-identity status. *Journal of Personality and Social Psychology, 3*, 551–558.

Marcuse, H. (1964). *One-dimensional man: Studies in the ideology of advanced industrial society.* Beacon.

Marx, K. (1859). *Zur Kritik der Politischen Ökonomie*, Franz Duncker, Berlin, Available online from https://www.marxists.org/archive/marx/works/1859/critique-pol-economy/index.htm. Accessed on June 20, 2022.

Marx, K. (1976). Capital: A critique of political economy. In *The process of production of capital* (vol. 1), in particular Chaps. 14 and 15. Penguin in association with New Left Review.

Marx, K., & Engels, F. (1848). *The Communist Manifesto*, Available from https://www.marxists.org/archive/marx/works/1848/communist-manifesto/ch01.htm#007. Accessed on June 20, 2022.

Marx, L. (1994). The Idea of "Technology" and postmodern pessimism. In M. R. Smith, & L. Marx, (Eds.), *Does technology drive history?*. The MIT Press.

Maslow, A. (1968). *Toward a psychology of being* (2nd edn, , p. 161), Van Nostrand Reinhold Company.

Maslow, A. H. (1943). A theory of human motivation. *Psychological Review, 50*(4), 370–96. Available online at http://psychclassics.yorku.ca/Maslow/motivation.htm. Accessed on June 20, 2022.

Matthews, J. (2013). The educational imagination and the sociology of education in Australia. *Australian Educational Researcher, 40*, 155–171.

Mayhew, B. H., & Levinger, R. (1976). Size and density of interaction in human aggregates. *American Journal of Sociology, 82*(July), 86–110.

Mayhew, B. H., Levinger, R. L., McPherson, J. M., & James, T. F. (1972). System size and structural differentiation in formal organizations. *American Sociological Review, 37*(October), 629–633.

Medawar, P. B. (1959). The future of man. *Reith Lectures 1959*, in particular Lecture 6. Available online from BBC as a recording, https://www.bbc.co.uk/programmes/p00h9lc9. Accessed on June 20,2022. As a book: Methuen, 1960.

Mendel, T., Castillejo, A. G., & Gómez, G. (2017). *Concentration of media ownership and freedom of expression: Global standards and implications for the Americas.* Published by the United Nations Educational, Scientific and Cultural Organization (7, place de Fontenoy, 75352 París 07 SP, Francia) and the Regional Bureau for Sciences in Latin America and the Caribbean/UNESCO Montevideo Office, (Luis Piera 1992, Piso 2, 11200 Montevideo, Uruguay).

Meyer, H., & Rutherford, J. (Eds.). (2012). *The future of European social democracy*. Palgrave Macmillan.

Michelfelder, D. P., Newberry, B., & Zhu, Q. (Eds.). (2017). *Philosophy and engineering*. Springer Nature.

Miller, G. A. (1956). The magical number seven, plus or minus two: Some limits on our capacity for processing information. *The Psychological Review, 63*, 81–97.

Mills, C. W. (1956). *The power elite*. Oxford University Press.

Mitcham, C., et al. (Eds.). (2018). *Philosophy of engineering*. Springer.

Montero, J. R., & Gunther, R. (2003). *The literature on political parties: A critical reassessment*. WP núm. 219, Institut de Ciències Polítiques i Socials, Barcelona. https://www.icps.cat/archivos/WorkingPapers/wp219.pdf?noga=1. Accessed on June 20, 2022.

Mueller, D. (2016). Free speech and the media are often in a marriage of convenience. *The Conversation*, 8 September, Available at https://theconversation.com/free-speech-and-the-media-are-too-often-in-a-marriage-of-convenience-64804. Accessed on June 20, 2022.

Munn, K., & Smith, B. (2008). *Applied ontology—An introduction*. Ontos Verlag.

Naas, M. B. (1990). *Turning: From persuasion to philosophy*. PhD dissertation, State University of New York at Stoney Brook.

Nye, J. S. (1990). *Bound to lead: The changing nature of American power*. Basic Books.

O'Keefe, D. J. (2002). *Persuasion: Theory and research* (2nd edn). Sage Publications, Inc.

OECD. (2021). *Key indicators on the distribution of the household disposable income and poverty*, version 9, December 2021.

Ogburn, W. F. (1922). *Social change with respect to culture and original nature*. Huebsch Inc.

Oxfam. (2016). *An economy for the 1%*. Available at https://www.oxfam.org.au/wp-content/uploads/2016/01/An-economy-for-the-1-percent.pdf. Accessed on June 20, 2022.

Parsons, T. (1971). *The system of modern societies*. Prentice-Hall.

Passmore, J. (1967). On teaching to be critical, Chapter 12. In R. S. Peters, (Ed.). *The concept of education*. Rutledge & Kegan Paul.

Patterson, O. (1991). *Freedom* (vol. 1). I.B. Tauris & Co Ltd.

Pentland, A. (2014). *Social physics*. The Penguin Press.

Peters, R. S. (1967). *What is an educational process?*, Chapter 1. In Peters, R. S. (Ed.), *The concept of education*. Rutledge & Kegan Paul.
Piketty, T. (2014). *Capital in the twenty-first century*. Harvard University Press.
Piketty, T. (2021). *Time for socialism*. Yale University Press.
Plato, *Dialogues*, B. Jowett, transl, Available from https://webs.ucm.es/info/diciex/gente/agf/plato/The_Dialogues_of_Plato_v0.1.pdf. Accessed on June 20, 2022.
Pondy, L. R. (1969). Effects of size, complexity, and ownership on administrative intensity. *Administrative Science Quarterly, 14*(1), 47–60.
Pope, W. (1975). Durkheim as a functionalist. *The Sociological Quarterly, 16*, 361–379.
Popper, K. R. (1959). *The logic of scientific discovery*. Hutchinson.
Raven, B. (1965). Social influence and power. In I. D. Seiner, M. Fishbein, (eds.), *Current studies in social psychology* (pp. 371–382). Holt, Rinehart, Winston.
Reanney, D. (1991). *The death of forever. A new future for human consciousness*. Longman Cheshire.
Reichenbach, R. (2003). On irritation and transformation: A-telelogical Bildung and its Significance (for the Democratic Form of Living, Ch. 6 in Lövlie (*op.cit*).
Rice, J. M., Edwards, D., & McMillan, J. (2019). *Education expenditure in Australia*, Aust. Council for Education Research, https://research.acer.edu.au/cgi/viewcontent.cgi?article=1029&context=policy_analysis_misc. Accessed on June 20, 2022.
Richerson, P. J., & Boyd, R. (2005). *Not by genes alone*. The University of Chicago Press.
Ridley, M. (1999). *Genome—The autobiography of a species in 23 chapters*. Fourth Estate Limited.
Riker, W. H. (1982). The two-party system and duverger's law: An essay on the history of political science. *The American Political Science Review, 76*(4), 753–766.
Robins, K. (1982). *New technology—The political economy of general Ludd*, Chapter 4. In L. Barry, U. Bannon, & O. Holst (*op.cit.*).
Robinson, W. I. (2014). *Global capitalism and the crisis of humanity*. Cambridge University Press.
Rooduijn, M. (2015). The rise of the populist right in Western Europe. *European View, 14*(3).
Rogers, E. M. (2003). *Diffusion of innovation* (5th edn, pp. 201–213). The Free Press.
Scanlon. (2021). Mapping social cohesion 2021, report by the Scanlon Foundation, https://scanloninstitute.org.au/sites/default/files/2021-12/Mapping_Social_Cohesion_2021_Report_0.pdf. Accessed June 20, 2021
Scheler, M. (1924). *Probleme einer Soziologie des Wissens*. Duncker & Humblot.
Schiller, H. I. (1972). *The mind managers*. Beacon Press.
Schiller, H. I. (1981). *Who knows: Information in the age of fortune 500*. Ablex.
Schmidt, B. C. (2019). *The debate on American hegemony*. Dialogue of Civilizations Research Institute GmbH.
Schopenhauer, A. (1969). *Die Welt als Wille und Vorstellung*, 1819/44, *The World as Will and Representation* (vol. I and II, translated by Payne, E. F. J.). Dover Publications.
Schrödinger. A very interesting (and famous) publication is the little book by Erwin Schrödinger, *What is Life? The Physical Aspect of the Living Cell*, based on lectures delivered at Trinity College, Dublin, in 1943, and published by Cambridge University Press, 1944. A version can be downloaded from http://whatislife.stanford.edu/LoCo_files?What-is-Life.pdf. Accessed on June 20, 2022.
Schwartz, S. J. (2001). The evolution of eriksonian and neo-eriksonian identity theory and research: A review and integration. *Identity, 1*(1), 7–58.
Selwyn, N. (2016). *Are we being too quick to embrace technology in education*. Radio National. Retrieved 13.01.2018 from www.abc.net.au/radionational/programs/futuretense/are-we-being-too-quick-to-embrace-technology-in-education/7211366. Accessed on June 20, 2022.
Serafini, T. E., & Adams, G. R. (2002). Functions of identity: Scale construction and validation. *Identity, 2*(4), 361–389.
Shafer, G. (1976). *A mathematical theory of evidence*. Princeton University Press.
Shannon, C. E. (1948). A mathematical theory of communication. *Bell System Technical Journal, 27*, 379–423, 623–656.
Shiller, R. J. (2012). *Finance and the good society*. Princeton University Press.

Shiller, R. J. (2013). Reflections on finance and the good society. *American Economic Review, 103*(3), 402–405.

Shirras, G. F. (1949). The measurement of national wealth. *Econometrica, 17*, 247–254.

Simpkins, B., Sieck, W., Smart, P., & Mueller, S. (2010). *Idea propagation in social networks: The role of 'cognitive advantage. In 1st ITA Workshop on Network-Enabled Cognition: The Contribution of Social and Technological Networks to Human Cognition.* Maryland.

SMH. (2017). Money transfer hack storm. *Sydney Morning Herald*, 17 April, p. 23.

SMH. (2021). All respect to Kristina Keneally, but my Labor Party is wrong to impose her on Western Sydney. *Sydney Morning Herald.*

SMH. (2022). Move to protect sitting MPs angers party. *Sydney Morning Herald*, p. 6.

Social Europe. (2017). Various articles at the web site https://www.socialeurope.eu/2017/04 (e.g. Corbey, D., *The sixth way: Devolution (/*sixth-way-devolution*)*; Schmidt, V., *How do progressives fight back against populism?* (/progressives-fight-back-populism); and N. Kowall, N., & Lindner, F., *Right-Wing populism and the social question* (/right-wing-populism-social-question)).

Solow, R. (1956). A contribution to the theory of economic growth. *Quarterly Journal of Economics, 70*(1), 65–94.

Spencer, H. (1899), *First principles*, D. Appleton and company. Available from https://openlibrary.org/works/OL1067709W/First_principles

Stahl, T. (2011). Institutional power, collective acceptance, and recognition. In H. Ikäheimo, & A. Laitinen, (Eds.), *Recognition and social ontology*. Brill.

Stevenson, R. L. (1973). In A. G. Day, (Ed.), *Travels in Hawaii* (p. 155). University of Hawaii Press.

Stiglitz, J. (2021a). A call to arms to change capitalism. *The Lancet*.

Stiglitz, J. (2021b). https://www8.gsb.columbia.edu/faculty/jstiglitz/. Accessed on June 20, 2022.

Storey, L. (2006). *Humanity and sovereignty—A political roadmap for the 21st century*. Peter Lang.

Strom, K. (1990). A behavioural theory of competitive political parties. *American Journal of Political Science, 34*(2), 565–598.

Suarez-Villa, L. (2009). *Technocapitalism: A critical perspective on technological innovation and corporatism*. Temple University Press.

Summers, L. H. (2014). The Inequality Puzzle. *Democracy*, no. 33, Available at http://democracyjournal.org/magazine/33/the-inequality-puzzle/. Accessed on June 20, 2022.

Thatcher, M. (1987). There is no such thing as society; there are individual men and women, and there are families. As quoted by Douglas Keay in interview 23 September 1987. *Woman's Own, 31*, 8–10.

TRIZ. No publication by the inventor of TRIZ, Genrich S. Altshuller, is known to the author, but a translation, by Shulyak, L. of a book by his son, Henry Altshuller, is available as *The Art of Inventing (And Suddenly the Inventor Appeared)*, Technical Innovation Center, Worchester, MA, 1994, ISBN 0-96-40740-1-X. The main promoter of TRIZ is Ideation International (www.ideationtriz.com. Accessed on June 20, 2022), and a good introduction is available online at www.mazur.net/triz (Accessed on June 20, 2022.

Turchin, P. et al. (2017). Quantitative historical analysis uncovers a single dimension of complexity that structures global variation in human social organization. *Proceeding of National Academy of Science*, E144–151, Available at www.pnas.org/cgi/doi/https://doi.org/10.1073/pnas.1708800115. Accessed on June 20, 2022.

Turner, R., & Golden, S. L. (Eds.). (1981). *Great engineers and pioneers in technology*. St. Martins Press.

Uhlíř, V., & Stella, M. (2012). Who needs memetics? Possible developments of the meme concept and beyond. *Anthropologie, 1*, 127–142.

Usó-Doménech, J. L., & Nescolarde-Selva, J. (2016). What are belief systems. *Journal Foundation of Science, 21*, 147.

van Schaik, C. P., Pradhan, G. R., & Tennie, C. (2019). Teaching and curiosity: Sequential drivers of cumulative cultural evolution in hominin lineage. *Behavioural Ecology and Sociobiology, 73*(2).

Veblen, T. (1921). The captains of industry and the engineers. In *The engineers and the price system* (p. 61). B.W. Huebsch Inc..

von Clausewitz. (1832). *Vom Kriege*. Dümmlers Verlag.
von der Schulenburg, M. (2017). *On building peace—Rescuing the nation-state and saving the United Nations*. Amsterdam University Press.
Walzer, M. (2009). What is "The Good Society"? *Dissent, 56*, 74–78.
Weber, M. (1930). *The protestant ethic and the spirit of capitalism*, (T. Parsons, transl.). Scribner's Sons, and George Allen & Unwin Ltd.
Weeranmanty, C. G. (1994). Access to information: A human right. In *4th Asian Year Book of International Law*. Brill.
Welch, S. R. (2010). *Social freedom: Choice, consent, commitment, and trust*. Doctoral Dissertation, State University of New York at Binghamton, ProQuest Dissertations 3413914.
Winner, L. (1977). *Autonomous technology: Technics out-of-control as a theme in political thought*. MIT Press.
Woolley, A. W, Chabris, C. F., Pentland A., Hashmi, N., & Malone, T. W. (2010). Evidence for a collective intelligence factor in the performance of human groups. *Science, 330*, 686–688.
World Bank. (2013). *Spending more or spending better: Improving education financing in Indonesia*, Report No. 73050-ID.
World Bank. (2018). https://datacatalog.worldbank.org/search/dataset/0038130/GDP-ranking. Accessed on June 20, 2022.
Yan, X. (2021). From keeping China's true strength hidden to striving to shine, Chapter 4. In Y. Zhang, & B. Shao, (Eds.), *China's international relations—Evolving landscapes and strategic adaptions*. Springer Nature.
Yang, J. (2021). Constructing theories of diplomacy: Preparing for China's transition to global great power, Chapter 1. In Y. Zhang, & B. Shao, (Eds.), *China's international relations—Evolving landscapes and strategic adaptions*. Springer Nature.
Yuan, C., & Zhang, L. (2015). Public education spending and private substitution in urban China. *Journal of Development Economics, 115*, 124–139.

Referenced Authors

E. W. Aslaksen, *The Evolution of Society*,
https://doi.org/10.1007/978-3-031-23054-7